ALLE ZEIT WACH
1842

Current Concepts of Diagnosis and Treatment of Bone and Soft Tissue Tumors

Edited by Hans K. Uhthoff

Associate Editor Elvira Stahl

With 143 Figures

Springer-Verlag
Berlin Heidelberg New York Tokyo 1984

Editor:
HANS K. UHTHOFF MD, FRCS(C)
Prof. and Head, Division of Orthopaedics
Faculty of Medicine of the University of Ottawa, Canada

Associate Editor:
ELVIRA STAHL, BA, FAAAS, FAMWA
2021 Atwater, Suite 1610
Montréal, Qué. H3H 2P2, Canada

ISBN-13:978-3-642-69212-3 e-ISBN-13:978-3-642-69210-9
DOI: 10.1007/978-3-642-69210-9

Softcover reprint of the hardcover 1st edition 1984

2124/3130-543210

Preface

During the past decade, the prognosis of bone and soft tissue tumors has improved considerably. This progress has not only been due to improved therapeutics and diagnostic methods but also due to a team approach now universally adopted.

Comparison of results, possible today, were hampered in the past by lack of a clinically relevant staging system and definitive treatment protocols. The complexity of the therapeutic management of bone tumors requires a constant exposure to tumors, highly skilled diagnostic techniques, and specialized expertise. It is therefore not surprising to learn that small and medium-sized hospitals embarking on the diagnosis and treatment of bone and soft tissue tumors are generally less successful than better equipped major centers. This fact has already been recognized by most physicians and surgeons who now refer their patients to specialized oncologic centers. Nevertheless, they all have to know the present state of the art. This symposium was therefore mostly addressed to them.

This international symposium, organized by the Division of Orthopaedic Surgery, University of Ottawa, and cosponsored by the Association des Chirurgiens Orthopédistes du Québec, was held in Ottawa from May 12-14, 1983. Each of the 40 presentations clearly demonstrates that disregard for proper staging before initiation of treatment and omission of a clearly defined treatment protocol is definitely unacceptable in view of the knowledge we have acquired.

The fast publication of these Proceedings has been made possible by the diligence of all contributors, the extraordinary editorial skills of Mrs. Elvira Stahl and her team, and the efficient collaboration of Mrs. Marianne Kalow and Dr. Heinz Götze, Springer Verlag, Heidelberg.
I sincerely hope that a comprehensive review of such a vitally important topic by a distinguished international panel of investigators, pathologists, and surgeons will find acceptance by a wide readership.

Hans-K. Uhthoff
University of Ottawa
Ottawa, Canada

IN MEMORIAM

Crawford Jennings Campbell, MD

Crawford Jennings Campbell, a distinguished academic physician, educator and medical investigator, died on June 24, 1983, at the age of 69, in Laconia, New Hampshire, after a prolonged illness.

Crawford Campbell, the son of James S. and Marian J. Campbell, was born in Cooperstown, New York, December 15, 1913, and lived during his early years in the Panama Canal Zone where his father was employed as a lawyer with the Canal Authority. In 1932, he entered Yale University, receiving his Bachelor of Arts degree in 1936 and his Doctorate in Medicine in 1940. He received his surgical and orthopedic education at the University of Chicago Clinics under Dallas B. Phemister and C. Howard Hatcher, two leading orthopedic scientists of the period. During the Second World War he served with distinction as a Major in the United States Army Medical Corps with the 31st Field Hospital in Hawaii, the Philippines, and in Okinawa.

After the war he returned to Chicago to complete his education in orthopedics and then joined the faculty of the Albany Medical College. He became Chairman of the Section of Orthopedics and Professor of Orthopedics in 1961, a position he held until 1975 when he joined the faculty at Harvard Medical School as Lecturer and Visiting Staff at the Massachusetts General Hospital.

Crawford Campbell became widely recognized for his special skills in musculoskeletal pathology and his ability in the management of patients with bone tumors. He was insatiably interested in all aspects of orthopedics and maintained a close relationship with a group of academic orthopedists, spread widely across the United States, who shared a profound interest in musculoskeletal pathology. Under Crawford's gentle prodding, they continuously exchanged ideas, techniques, interesting case reports and pathologic materials.

Crawford assembled a large collection of rare and unusual pathologic findings which served as the focus of his research, chiefly in the description of a large variety of disease states and their effects on the skeletal system. In all, he contributed over 50 scholarly publications to the literature. He was frequently sought out as guest lecturer, visiting professor, and special essayist in his area of expertise. Up to the time of his death, he served as director of a series of Harvard Orthopedic Basic Science review courses.

His expertise in musculoskeletal pathology was so outstanding that any bone tumor conference in North America remained incomplete without his contribution. It is therefore not surprising that in the preliminary program of the symposium published in this book his name was associated with six different topics.

Crawford devoted much time to the education of medical students and residents. In recognition of his interest in and contribution to the education of foreign students, he was elected to honorary membership in the Japanese and Thai Orthopedic Associations. He was a diplomate of the American Board of Orthopedic Surgery and served as an examiner and member of the Board for many years. He was an associate editor and trustee of The Journal of Bone and Joint Surgery and an associate editor of The Journal of Trauma. He served with the National Institutes of Health, Surgical Study Section, and the National Academy of Science-National Research Council Committee on the Skeletal System.

Crawford will stand as an example of independence, integrity, and selfless service to the entire orthopedic community. All of us who were fortunate to have known Crawford Campbell will always remember him as an outstanding surgeon, a compassionate physician, a deeply respected educator, a true scientist and, above all, a warm and devoted friend.

H.J.M. W.F.E. H-K.U.

Contents

List of Contributors

ABOU ZAHR, K.	Memorial Sloan-Kettering Cancer Center, New York, NY, 10021, USA.
APARISI, T.	Department of Orthopaedic Surgery, Karolinska Hospital, 104 01 Stockholm 60, Sweden.
BELL, M.E.A.	University of Ottawa, Ottawa General Hospital, 501 Smyth Road, Ottawa, Ont. K1H 8L6, Canada.
BOLAND, P.	Memorial Sloan-Kettering Cancer Center, New York, NY, 10021, USA.
BOUCHARD, H-L.	Department of Orthopedics, L'Hôtel-Dieu de Québec, 11, côte du Palais, Québec, Qué. G1R 2J6, Canada.
BOWMAN, W.E. Jr.	Department of Orthopedics, Mayo Clinic and Mayo Foundation, Rochester, Minn. 55905, USA.
BROOKS, C.E.	Department of Orthopaedic Surgery, Montreal General Hospital, 1650 Cedar Ave., Montréal, Qué. H3G 1A4, Canada.
BROSTRÖM, L.-Å	Department of Orthopaedic Surgery, Karolinska Hospital, 104 01 Stockholm 60, Sweden.
BROWER, T.D.	Division of Orthopedic Surgery, University of Kentucky, Lexington, KY 40511, USA.
CAMPANACCI, M.	Clinica Ortopedica, Instituto Ortopedico Rizzoli, Via Codivilla 9, 40100 Bologna, Italy.
CAPANNA, R.	Clinica Ortopedica, Instituto Ortopedico Rizzoli, Via Codivilla 9, 40100 Bologna, Italy.
CAPARROS, B.	Memorial Sloan-Kettering Cancer Center, New York, NY 00221, USA
CARTER, S.K.	Anti-Cancer Research, Bristol-Myers Company, New York, NY, USA.
CAULKINS, E.	Division of Surgical Oncology, John Wayne Clinic, Jonsson Comprehensive Cancer Center, UCLA School of Medicine, Los Angeles, CA 90024, USA

CHAO, E.Y.S.	Department of Orthopedics, Mayo Clinic and Mayo Foundation, Rochester, Minn. 55905, USA.
CSATÓ, Z.	Department of Orthopaedics, Semmelweis Medical University, P.O. Box 45, H-1502, Budapest, Hungary.
CZITROM, A.	Department of Orthopedic Surgery, Mt. Sinai Hospital, 600 University Ave., Toronto, Ont. M5G 1X5, Canada.
DANIEL, R.K.	Department of Surgery, McGill University and Microsurgical Laboratories, Royal Victoria Hospital, 687 Pine Ave., Montréal, Qué. H3A 1A1, Canada
D'ASTOUS, J.	Division of Orthopaedics, Children's Hospital of Eastern Ontario, 401 Smyth Road, Ottawa, Ont. K1H 8L1, Canada.
DELAGE, C.	Department of Pathology, L'Hôtel-Dieu de Québec and Université de Laval, Québec, Qué. G1R 2J6, Canada
ECKHARDT, J.	Division of Surgical Oncology, John Wayne Clinic, Jonsson Comprehensive Cancer Centre, UCLA School of Medicine, Los Angeles, CA 90024, USA
EDWARDS, C.C.	Division of Orthopedic Surgery, School of Medicine, University of Maryland, Baltimore, MD 21201, USA.
EILBER, F.R.	Division of Surgical Oncology, John Wayne Clinic, Jonsson Comprehensive Cancer Centre, UCLA School of , Medicine, Los Angeles, CA 90024, USA
EINHORN, S.	Radiumhemmet, Karolinska Hospital, 104 01 Stockholm 60, Sweden.
ENNEKING, W.F.	Division of Orthopedic Surgery, University of Florida College of Medicine, Gainesville, FLA 32666, USA.
FARINE, I.	Department of Orthopedic Surgery, Mt. Sinai Hospital, 600 University Ave., Toronto, Ont. M5G 1X5, Canada.
FORNASIER, V.L.	Princess Margaret Hospital, 500 Sherbourne St., Toronto, Ont. M4X 1K9, Canada.
GALENTE, J.O.	Department of Orthopedics, Rush-Presbyterian-St. Luke's Medical Center, Chicago, ILL 60612, USA
GHERLINZONI, F.	1st Orthopaedics Clinic, University of Bologna, Instituto Ortopedica Rizzoli, Via Codivilla 9, 40100 Bologna, Italy.
GROSS, A.	Department of Orthopedic Surgery, Mt. Sinai Hospital, 600 University Ave., Toronto, Ont. M5G 1X5 Canada.

HADJIPAVLOU, A.	Paget's Clinic, Department of Orthopaedics, The Sir Mortimer B. Davis-Jewish General Hospital, 3755 Côte Ste-Catherine Road, Montréal, Qué. H3T 1E2, Canada.
HARRIS, J.E.	Section of Medical Oncology, Rush-Presbyterian-St. Luke's Medical Center, Rush Medical College, 1753 West Congress Parkway, Chicago, ILL 60612, USA.
HERNANDEZ, B.	Section of Medical Oncology, Rush-Presbyterian-St. Luke's Medical Center, Rush Medical College, 1753 West Congress Parkway, Chicago, ILL 60612, USA.
HURSON, B.	Memorial Sloan-Kettering Cancer Center, New York, NY 10021, USA.
HUVOS, A.	Memorial Sloan-Kettering Cancer Center, New York, NY 10021, USA.
INGIMARSSON, S.	Radiumhemmet, Karolinska Hospital, 104 01 Stockholm 60, Sweden.
JAWORSKI, Z.F.G.	Department of Medicine, Ottawa General Hospital, 501 Smyth Road, Ottawa, Ont. K1H 8L6, Canada.
JENKIN, R.D.T.	Department of Radiology, University of Toronto, Ontario Cancer Foundation, Toronto-Bayview Clinic, 2075 Bayview Ave., Toronto, Ont. M4N 3M5, Canada.
KÉRY, L.	Department of Orthopaedics, Semmelweis Medical University, P.O. Box 45, H-1502, Budapest, Hungary.
KINNARD, P.	Department of Orthopaedics, Centre Hospitalier Universitaire de Sherbrooke, Sherbrooke, Qué. J1H 5N4, Canada.
KLOIBER, R.	Division of Nuclear Medicine, Foothills Hospital, 1403 - 29th St.N.W., Calgary, Alta. T2N 2T9, Canada.
KOLAR, J.	Radiodiagnostic Clinic, Institute of Postgraduate Studies, Bulovka, 180 81 Prague, Czechoslovakia.
KUETTNER, K.E.	Departments of Biochemistry, Orthopedic Surgery and Pathology, Rush Medical College, 1753 West Congress Parkway, Chicago, ILL 60612, USA.
LAGACÉ, R.	Department of Pathology, L'Hôtel-Dieu de Québec and Université de Laval, Québec, Qué. G1R 2J6, Canada.
LANE, J.M.	Memorial Sloan-Kettering Cancer Center, New York, NY 10021, USA.
LANGER, F.	Mt. Sinai Hospital, 600 University Ave., Toronto, On. M5G 1X5, Canada.

LIVER, J.A.	Department of Radiology, University of Ottawa and Ottawa General Hospital, 501 Smyth Road, Ottawa, Ont. K1H 8L6, Canada.
MANKIN, H.J.	Orthopedic Oncology Service, Massachusetts General Hospital, Harvard Medical School, Boston, MASS 02114, USA.
MAROUN, J.A.	Ontario Cancer Treatment and Research Foundation, Department of Medicine, Ottawa General Hospital, 501 Smyth Road, Ottawa, Ont. KLH 8L6, Canada.
MATEJOVSKY, Z.	Division of Orthopaedic Oncology, Orthopaedic Clinic, Institute for Postgraduate Studies, Bulovka, 180 81 Prague, Czechoslovakia.
McCORMACK, R.R.	Memorial Sloan-Kettering Cancer Center, New York, NY 10021, USA.
McKEE, N.	Department of Orthopedic Surgery, Mt. Sinai Hospital, 600 University Ave., Toronto, Ont. M5G 1X5, Canada.
MINDELL, E.R.	Department of Orthopaedics, State University of New York at Buffalo, 462 Grider Street, Buffalo, NY, USA.
MIRRA, J.	Division of Surgical Oncology, John Wayne Clinic, Jonsson Comprehensive Cancer Center, UCLA School of Medicine, Los Angeles, CA 90024, USA
NILSONNE, U.	Department of Orthopaedic Surgery, Karolinska Hospital, 104 01 Stockholm 60, Sweden.
ORIZAGA, M.	Division of Anatomical Pathology, Department of Laboratory Medicine, Ottawa Civic Hospital, 1053 Carling Ave., Ottawa, Ont. K1Y 4E9, Canada.
OTIS, J.	Memorial Sloan-Kettering Cancer Center, New York, NY 10021, USA.
PAULI, B.U.	Departments of Biochemistry, Orthopedic Surgery and Pathology, Rush Medical College, 1753 West Congress Parkway, Chicago, ILL 60612, USA.
PICCI, P.	1st Orthopaedics Clinic, University of Bologna, Instituto Orthopedica Rizzoli, Via Codivilla 9, 40100 Bologna, Italy.
PRITCHARD, D.J.	Department of Orthopedics, Mayo Clinic, Rochester, Minn. 55905, USA.
RENNIE, W.	Department of Orthopaedic Surgery, McGill University and Royal Victoria Hospital, 687 Pine Ave.W., Montréal, Qué. H3A 1A1, Canada.

RIVARD, C.H.	Department of Surgery, Université de Montréal, Montréal, Qué., Canada.
ROSEN, G.	Memorial Sloan-Kettering Cancer Center, New York, NY 10021, USA.
SIM, F.H.	Department of Orthopedics, Mayo Clinic and Mayo Foundation, Rochester, Minn. 55905, USA.
SÖDERBERG, G.	Department of Tumor Pathology, Karolinska Hospital, 104 01 Stockholm 60, Sweden.
STENER, B.	Department of Orthopaedic Surgery II, Sahlgren Hospital, S-413 45 Göteborg, Sweden
STRANDER, H.	Radiumhemmet, Karolinska Hospital, 104 01 Stockholm 60, Sweden.
STÜRZ, H.	Oberarzt der Orth. Klinik der Med. Hochschule Hannover, im Annastift (Klinik III), Heimschenstr. 1-7, D,3000 Hannover 61, Germany.
SUNDARESAN, N.	Memorial Sloan-Kettering Cancer Center, New York, NY 10021, USA.
TELFER, C.D.	Division of Orthopaedic Surgery, University of Ottawa, Ottawa General Hospital, 501 Smyth Road, Ottawa, Ont. K1H 8L6, Canada.
UHTHOFF, H.-K.	Division of Orthopaedic Surgery, University of Ottawa, Ottawa General Hospital, 501 Smyth Road, Ottawa, Ont. K1H 8L6, Canada.
VIZKELETY, T.	Department of Orthopaedics, Semmelweis Medical University, P.O. Box 45, H-1502 Budapest, Hungary.
WEISENBURGER, T.	Division of Surgical Oncology, John Wayne Clinic, Jonsson Comprehensive Cancer Center, UCLA School of Medicine, Los Angeles, CA 90024, USA
ZENKER, H.	Orthopädische Universitäts-Klinik, Harlachingerstr. 51, D-8000 München 90, Germany.
ZIDKOVA, H.	Radiodiagnostic Clinic, Institute for Postgraduate Studies, Bulovka, 180 81 Prague, Czechoslovakia.
ZUCKER, J.	Department of Orthopaedics, Royal Victoria Hospital, 687 Pine Ave. W., Montréal, Qué. H3A 1A1, Canada.

TENTH ANNUAL APPLIED BASIC SCIENCES COURSE

UNIVERSITY OF OTTAWA - FACULTY OF HEALTH SCIENCES

12 - 14 MAY 1983

LEFT TO RIGHT: 1 - F.R. Eilbel; 2 - C.C. Edwards; 3 - E. Brooks; 4 - J. Harris; 5 - M. Orizaga; 6 - R.D.T. Jenkin; 7 - H.K. Uhthoff; 8 - H. Zenker; 9 - H. Mankin; 10 - A. Gross; 11 - B. Stener; 12 - M. Campanacci; 13 - A. Liver; 14 - T. Brower; 15 - D. Pritchard; 16 - Z. Matejovsky; 17 - E. Mindell; 18 - J.M. Lane; 19 - W. Rennie; 20 - F. Langer; 21 - A. Hadjipavlou.

Staging of Musculoskeletal Neoplasms

W. F. ENNEKING

Introduction

The purposes of a staging system for musculoskeletal neoplasms are to:

1) incorporate the significant prognostic factors into a system which describes progressive degrees of risk of local recurrence and distant metastases to which a patient is subject,

2) stratify the stages so they have specific implications for surgical management, and

3) provide guidelines for adjunctive therapies.

Over a number of years, staging systems for various classes of malignant tumors have been developed under the auspices of the American Joint Committee for Cancer Staging and End Results Reporting (AJC). The systems vary among cancers related to the natural course of a particular type of cancer. In 1980, a system for the surgical staging of musculoskeletal sarcoma was proposed, studied, and adopted by the Musculoskeletal Tumor Society (Enneking et al, 1980) and subsequently adopted by the AJC.

In this review, I will report the natural evolution of benign and malignant lesions of connective tissue derivation that led to the staging system, the system for both benign and malignant lesions, its articulation with surgical treatment, and early experience with its use.

Natural Evolution

The natural course in progression from the most benign to the most malignant connective tissue tumor is the same, lesion for lesion, whether the tumor arises in bone or somatic soft tissue. A fibrosarcoma behaves as a fibrosarcoma whether it arises in soft tissues and invades bone or vice versa. Aggressive benign fibrous lesions in soft tissue (fibromatosis) behave the same as their counterparts arising in bone (desmoplastic fibroma). Therefore, a common staging system was devised for bone and soft tissue as opposed to separate systems for bone and soft tissue lesions. The system, as befits the natural history, applies only to lesions of connective tissue histogenesis and not to primary lesions of round cell origin (leukemias, lymphomas, myeloma, Ewing's sarcoma) or metastatic lesions.

Current Concepts of Diagnosis and Treatment
of Bone and Soft Tissue Tumors.

The significant progressive changes in the biologic behavior of musculoskeletal lesions are: 1) localized, latent or static, inactive, benign; 2) localized, active, benign; 3) aggressive, invasive, but still benign; 4) indolent, invasive, malignant, low risk of regional lymphatic or distant metastases; 5) rapidly, destructive, malignant, high risk of local, regional, and distant metastases; and 6) regional and/or distant metastases. Each of these progressions has distinctive clinical, radiographic, and histologic features that form the basis for the staging system and will be presented in some detail. The radiographic features that express these evolutionary changes in skeletal lesions have been previously studied and classified by their probability of occurrence by Lodwick et al (1980).

Inactive benign lesions are usually asymptomatic, discovered incidentally, and seldom related to pathologic fracture or mechanical dysfunction. They may slowly attain large size but eventually reach a steady state where they no longer grow. They appear quite responsive to contact inhibition and remain completely encapsulated. They remain intracompartmental and seldom deform the compartmental boundaries of cortical bone, articular cartilage, or dense fascial septae. When palpable in soft tissue, they are often small, movable, nontender, with little or no significant enlargement on subsequent clinical observation. Radiographic characteristics are lesions that are well marginated by a mature shell of cortical-like reactive bone without deformation or expansion of the encasing bone (Lodwick IA). Angiographic staging studies show little or no increase in isotope uptake and no significant reactive neoangiogenesis about the lesion or intralesional neoangiogenesis. CT scanning shows a homogenous density, well marginated, with no cortical broaching or cross-fascial extension.

The histologic characteristics of the lesion are: 1) low cell-to-matrix ratio; 2) mature, well-differentiated matrices; 3) benign cytologic characteristics (no hyperchromasia, anaplasia, or pleomorphism); 4) encapsulation by mature fibrous tissue or cortical bone; and 5) little or no reactive mesenchymal proliferation, inflammatory infiltrate, or neoangiogenesis about the lesion.

Active benign lesions are mildly symptomatic, discovered because of discomfort, and occasionally associated with pathologic fracture or mechanical dysfunction. They grow steadily and continue to enlarge during observation. They appear responsive to contact inhibition but not at normal levels as they can expand by deformation of overlying cortical bone, articular cartilage, or fascial septae. They remain encapsulated and have only a thin layer of filmy areolar tissue forming the reactive zone between the lesion and surrounding normal tissue. When palpable in soft tissue, they are usually small, movable, moderately tender, and grow slowly during clinical observation. Radiographic characteristics of active lesions in bone are well, but irregularly, marginated defects. The margin is a mature cancellous ring rather than a cortical shell, and the inner aspect is often irregular or corrugated giving a septated appearance. Expansion, bulging, or deformation of the combination of overlying cortex/reactive bone (Lodwick IB) is frequently observed.

Staging studies show increased isotope uptake that conforms closely to the limits of the radiographic defect and reactive changes. A thin but discernible rim of reactive neoangiogenesis about the lesion is frequently observed but seldom any significant intralesional neoplastic neoangiogenesis. CT scanning shows a homogenous density, irregular but intact reactive bone, "expansion" of the overlying cortex, and intracompartmental containment by bone or fascia.

The histologic characteristics of active benign lesions are: 1) a relatively balanced cell-to-matrix ratio with homogenous distribution of the matrix; 2) well-differentiated matrices; 3) benign cytologic characteristics; 4) an intact capsule of mature fibrous tissue and/or cancellous bone; 5) a narrow zone of mesenchymal, inflammatory, and vascular reactive tissue between the capsule and the surrounding normal tissue; 6) resorption of preexisting bone by osteoclasts rather than by neoplastic cells as the mechanism of expansion. Intermittent areas of resorption often produce an irregular, serrated, sometimes corrugated interface between the capsule and the adjacent reactive bone.

Aggressive benign lesions are often symptomatic, discovered because of discomfort and/or a growing mass, and when in a stress-bearing bone associated with a pathologic fracture. When palpable in the soft tissues, they grow rapidly, sometimes alarmingly. These lesions are frequently tender and may have an inflammatory-like appearance. They are little affected by contact inhibition and readily penetrate or permeate the natural barriers to tumor growth: cortical bone, fascial septae, and in some cases, articular cartilage or joint capsules. They penetrate the capsule with finger-like extensions protruding directly into the surrounding zone. The reactive zone is thick, edematous, and often appears inflammatory. These aggressive lesions invade by destroying or resorbing the restraining bone or fascia and permeate into adjacent tissues or compartments rather than expanding by concomitant endosteal resorption and subperiosteal apposition.

When aggressive benign lesions involve unrestrained areas (medullary canal, cancellous bone, advential intermuscular planes within muscle bellies, pararticular tissues), they extend rapidly although usually preceded by a pseudocapsule of reactive tissue. Aggressive soft tissue lesions are usually firm, fixed, tender, and have a rapid growth history. The radiographic features of aggressive benign lesions are a ragged permeative interface with adjacent bone, incomplete attempts at containment by reactive bone, cortical destruction, endosteal buttresses and periosteal Codman's triangles, and rapid soft tissue extension (Lodwick IC). Staging studies often reflect the aggressive nature and behavior of these lesions. Isotope scans show increased uptake in both the early vascular phase and the late bone phase. The extent of the increased uptake is often well beyond the apparent radiographic limits. Angiograms show a distinct reactive zone of neovasculature on the early arterial phase and an intralesional hypervascular blush on the late venous phase of the study. CT scans show nonhomogenous mottled densities with defects in attempts at reactive containment, early extracompartmental extension from bone, and indistinct margins in soft tissues. Otherwise, occult involvement of major neurovascular bundles is often shown by either angiography or CT scans in aggressive soft tissue lesions.

Histologic features of aggressive lesions are: 1) high cell-to-matrix ratio; 2) clearly differentiated matrices of varying maturity; 3) predominantly benign cytologic characteristics without anaplasia or pleomorphism, but with frequent hyperchromatic nuclei. Mitoses are occasionally encountered. Vascular invasion within the lesion may also be found; 4) finger-like projections of tumor extend through gaps in the capsule and grow into the surrounding reactive zone. These extensions usually maintain continuity with the main mass, although occasional apparently isolated satellites are seen; 5) a thick, succulent zone of reactive tissue is interposed between the penetrated capsule and the more peripheral normal tissue. This zone or pseudocapsule encircles but does not inhibit growth of an aggressive tumor. The reactive zone prevents tumor nodules from extending directly into normal tissue. Destruction of surrounding bone is by reactive osteoclasts rather than by tumor cells, although fingers of tumor grow rapidly into the reactive bone.

Despite the benign cytologic characteristics, the invasive behavior of these lesions is more like that of a low-grade malignancy than of an active benign process. The reliability of the benign cytologic features is occasionally challenged by the occurrence of distant, usually pulmonary, metastases. These metasteses look histologically as benign as the primary tumor and have a much better prognosis than the metastases of frankly malignant lesions.

Low-grade sarcomas have all the invasive growth mechanisms of local malignancy, but have a low risk of distant metastases and an indolent rate of evolution. With time, and particularly with repeated cycles of unsuccessful excisions and recurrences, the risk of dedifferentiating into high-grade and distant metastases is increased.

Low-grade sarcomas often present as a slow-growing, painless mass with an indolent but steady growth rate. They are seldom symptomatic. They stimulate generous amounts of reactive bone or fibrous tissue that often impart a false impression of encapsulation. They are not inhibited by the natural barriers to tumor growth, but their extension through them is one of gradual erosion rather than rapid destruction. Their relentless progression often produces extraosseous or extracompartmental extension and neovascular bundle involvement. They seldom traverse articular cartilage or joint capsule to extend intra-articularly. They may, however, stimulate reactive synovitis and effusion when adjacent to joints in either bone or soft tissue. Similarly, they do not extend through tendon sheath, nerve sheath, or the outer adventitial layer of major arteries but tend to push these structures aside. Tendons, nerves, and vessels, however, are often involved by the reactive zone about the lesion.

When low-grade sarcomas arise in the soft tissue, they are apt to be small, superficial, fixed, nontender, and without inflammatory signs from the reactive zone. The radiographic features of low-grade skeletal sarcomas are similarly apt to be less ominous than many aggressive benign lesions (Lodwick II). They often have a generous reactive rim of cancellous bone admixed with defects of extracapsular and/or soft tissue extension. Buttressing in the meduallary canal, Codman's triangles externally, and especially endosteal scalloping are features of these lesions. Staging studies again accurately reflect the behavior of these lesions. Isotope scans in both bone or soft tissue show increased early and late uptake that is more extensive than the radiographic limits would suggest. The angiographic findings show little or no reactive neovasculature or intralesional neoangiogenesis and often appear deceptively benign. CT findings are of nonhomogenous density, a thick perforated ring of reactive bone, and occult soft tissue or intraosseous extension.

The histologic features of locally invasive low-grade malignancies are: 1) a relatively even proportion of cells to matrix; 2) well-differentiated and usually mature matrices; 3) malignant cytologic characteristics -- anaplasia, pleomorphism, and hyperchromasia -- with a modest number of mitoses. These features are consistent with Broder's grades 1 and occasionally 2 in other malignances (Broders et al, 1939); 4) varying amounts of necrosis, hemorrhage, and vascular invasion are found in these lesions that are seldom seen in benign lesions; 5) there are numerous interruptions in the continuity of the capsule where the tumor extends directly into the reactive zone which, in turn, forms a pseudocapsule about the lesion. In the reactive zone, nodules of isolated tumor-forming satellites are almost universally found. Skip lesions in normal tissue beyond the reactive zone are rarely if ever seen about low-grade lesions.

Due to the indolent growth rate of low-grade locally invasive sarcomas, they remain contained by the natural barriers to tumor growth within their compartments of origin for long periods. With their ability to destroy or pervade normal tissue, however, they eventually extend through their barriers to involve adjacent extracompartmental tissues. Metastasis to regional lymph nodes is unusual. Distant pulmonary metastases occur late in the course and are often solitary.

High-grade sarcomas usually appear as destructive symptomatic masses that are often associated with pathologic fractures when they involve bone. They stimulate generous amounts of reactive tissue but overgrow it so rapidly that they appear to have little or no pseudoencapsulation. They are uninhibited by the natural barriers to tumor growth and rapidly extend into adjacent tissues by destruction of cortical bone, fascial septae, articular cartilage, or joint capsules. They quickly extend extracompartmentally, involve adjacent neurovascular bundles, and extend proximally along ill-defined extracompartmental fascial plane and spaces. They often cross epiphyseal growth plates, and although they respect articular cartilage, will extend intra-articularly at sites of capsular or ligamentous attachment.

High-grade sarcomas of soft tissue origin are usually deep, large, fixed, tender, and stimulate a soft, edematous, succulent inflammatory-like reaction about them. The radiographic features of high-grade sarcomas arising in bone are quite predictive of their behavior (Lodwick III). The reactive response is so rapidly destroyed that the interface between the lesion and the surrounding bone is poorly marginated with a diffuse permeative border. Patchy cortical destruction, early occult soft tissue extension, obliteration of periosteal reaction with only small Codman's triangles remaining, and ill-defined intramedullary extension beyond the extent suggested by the periosteal reaction are all features of high-grade primary skeletal sarcomas.

Staging studies are usually accurate in suggesting their high-grade nature. Isotope scans show increased uptake both early and late that extends beyond the radiographic limits of the lesion. Increased isotope uptake in radiographically normal bone adjacent to high-grade soft tissue sarcomas may be the only clue to occult reaction to the soft tissue lesion. They may also be the first hint of a "skip" metastasis. Angiograms show a vigorous zone of reactive neovasculature about the lesion on the early arterial phase and often a hypervascular intralesional "blush" on the late venous phase. The extent of involvement of neurovascular bundles is often best shown on angiography: CT scans show occult intraosseous extension, skip lesions in the medullary canal, occult soft tissue extension, and extra-compartmental extension through cortical bone and across fascial septae.

The histologic features of high-grade sarcomas are: 1) high cell-to-matrix rate; 2) poor differentiations of immature matrices; 3) all the cytologic characteristics of high-grade malignancy. Mitoses are abundant. Vascular invasion, necrosis, hemorrhage, and direct destruction of normal tissue by tumor cells are part of this picture. These findings are those of Broder's Grade 2, 3, or 4 lesion (Broders et al, 1939); 4) little or no encapsulation is apparent and isolated satellites are found in the pseudocapsule of reactive tissue about the lesion; 5) skip metastases -- isolated nodules of tumor in the normal tissue well beyond the reactive zone -- are found in a significant proportion (approx. 25%) in both bone and soft tissue sarcomas (Enneking and Kagan, 1975). In bone, they are either in the medullary canal or occasionally transarticularly in an adjacent metaphysis. In soft tissue, they may occur more proximally in either the ill-defined extracompartmental spaces and planes or within skeletal muscle.

High-grade sarcomas quickly cross barriers to tumor extension and a relatively small proportion (approx. 10%) are still intracompartmental at the time of presentation. The majority extend extracompartmentally; an occasional lesion, usually of soft tissue origin, will develop regional lymphatic metastasis; and a significant proportion (approx. 10%) will present with distant (usually pulmonary) metastases.

These behavioral changes (latent, active, aggressive, invasive, destructive, and metastatic) that form the basis of the staging system together with their clinical, radiographic, and staging studies are summarized in Tables I (benign) and II (malignant).

The Staging System

The system is based on the interrelationship of three factors: 1) Grade (G), 2) Site (T), and 3) metastases (M). Each of these in turn is stratified by components that influence both prognosis and response to treatment.

Grade

The grade is an assessment of the biologic aggressiveness of the lesion. It is not a purely histologic assessment (as in Broders' 1, 2, 3, 4 grading of malignancies [Broders et al, 1939]), nor a purely radiographic assessment (as in Lodwick's IA, IB, IC, II, and III radiographic classification of probabilities [Lodwick et al, 1980]), nor a purely clinical reflection of growth rate, doubling time, size, temperature, tissue pressure, or bio-

Table I. Stages of benign musculoskeletal lesions

	1	2	3
Grade	G_0	G_0	G_0
Site	T_0	T_0	T_{1-2}
Metastases	M_0	M_0	M_{0-1}
Clinical course	Latent, static, self-healing	Active progressing expands bone or fascia	Aggressive, invasive, breaches bone or fascia
Radiographic grade	I_A	I_B	I_C
Isotope scan	Background uptake	Increased uptake in lesion	Increased uptake beyond lesion
Angiogram	No neovascular reaction	Modest neovascular reaction	Moderate neovascular reaction
CT	Crisp, intact margin - well defined capsule homogeneous	Intact margin "expansile" - thin capsule homogeneous	Indistinct broached margin - extracapsular and/or extracompartmental extension nonhomogeneous

chemical markers. It is a blending of all of these into patterns. The three stratifications of grade are G_0, G_1, and G_2. Their identifying characteristics are:

G_0 - Benign

Histologic - benign cytology, clearly differentiated, low to moderate cell-to-matrix ratio.

Radiographic - Lodwick IA, IB, or IC ranging from clearly marginated to those with capsular broaching and soft tissue extensions.

Clinical - Distinct capsule, no satellites, no skips, rare metastases, variable growth rate, predominantly in adolescents and young adults.

G_1 - Low-grade malignant

Histologic - Broders' Grade 1 and some 2. Few mitoses, moderate differentiation, distinct matrix.

Table II. Stages of malignant musculoskeletal lesions

	I_A	I_B	II_A	II_B	III_A	III_B
Grade	G_1	G_1	G_2	G_2	G_{1-2}	G_{1-2}
Site	T_1	T_2	T_1	T_2	T_1	T_2
Metastases	M_0	M_0	M_0	M_0	M_1	M_1
Clinical course	Symptomatic indolent growth	Symptomatic mass indolent growth	Symptomatic rapid growth	Symptomatic rapid growth fixed mass path. fracture	Sytemic symptoms palpable nodes pul. symptoms	
Isotope scan	Increased uptake	Increased uptake	Increased uptake beyond radio-graphic limits	Increased uptake beyond radio-graphic limits	Pul. lesions no increased uptake	
Radiographic grade	II	II	III	III	III	
Angiogram	Modest neo-vascular react. involvement of neurovascular bundle	Modest neo-vascular react. involvement of neurovascular bundle	Marked neovascular reaction - no involvement of neurovascular bundle	Marked neovascular reaction - involve-ment of neurovascu-lar bundle	Hypervascular lymph nodes	
CT	Irregular or broached capsule but intracompartmental	Extracompartmental extension or location	Broached (pseudo) capsule - intra-compartmental	Broached (pseudo) capsule - extra-compartmental	Pulmonary lesions or enlarged nodes	

Radiographic - Lodwick II with indolent invasive features.

Clinical - Indolent growth, extracapsular satellits in the reactive zone, no skips and only occasional distant metastases.

G_2 - High-grade malignant

Histologic - Broders' Grades 2, 3, and 4. Frequent mitoses, poorly differentiated, sparse and immature matrix. High-grade cytologic features: anaplasia, pleomorphic, and hyperchromatic.

Radiographic - Lodwick III: destructive, invasive.

Clinical - Rapid growth, symptomatic, both satellites and skips, occasional regional and frequent distant metastases.

The behavior of G_0 benign lesions may be latent, active, or aggressive. Their histologic features are often poor indicators of their behavior, and within this spectrum G_0 lesions are often better predicted by their radiographic, staging, and clinical features (Table I). The histologic characteristics of G_1 low-grade sarcomas make their distinction from G_2 high-grade lesions on histologic grounds predictably accurate, and their radiographic staging and clinical features are supportive and confirmatory of the histologic distinction. However, it may be difficult to distinguish G_0 from G_1 lesions on purely histologic features and, in many instances, the radiographic and particularly the staging studies may be of more value than the histologic findings (Table II).

A promising new method of assessing grade is the determination of the nuclear DNA concentration (ploidy) by flow cytometry. Individual cell nuclei are stained with a specific fluorescent DNA dye and the concentration assessed rapidly by fluorometric assay of the cells as they pass through a focussed laser beam. Normal cells are euploid and so are most G_0 lesions. G_1 lesions have increased amounts of cells in normal replicative activity (tetraploidy). G_2 lesions have both abnormal numbers of cells in tetraploidy but also may show an abnormal cell line (aneuploid) quite distinctive for high-grade neoplasms. These correlations between ploidy and prognosis have been shown to be valid for other classes of neoplasia -- particularly myelomas and lymphomas -- and preliminary results suggest that this technique may be quite helpful in connective tissue lesions (Kreicbergs, et al, 1982).

In summary, surgical grading into G_0, G_1, or G_2 requires histologic, radiographic, and clinical correlation to achieve accuracy and reliability. Although certain histogenic types of sarcomas may have a preponderance of their lesions G_1 or G_2 (Table III), each lesion must be assessed on its own characteristics before a grade is assigned. For example, most parosteal osteosarcomas are G_1, but a few dedifferentiate into G_2 lesions and accordingly have a much more ominous prognosis. Conversely, although most classic osteosarcomas are G_2, occasionally one will be G_1 with a much more favorable prognosis.

Table III. Surgical grade (G)

Low (G_1)	High (G_2)
Parosteal osteosarcoma Endosteal osteosarcoma	Classic osteosarcoma Radiation sarcoma Paget's sarcoma
Secondary chondrosarcoma	Primary chondrosarcoma
Fibrosarcoma, Kaposi's sarcoma Atypical mal. fib. hist.	Fibrosarcoma Malig. fib. histiocytoma Undiff. primary sarcoma
Giant cell tumor, bone	Giant cell sarcoma, bone
Hemangioendothelioma Hemangiopericytoma	Angiosarcoma Hemangiopericytoma
Myxoid liposarcoma	Pleomorphic liposarcoma Neurofibrosarcoma (Schwannoma) Rhabdomyosarcoma Synovioma
Clear cell sarcoma, tendon sheath Epitheloid sarcoma	
Chordoma	
Adamantinoma	
Alveolar cell sarcoma	Alveolar cell sarcoma
Other and undifferentiated	Other and undifferentiated

Site

The anatomic setting of the lesion has a direct relationship to the prognosis and the choice of surgical procedure. The three strata of anatomic settings are T_0, T_1, and T_2. These are determined primarily by clinical and radiographic techniques. Staging studies (isotope scanning, angiography, CT, and NMR imaging, ultrasonography, myelography, etc.) can make valuable contributions in preoperatively assessing the anatomic setting.

T_0 - The lesion remains confined within the capsule and does not extend beyond the borders of its compartment of origin. While the boundaries of the capsule and/or the compartment of origin may be distorted or deformed, they both remain intact.

T_1 - The lesion has extracapsular extensions, either by continuity or isolated satellites, into the reactive zone, but both the lesion and the reactive zone about it are contained within the anatomic compartment bounded by the natural barriers to tumor extension: cortical bone, articular cartilage, joint capsule, or the dense fibrous tissue of fascial septa, ligaments, or tendon (sheath). To be classified as T_1, both the lesion and its (pseudo)capsule must be within the

Table IV. Surgical sites (T)

Intracompartmental (T_1)		Extracompartmental (T_2)
Intraosseous	--->	Soft tissue extension
Intra-articular	--->	Soft tissue extension
Superficial to deep fascia	--->	Deep fascial extension
Parosseous	--->	Intraosseous or extrafascial
Intrafascial compartments		Extrafascial planes or spaces
Ray of hand or foot		Mid- and hindfoot
Post. calf		Popliteal space
Anterolat. leg		Groin - femoral triangle
Ant. thigh		Intrapelvic
Midthigh		Midhand
Post. thigh		Antecubital fossae
Buttocks		Axilla
Volar forearm		Periclavicular
Dorsal forearm		Paraspinal
Ant. arm		Head and neck
Post. arm		
Periscapular		

compartment. If the reactive zone extends outside the compartment while the tumor remains within, the lesion is classified as extracompartmental. The anatomic compartments of bone bone and soft tissue are shown in Table IV.

Three particular points about compartmentalization require elaboration. The skin and subcutaneous tissue are classified as a compartment even though there are no longitudinal boundaries. In the transverse dimension, however, the deep fascia forms an effective barrier between the subcutaneous and deeper tissues. The parosseous compartment is a potential compartment between cortical bone and overlying muscles. Lesions on the surface of bone that have not invaded either the underlying cortical bone or the overlying muscle but have pushed them apart are defined as intracompartmental. Lesions within muscular compartments that contain more than one muscle (eg, the volar compartment of the forearm) are considered intracompartmental despite involving more than one muscle.

T_2 - Lesions extending beyond compartmental barriers into the loosely bounded fascial planes and spaces that have no longitudinal boundaries are extracompartmental or T_2. Extracompartmental involvement may be either by virtue of extension of a previously intracompartmental lesions, by arising de novo in the extracompartmental tissues, or by inadvertent transmission by trauma or surgical excision. The various sites that are extracompartmental are shown in Table IV. Almost without exception, lesions (or their reactive zones) that abut or involve major neurovascular bundles are extracompartmental by virtue of the extracompartmental location of these structures.

Metastasis

In most staging systems for carcinomas, metastases are stratified by virtue of being regional (N for nodes) or distant (M) since the prognosis and treatment is significantly different for these two sites of metastasis.

For sarcomas, metastatic involvement of either regional lymph nodes or distant organs has the same ominous prognosis, and both are designated by M. There are only two strata of metastasis - M_0 and M_1. M_0 indicates no evidence of regional or distant metastases while M_1 signifies either regional or distant metastases.

These three factors, G, T and M, are combined to form the criteria for the progressive stages of benign and malignant lesions (Tables I and II).

Benign lesions are designated by Arabic numerals 1, 2, or 3 that are synonymous with latent, active, or aggressive. The characteristics of Stage 1 latent, Stage 2 active, and Stage 3 aggressive lesions are shown in Table I, and were described in detail as the latent, active, and aggressive progressions in the proceding section on evolution. As indicated, Stages 1, 2, and 3 correspond closely to the Lodwick classification of radiographic features as IB, and IC.

Malignant lesions are designated by the Roman numerals I, II, or III that are synonymous with low-grade, high-grade, and metastatic. These three stages of sarcomas are further stratified into A or B depending on whether the lesion is anatomically intracompartmental (A) or extracompartmental (B). The characteristics of these malignant lesions are shown in Table II, and were described in detail as low-grade invasive and high-grade destructive lesions in the evolution of connective tissue lesions. Their radiographic characteristics correspond closely to Stages II and III in the Lodwick classification. Only after each lesion has been studied clinically, radiographically, and biopsied for histogenic typing and cytologic grading, can it be staged according to its characteristics. Although particular lesions tend to cluster in particular stages (ie, >90% of classic osteosarcomas present as Stabe IIB) others tend to be more evenly distributed (ie, giant cell tumors of bone approximately 10% Stage 1, 75% Stage 2, and 15% Stage 3).

Clearly, a particular lesion may undergo transition from one stage to another. Benign lesions that are Stage 2 active or even Stage 3 aggressive during adolescence frequently undergo involution into Stage 1 latent lesions after growth has ceased. On the other hand, certain benign lesions of any stage may undergo transformation into Stage I, II or even III sarcomas. Obviously, high-grade Stage II and occasionally low-grade Stage I lesions become Stage III lesions after presentation by virtue of either regional or distant metastases. Certain factors have been implicated directly or by inference in the upstaging of benign or malignant lesions. Radiation has been held responsible for transition of giant cell tumor, chondroblastoma, and other benign lesions to sarcomas. Repeated inadequate surgical interventions have been implicated in the evolution of low-grade

Table V. Surgical Margins

Type	Plane of dissection	Microscopic appearance
Intracapsular	Within lesion	Tumor at margin
Marginal	Within reactive zone - extracapsular	Reactive tissue ± microsatellites tumor
Wide	Beyond reactive zone through normal tissue within compartment	Normal tissue ± "skips"
Radical	Normal tissue extracompartmental	Normal tissue

fibrous lesions into high-grade fibrosarcomas and in the dedifferentiation of Stage I parosteal osteosarcoma into Stage II or III high-grade osteosarcoma

Articulation with Surgical Treatment

Articulating the staging system with the surgical treatment of connective tissue tumors requires precise definitions of the procedures as well as the stages. The traditional terms of incisional biopsy, excisional biopsy, resection, and amputation are difficult to define in either biologic, anatomic, or physical terms. After a number of physical and surgical criteria were postulated, a method of definition was devised based on the margin the procedure obtained in relation to the lesion and the barriers of its extension.

The four oncologic surgical margins, the plane of dissection that achieves them, and the microscopic appearance of the tissue at the margin of the wound are shown in Table V. The four margins are described in surgical terms (intracapsular, marginal, wide, and radical), and they reflect the progressive barriers to tumor extension in their natural evolution, eg, the (pseudo)capsule, reactive zone, intracompartmental normal tissue, and compartmental boundaries. Although marginal, wide, and radical margins may all be tumor-free, the residual reactive tissue at a marginal margin often contains extensions or satellites, and the residual normal intracompartmental tissue beyond a wide margin occasionally contains skip lesions. For high-grade sarcomas only a radical margin with an intact barrier of normal tissue between the margin and the reactive zone consistently and reliably can be called tumor-free.

Determinations of margins may be estimated by inspection of the cut surface of either bone or soft tissue. Tetracycline-labeling may be quite helpful in visually identifying the type of osseous margin as it distinguishes reactive from normal bone. Often specimens will have to be taken for histologic study from questionable areas to verify whether non-neoplastic tissue at a margin is reactive or normal. The microscopic appearance of wide and radical margins is histologically identical (ie, normal) and the distinction as to the type of margin obtained is made by identifying whether

Table VI. Musculoskeletal oncologic surgical procedures

Margin	How margin achieved	
	Limb-salvage	Amputation
Intracapsular	Intracapsular piecemeal excision	Intracapsular amputation
Marginal	Marginal en bloc excision	Marginal amputation
Wide	Wide en bloc excision	Wide through-bone amputation
Radical	Radical en bloc resection	Radical exarticulation

or not the margin is beyond a compartmental barrier. This is usually done by gross inspection or radiographic examination of the specimen. As shown in Table VI, each of the four margins can be achieved by a local or limb-salvaging procedure or an amputation making eight possible oncologic procedures.

The four types of limb-salvaging procedures are: 1) intracapsular excision, ie, debulking, cytoreductive excision, etc, done piecemeal within the (pseudo)capsule; 2) marginal (local) excision, ie, en bloc excisional biopsy, shell-out, etc, done en bloc extracapsularly within the reactive zone; 3) wide (local) excision, ie, en bloc excision done through normal tissue beyond the reactive zone, but within the compartment of origin leaving *in situ* some portion of that compartment; or 4) radical (local) resection, ie, en bloc excision of the lesion and the entire compartment of origin leaving no remnant of the compartment of origin.

The terms excision and resection are coupled with wide and radical to emphasize the biologic differences between the two procedures. Wide excision and radical resection are correct; by definition wide resection or radical excision become incompatible terms. This is important conceptually, if not semantically, because in Europe, the term "radical" in terms of margin is synonymous with tumor free and can in the above terms be either marginal, wide, or radical. Therefore, in the European literature, excision and resection are used interchangeably with "radical" taken to mean any local procedure with a tumor-free margin.

The other four types of oncologic procedures are amputations that achieve various margins; 5) an amputation whose level passes within the (pseudo)-capsule is an intracapsular amputation; 6) through the reactive zone is a marginal amputation; 7) through normal tissue proximal to the reactive zone, but with the compartment of involvement (usually a through-bone) is a wide amputation, and 8) proximal to the involved compartment (usually a disarticulation) is a radical amputation, as it removes the entire compartment at risk.

Table VII. Articulation of benign stages with surgical margins

Stage	Grade	Site	Metastases	Margin for control
1	G_0	T_0	M_0	Intracapsular
2	G_0	T_0	M_0	Marginal or ? intracapsular plus effective adjuvant
3	G_0	T_{1-2}	M_{0-1}	Wide or ? marginal plus effective adjuvant

Table VIII. Articulation of malignant stages with surgical margins

Stage	Grade	Site	Metastases	Margin for control
I_A	G_1	T_1	M_0	Wide - usually excision
I_B	G_1	T_2	M_0	Wide - consider amputation vs joint or neurovascular deficit
II_A	G_2	T_1	M_0	Radical - usually resection or wide excision plus effective adjuvant
II_B	G_2	T_2	M_0	Radical - consider exarticulation or wide excision or amputation plus effective adjuvant
III_A	G_{1-2}	T_1	M_1	Thoracotomy - radical resection or palliative
III_B	G_{1-2}	T_2	M_1	Thoracotomy - radical exarticulation or palliative

In terms of these definitions, articulation of the stages with the surgical margins and procedures can be done with anatomic and biologic meaning rather than in less significant physical dimensions. The articulation for benign lesions is shown in Table VII, and the articulation for malignant lesions is summarized in Table VIII.

Benign lesions

Stage 1 latent lesions have a negligible recurrence rate following intracapsular excision (ie, curettage, piecemeal removal) as their natural history is to heal spontaneously. Examples would be a ganglion, nonossifying fibroma, solitary eosinophilic granuloma, simple cyst, giant cell tumor of tendon sheath, etc, whose behavior was latent. En bloc marginal or even wide excision, if feasible, without additional morbidity or disability, while desirable for insurance, is unnecessary to achieve a low-risk margin for Stage 1 lesions.

Stage 2 active lesions have a significant recurrence rate after intracapsular procedures and negligible recurrence rates after marginal en bloc excision. Since active lesions are by definition intracapsular (T_0), dissection through the extracapsular reactive zone carries little risk of leaving residual neoplastic tissue with a marginal margin. When obtaining a marginal margin by an en bloc excision carries a significant risk of morbidity or disability, then either these risks or the risks or recurrence after an intracapsular procedure must be assumed or consideration be given to extending the margin of an intracapsular procedure to the equivalent of a marginal (or better) margin by the use of nonsurgical adjuvants. Since both chemotherapy and radiation therapy are effective by virtue of their effect on mitotically active cells, their effectiveness on benign G_0 lesions is limited, and the side effects make their use inappropriate. Physical adjuvants such as phenol, hypertonic saline, merthiolate, methylmethacrylate, and repeated freezing and thawing (cryosurgery) have all had trials and advocates. Only cementation with thermal (and perhaps chemotoxic necrosis) and cryosurgery producing cell rupture have been documented to produce significant extensions of surgical margins (Feith, 1975; Gage et al, 1966). Both have been shown to produce millimeters of necrosis, and when used appropriately and judiciously can reliably extend an intracapsular margin to become the equivalent of a marginal margin with significant reduction in recurrence rates for active Stage 2 benign lesions.

Stage 3 aggressive benign lesions with their extracapsular extensions (T_1 or even T_2) have high recurrence rates after either intracapsular or marginal procedures. Wide surgical margins beyond the extensions in the reactive zone are the least that provide a low-risk procedure for these aggressive lesions. Marginal procedures coupled with effective adjuvants may be clinically prudent when location makes wide surgical procedures impractical. Thermotherapy has been ineffective, albeit largely untried; but radiation therapy, when coupled with marginal excision, has considerably reduced the risk of recurrence. Surgically inaccessible lesions such as aggressive fibromatosis, aggressive recurrent aneurysmal bone cysts, and aggressive giant cell tumors are in this category. The effectiveness of physical adjuvant for Stage 3 lesions has not been as extensively investigated as for Stage 2 lesions, but may be rational in lieu of disabling wide procedures. Despite their appeal, to date no combination of adjuvant/marginal surgery has been as effective as wide surgery. When prior recurrences after inadequate excisions have occurred, the pattern of local dissemination is often so diffuse that only amputation offers a practical method of achieving a low-risk wide margin.

Malignant lesions

Stage IA - Low-grade G_1 locally invasive tumors with predilection for occult extracapsular satellites in the enveloping reactive zone have high recurrence rates after either intracapsular or marginal procedures. Wide procedures are low-risk and because of their intracompartmental location, Stage IA lesions are usually excellent candidates for excision rather than amputation. Marginal excision plus an effective adjuvant is less risky than marginal excision alone, but identification of an effective adjuvant is difficult. Adjuvant radiation therapy has been effective in reducing the incidence of recurrence after marginal excision in soft tissue lesions, but of limited value in skeletal lesions. Chemotherapy has been ineffective in G_1 sarcomas. Physical adjuvants are largely untested, but as their effectiveness is difficult to predict and usually measured in millimeters, they would not seem rational for G_1 sarcomas that have a low incidence of recurrence after wide procedures.

Stage IB - Extracompartmental G_1 lesions require the same margins as their intracompartmental counterparts. The extracompartmental location often makes a wide margin unattainable by anything short of amputation or wide excision with sacrifice of significant neurovascular or articular structures. Marginal surgery coupled with an adjuvant is no more effective in IB than in IA lesions. When the extracompartmental status has occurred as a result of recurrences after prior inadequate excision, achieving a wide margin in the face of widely disseminated occult disease by limb-salvaging excision becomes less and less practical, and amputation more and more a serious consideration.

Stage IIA - High-grade, high-risk destructive lesions are seldom intracompartmental and have a significant incidence of skips. Low-risk control is offered by either radical resection (frequently practical in the unusual circumstance of intracompartmental confinement), radical amputation, or by the combination of a wide margin and an effective adjuvant(s). Soft tissue sarcomas are more often still intracompartmental and radiation therapy is an effective adjuvant for local control of the majority of the various histogenic types (exception: chemotherapy is more effective in rhabdomyosarcoma) of soft tissue sarcomas. Chemotherapy may be effective in assisting local control in some skeletal lesions, ie, osteosarcoma and perhaps malignant fibrous histiocytoma; while in others, ie, chondrosarcoma and fibrosarcoma, it has little benefit. It is quite clear, however, that all combinations of wide surgery with adjuvant therapy carry a significantly greater risk of local recurrence than a radical procedure (approx. 25% versus < 5%), whether achieved by limb salvage or amputation.

Stage IIB - Radical margins are the most effective way of assuring local control and in IIB lesions are often attainable only by disarticulation. Wide margins, coupled with adjuvant therapy, have significant risks of recurrence in their own right and often achieve only marginal margins because of the proclivity of these lesions to have occult proximal microextensions along the major neurovascular bundles. The risk of local recurrence in Stage IIB sarcomas treated by wide excision without adjuvant therapy is 40-60%, and with adjuvant therapy in responsive lesions is 20%. When the wide margin is achieved by through-bone amputation, the risk of recurrence with effective adjuvant is less than with wide excision -- approximately 10% rather than 20%.

Stage III - Control of the disease requires both the appropriate surgical management of the primary plus control of the pulmonary or other distant metastases. When the appropriate wide or radical procedure entails significant morbidity or disability, a lesser palliative procedure may be rational unless control of the metastasis can be reasonably anticipated.

Both preliminary aggressive chemotherapy (with decision as to subsequent thoracotomy) and definitive surgical treatment of the primary tumor -- based on the response of the metastases to the chemotherapy or, alternatively, aggressive thoracotomy and definitive radical surgery followed by postoperative adjuvant chemotherapy in the clinically tumor-free state -- are recommended by some authors, although long-term results are not yet known. Either procedure appears preferable to palliation alone as each has significant salvage rates at five years.

Discussion

In its preliminary trials by the Musculoskeletal Tumor Society, this staging system was shown to be practical, reproducible, and of significant prognostic value for sarcomas of both bone and soft tissue origin (Enneking et al, 1980). Subsequent reports have shown its value in surgical planning and treatment evaluation (Bononi et al, 1983; Eriksson et al, 1980; Gherlinzoni et al, 1983; Gitelis et al, 1981; Sim, 1983).

Since the original presentation of the definitions of surgical margins and procedures in 1980, some misperceptions need clarification. The common ones are how to describe the margins (and procedures) about superficial lesions, extracompartmental lesions, and lesions that are inadvertently entered but subsequently reexcised.

A superficial lesion in the skin and/or subcutaneous tissue that has not penetrated the deep fascia is intracompartmental. En bloc removal with a plane of dissection deep to the deep fascia and through normal tissue well around the lesion obtains an extracompartmental radical margin depth (on the other side of the deep fascia -- a natural barrier), but only a wide margin circumferentially (there are no natural barriers within skin and subcutaneous tissue, and so an extracompartmental radical margin in the defined sense is not possible). This ambiguity has been resolved by arbitrarily calling a margin less than 5 cm about the reactive zone wide, and a margin more than 5 cm radical. This dimension was chosen in conformity with the melanoma experience. Thus, a superficial IA lesion excised en bloc deep to the deep fascia with a surrounding margin of 2 cm has been widely excised, while the same lesion with a 6 cm margin about it has been radically resected. Whether or not these physical dimensions are appropriate for the articulation of margin and stage remains to be seen.

Extracompartmental "B" lesions -- whether by extension or origin -- by definition cannot be radically resected since the extracompartmental spaces and planes have no longitudinal barriers. For such lesions, a local procedure that removes en bloc a lesion with a margin of normal tissue is a wide excision for an en bloc. Procedure of an extracompartmental lesion that is beyond natural barriers in the transverse plane but by definition cannot be radical in the longitudinal sense is arbitrarily defined as a radical resection when the longitudinal margin is at the same level as the origin or insertion of the adjacent muscles. For example, a lesion in the subsartorial canal abutting the femoral neurovascular bundle that was removed en bloc, including the bundle with a plane of dissection beyond the fascial boundaries of the canal (ie, radical transversely), but with a proximal and distal margin less than the musculotendinous junctions of the sartorius would have been widely excised. The same procedure with the proximal and distal margins at or beyond the musculotendinous junctions of sartorius would be a radical resection. If the lesion were removed en bloc by dissection within the canal, sacrificing the bundle, the procedure would be a marginal excision. If the lesion were dissected away from the bundle, preserving the bundle, the procedure would be designated as either an intracapsular or marginal excision depending on whether the dissection was within the (pseudo)-capsule or extracapsular reactive zone.

If a lesion involves two compartments, ie, a lesion arising in bone extending into the adjacent soft tissues, then, to achieve a radical margin, both compartments would have to be removed en bloc in toto. For example, to achieve a radical resection of a lesion of the distal femoral metaphysis extending into the posterior thigh would require removal of the entire femur, hamstrings, and sciatic nerve en bloc. From the above, it is evident that in certain instances the only practical way of achieving a radical margin is by amputation. This may be particularly true in certain anatomic sites (ie, popliteal fossae, femoral triangle, axillae, antecubital fossae, flexor canal of the forearm) where a radical margin can be obtained by resection, but the virtually functionless salvaged limb hinders rather than aids rehabilitation.

When lesions are entered, the wound is being contaminated and all the tissues exposed are at risk for recurrence. If these at risk tissues are not removed, the margin is intracapsular. If the tissues are removed, the margin becomes what the subsequent removal achieves in relation to the lesion as if the exposure had not happened. The procedure is said to be a contaminated procedure. For example, if a lesion in the quadriceps were inadvertently entered exposing the rectus femoris muscle and the lesion was subsequently widely excised with a cuff of normal tissue, the procedure would be designated a contaminated wide excision since some of the exposed more proximal rectus femoris would remain in the wound.

If, under the same circumstance, the entire quadriceps compartment were removed en bloc by extracompartmental dissection, then the procedure would be a (uncontaminated) radical resection. This means that after incisional biopsy, the entire tract at risk must be appropriately excised en bloc with the lesion and tissues to achieve wide or radical margins. It also means that if the (pseudo)capsule is inadvertently entered during attempted excisional biopsy, a great deal more tissue will have to be removed to achieve an uncontaminated wide or radical margin than if such contamination had not occurred. In certain instances, contamination may take place in such a way that the only way of achieving an uncontaminated wide or radical margin is by amputation (previously unnecessary to achieve an uncontaminated wide or radical margin), and in other circumstances (eg, the pelvis) inadvertent contamination may make obtaining an uncontaminated margin of any kind impossible.

It is evident that continuous refinement and classification of these terms, definitions, and concepts are needed for them to be of optimal value.

A serious consideration of stratification of Stage III is also in order. It is becoming clearer that the prognosis of a patient who develops a solitary pulmonary metastasis from a G_0 or G_1 primary tumor some years following local control is significantly different than multiple metastases from a G_2 lesion at the time of presentation or shortly after apparent local control of the primary lesion (Sim, 1983). It may well be that meaningful stratifications will offer guidelines for the management of these lesions.

The final objective of this staging system -- development of guidelines for adjunctive therapy -- has yet to be realized. The effectiveness of adjuvant therapy continues to be judged by survival rates of various histogenic types of sarcomas largely ignoring the influence of the stage, surgical margin, or adequacy of the surgical procedure on survival rates. This lamentable state is exemplified by the fact that one decade after the enthusiastic widespread adoption of prophylactic chemotherapy, serious doubt continues whether the increase in survival rates during this period is the result of adjuvant chemotherapy or improvement in staging techniques with resultant improvements in surgical control of the primary tumor. In the light of this, it would seem obvious that data concerning staging, surgical margins, and surgical procedures must be gathered to establish significant variables in the assessment of the current explosive proliferation of protocols for adjunctive management of musculoskeletal lesions.

References

1. Bononi, S., Bacchini, P., Bertoni, F. and Campanacci, M.: Periosteal chondroma. J. Bone Joint Surg. 65(A), 205, 1983.

2. Broders, A.C., Hargrave, R. and Meyerding, H.W.: Pathologic features of soft tissue fibrosarcoma. Surg. Gyn. Obstet. 69, 267, 1939.

3. Enneking, W.F., Spanier, S.S., and Goodman, M.A.: A system for the surgical staging of musculoskeletal sarcoma. Clin. Orthop. Rel. Res. 153, 106, 1980.

4. Enneking, W.F. and Kagan, A.: "Skip" metastases in osteosarcoma. Cancer 36, 2192, 1975.

5. Eriksson, A.I., Schiller, A. and Mankin, H.J.: The management of chondrosarcoma of bone. Clin. Orthop. Rel. Res. 153, 44, 1980.

6. Feith, R.: Side effects of acrylic cement implanted into bone. Acta Orthop. Scand. Suppl. 161, 1975.

7. Gage, A.A., Green, G.W., Neiders, M.E. and Emmings, F.G.: Freezing bone without excision - an experimental study of bone cell destruction and manner of regrowth in dogs. JAMA 196, 770, 1966.

8. Gherlinzoni, F., Rock, M. and Picci, P.: Chondromyxofibroma. J. Bone Joint Surg. 65(A), 198, 1983.

9. Gitelis, S., Bertoni, F., Picci, P., and Campanacci, M.: Chondrosarcoma of bone. J. Bone Joint Surg. 63(A), 1248, 1981.

10. Kreicbergs, A., Boquist, L. and Larsson, S.E.: Prognostic factors in chondrosarcoma. Cancer 50, 577, 1982.

11. Lodwick, G.S., Wilson, A.J., Farrell, C., Virtamma, P. and Dittrich, F.: Determining growth rates of focal lesions of bone from radiographs. Radiology 134, 577, 1980.

12. Sim, F.H.: Principles of surgical treatment. In, Diagnosis and Treatment of Bone Tumors - A Teaching Approach. 1983.

Classification of Bone Tumors

V. L. Fornasier

Bone tumors are rare. They make up less than 1% of all solid tumors, but are the most important single group of tumors in under 20-year old patients. Despite their rarity, they are frequently considered in the differential diagnosis of radiologically detected bone abnormalities and hold an irresistible fascination for those who deal with them. Tumors of bone may arise from the various mesenchymal elements present in and around bone, and have been assumed to recapulate the biologic and histomorphologic characteristics of the tissue of origin of the tumor cells. This histogenetic concept is the basis of all classifications of bone tumors. However, our incomplete knowledge of the cell of origin of several bone tumors (eg, Ewing's sarcoma) has resulted in confusion and difficulty.

Historically, Virchow made the first attempt at a classification of bone tumors in 1865. He divided bone sarcomata into three categories: the spindle-celled, the round-celled and the giant-celled types. Individual categories of bone tumors were really not categorized until chondrosarcoma was identified as a specific entity by Phemister in 1930. The giant cell tumor of bone was first separated from other lesions with giant cells by Coley and Higinbotham in 1938. They were also the first investigators to arbitrarily divide giant cell tumors of bone into a benign and malignant category in accordance with the clinical behavior exhibited by their patients. Parker and Jackson introduced a classification of malignant lymphomata in 1939. They were the first to identify a category of primary lymphoma of bone distinct from other malignant lymphomas and from Ewing's sarcoma.

In 1939, Ewing produced one of the first classifications of bone tumors. Based on his experience with 2,000 patients, he applied the histogenetic concept and produced a categorization of bone tumors which has been the basis of most classifications still in use today. This includes the classifications published by Geschickter and Copeland (1949), Johnson (1953), Lichtenstein (1952), Jaffe (1958) and the histologic classification of bone tumors by the World Health Organization (1972). In more recent years, classifications have been modified and updated in accordance with the evolution of newer knowledge and understanding of the cell types involved. This has resulted in a number of modifications based in part on the pattern of cellular differentiation of the neoplasm and in part on the experience of the author. While none of these represents entirely accurate rates of incidence across the population at large, they are valid analyses of the cases that had been studied within a centre, and provide a basis for com-

Current Concepts of Diagnosis and Treatment
of Bone and Soft Tissue Tumors.

parison between involved centres and interested individuals. Because of the difference in the experience of the centres publishing classifications based on their own experience, difficulties in terminology and statistical analyses are inevitable.

While the histogenesis of bone tumors has been and will continue to be the basis for classification, it is difficult to accept the statement that tumors arise from the type of tissue which they produce (eg, that bone tumors arise from bone, that cartilage tumors arise from cartilage, etc). How can primary chondrosarcoma of bone that usually arises in mature bone devoid of cartilage have arisen from cartilage? On occasion, neoplasms show a variety of differentiation (eg, classical mesenchymoma) and it is difficult to envisage each element arising from cells with different differentiation, while it is more probable that the differentiation expressed by a tumor may at least in part be the result of the combination of stimuli that lead to the oncogenetic transition that may permit the expression of pluripotentiality of differentiation within one neoplasm.

Any classification must assume that a similar histologic pattern of differentiation indicates a similar biologic potential. While newer knowledge may help us to further subclassify each existing category, it is important to the clinical management of patients and in the assessment of the efficacy of therapy, and thus to the prognosis of the tumors, that a classification be adopted that is accessible and easily understood by the user. In addition to the pattern of differentiation, bone tumors show a range of degrees of differentiation that in each category may extend from well differentiated and easily recognizable histomorphologic features to undifferentiated or purely mesenchymal patterns. For this reason, in addition to classification, the concept of grading and staging must be applied to the tumors in order that the implications of a tumor, particularly a malignant one, to an individual patient be more clearly defined, and that clinical results be more efficiently and reliably recorded.

From the therapeutic point of view, the simplest classification is the best and most successful. This same classification, however, may not necessarily be accurate in categorizing the biologic evolution and potential of a neoplastic cell at the purely biologic level. It is important, therefore, that these two separate and distinct approaches be retained, on the one hand to admit our ignorance and limitations while we treat patients, and on the other hand to permit continuing analysis for the improvement of our understanding of the pertinent cell biology.

In this era of easy international communication, the differences in terminology have begun to diminish. Referral patterns and geographic location may be significant factors in the recording of the incidence of individual bone tumors. This is illustrated in comparing published experiences. For example, Schajowicz (1981) reported a 10% incidence of multiple myeloma in 4,193 patients while Dahlin (1978) reported a 36% incidence of multiple myeloma in 6,221 patients. In contrast, in the same publication Dahlin has a far lower incidence of osteochondroma, enchondroma and giant cell tumor than Schajowicz.

The classification presented in Table I is a modification of that of Dahlin, incorporating some of the aspects of the classification presented in the

Table I. Bone tumor classification

Differentiation	Benign	Malignant
Bone	Ivory osteoma Medullary osteoma (bone island) Juxtacortical osteoma Osteoid osteoma Osteoblastoma	Osteosarcoma: primary - osteoblastic - chondroblastic -fibroblastic telangiectatic multicentric low grade central of jaws secondary - in Paget's - postradiation - benign lesion parosteal periosteal high grade cortical
Cartilage	Enchondroma - solitary or multiple Chondroblastoma Chondromyxoid fibroma Subperiosteal condroma Osteochondroma - solitary or multiple	Chondrosarcoma: primary ("differentiated") mesenchymal clear cell secondary to osteochondrom or enchondroma parosteal dedifferentiated
Fibrous tissue	(Fibrous cortical defect) Fibroma Desmoplastic fibroma Nonossifying fibroma (fibrous histiocytoma) Ossifying fibroma	Fibrosarcoma: Malignant fibrous histio cytoma - fibroblastic - histiocytic - pleomorphic
Blood vessels	(Telangiectasia) Angioma - cavernous or - capillary Hemangiopericytoma "Histiocytoid angiomatoses" (angiofollicular hyper-plasia with eosinophilia or Kimura's disease)	Angiosarcoma Hemangiopericytoma (Kaposi's sarcoma)
Fat	Lipoma (with or without other elements)	Liposarcoma
Nerve	Neurilemmoma (Schawannoma) Neurofibroma (Von Recklinghausen's)	Neurosarcoma
"Notochord"	(Rests)	Chordoma
Undetermined	Giant cell tumor Aneurysmal bone cyst Adamantinoma Mesenchymoma	Ewing's sarcoma Malignant mesenchymoma
Metastatic		From other bones From extraosseous tissues

Histologic Classification of Bone Tumors by the World Health Organization. While neoplasms are still generally categorized as either benign or malignant, one fully realizes that in many cases such a narrow classification is neither possible nor desirable for it does not necessarily reflect the biologic potential of a neoplasm.

There are still a number of difficult areas in any classification of bone tumors. Nomenclature has now been overcome as a major difficulty because of the greater communication between national and international centres, and a better understanding resulting from the more uniform use of terms by the various medical disciplines. It must be remembered that in our so-called developed world with the ease of access to good medical care and early treatment of bone tumors, the natural history is routinely modified since the patient is usually treated as soon as a lesion is identified. This may provide some difficulty in relating current experience with that published some years ago, and also from publications arising from less fortunate areas around the world where medical care is not easily available, and patients present in a far more advanced stage of their disease. Thus, we must not be confused by the variance from the expected or predicted potential of a specific neoplasm due only to the more advanced stage in the natural history of the disease.

Another difficulty is the apparent or real change in the biologic behavior of a neoplasm. This is apparent when a lesion, particularly a malignant lesion, is inadequately sampled and a diagnosis of a benign lesion is made. The change, especially from benign to malignant behavior, may occur as the result of therapy, in particular inappropriate irradiation. When inappropriate and unrepresentative biopsies lead to incorrect interpretation, inappropriate treatment may be given resulting in treatment failure. This sampling difficulty has been one of the major objections by many pathologists to the use of needle biopsies for the diagnosis of bone tumors.

A major stumbling block in the biologic assessment or grading of bone tumors has been the lack of reproducable histologic criteria for the assessment of the biologic behavior. This is best known in the case of giant cell tumors of bone. Mitotic activity, cellularity, presence of spindle cells, presence of necrosis, presence of hemorrhage, presence of cyst formation and even evidence of vascular intrusion within the tumor are all features that have been focused on with inconsistent and nonreproducable results in the prediction of the behavior of this neoplasm. Often, persistence of a lesion seems more related to inadequacy of removal rather than to an inherent malignant potential in the tumor.

Finally, since the majority of classifications is based almost exclusively on the histogenetic origin of the tumor cells, it is important that the histologic diagnosis be assessed together with the roentgenographic, clinical and laboratory investigation of the patient. It is crucial that all aspects of the investigation of a patient be consistent with each other before a diagnosis is assumed and therapy instituted. The planning of the biopsy with a view to obtaining the best preserved, viable and most representative portion of the lesion for interpretation cannot be overemphasized.

It is an unfortunate distortion of interdisciplinary cooperation that clinicians compose exhaustively comprehensive tomes, radiologists give nebulographic dissertations, the clinical laboratory ejects mysterious numerical hieroglyphs and that only the pathologist must provide the diagnosis.

References

1. Dahlin, D.: Bone Tumors, 3rd Edition. Chas. C. Thomas, Springfield, 1978.

2. Schajowitz, F.: Tumorlike Lesions of Bone and Joints. Springer-Verlag , Heidelberg, 1981.

3. World Health Organization: Histologic typing of bone tumors. International Histologic Classification of Tumors No. 6, 1972.

Grading of Bone Tumors

T. D. BROWER

Predicting the future has been the desire of man from the beginning of time. The experiences of the past are the only means with which we can forecast the future and such is the case with bone tumors. This attempt at predicting behavior is called grading. Several sources of information are available that allow us to predict with fair accuracy the course of bone tumors.

During the first several decades of microscopic pathology, observers spent almost all their efforts at classifying disease. In the 1920's, attempts were made to grade the degree of malignancy by certain histologic observations, and Broders' classification (Broders et al, 1939) was one of the giant steps toward developing an accepted grading system. Broders used the following histologic observations to assess the degree of malignancy of fibrosarcomas:

- degree of cellularity
- cellular anaplasia or pleomorphism
- mitotic activity
- evidence of invasion
- presence of necrosis

Most histologic grading of tumors today is based on Broders' classification with some variations.

For some bone tumors, such as classic osteosarcoma, histologic grading is not very helpful because such tumors fall into Grades III or IV, which means a bad or hopeless prognosis. Low grade osteosarcoma occurs rarely, but osteosarcoma presents other clinical features that assist us in predicting its behavior, such as the position of the tumor in relation to the intramedullary canal. Unni et al (1976a) have observed the rare incidence of an osteosarcoma occurring apparently beneath the periosteum. Although microscopically these appear to be classic chondroblastic osteosarcomas, their prognosis is far better than the classic intramedullary tumor.

Should the bone-forming neoplasm be entirely extracortical, the prognosis is even better. This tumor is known as parosteal osteosarcoma (Unni et al, 1976b). It is very dense, slow growing, and usually not painful. These clinical signs indicate a low-grade malignancy and, indeed, the histopathologic grading supports this opinion. The prognosis of this tumor is quite good after surgery.

Current Concepts of Diagnosis and Treatment
of Bone and Soft Tissue Tumors.

Another clinical observation assisting in the grading of the degree of malignancy of an osteosarcoma is its position in the skeleton. Osteosarcoma arising close to the axial skeleton, such as in the scapula, proximal humerus, proximal femur or pelvis, are fatal. Those occurring about the knee have a 20 to 25% five-year survival rate, while those rare tumors occurring in the distal tibia have a 40 to 50% survival rate. These prognoses are based on cases before the era of chemotherapy. Histopathologic findings alone do not predict such an outcome.

Chondrosarcoma, described by Phemister (1930), is the other major malignant primary bone tumor. He recognized this tumor to be clinically and histopathologically different from osteosarcoma. It occurs mostly in adults. Its clinical course is not as aggressive, and the survival rate, after complete excision, is far better than that with osteosarcoma. Histologic examination shows a totally cartilaginous stroma without formation of tumor bone or osteoid.

The histologic grading of chondrosarcoma is extremely difficult. Lichtenstein and Jaffe (1943) gave us the following criteria for assessing malignancy:

1) many cells with plump nuclei,
2) more than an occasional cell with two large nuclei,
3) giant cartilage cells with large single or multiple nuclei or with clumps of chromatin.

Although these criteria are specific, their application is fraught with difficulty. The presence of these histologic hallmarks is not diagnostic of a malignant cartilaginous tumor; for example, enchondromas of the hands can present areas of similar cytologic changes. Conversely, pelvic chondrosarcomas may appear histologically rather benign, yet be resistant to attempts at excision and they may later metastasize. The grading of cartilaginous tumors, therefore, depends on the clinical course, location of the tumor and, to some degree, the histologic profile.

Most chondrosarcomas fall into grade I and II, correlating with the clinical course, for the majority of patients do well who have had their tumors completely excised.

Some of the pitfalls in the diagnosis of chondrosarcoma can be shown by a few illustrations.

The first concerns a 38-year old farmer. He complained of spontaneous onset of pain in his right midarm 18 months before he was first seen in 1966. The pain occurred at night and was deep within the arm; it did not radiate and was unrelated to activity. Physical examination revealed a slight enlargement of the midshaft of the humerus. Roentgenogram revealed a radiolucent lesion within the midshaft which had caused a slow expansion of the cortex of the humerus. Within the lesion were a few flecks of radiopacity. Since the patient demanded immediate return to normal heavy work, this lesion was treated by curettage. The pathologic diagnosis was chondrosarcoma Grade I. Roentgenograms 14 years later show no significant change in the architecture of the humerus and the patient is asymptomatic. This is an example of a low-grade chondrosarcoma that, despite inadequate treatment, gave a good result.

A 17-year old girl noted swelling near her wrist and had some discomfort for six months. The only physical finding was an enlargement of the distal right radius. Roentgenographic examination revealed a lytic lesion enlarging the distal radius. The lesion was removed by curettage and the defect filled with bone chips. The pathologic examination revealed a Grade II chondrosarcoma. The defect healed well and the patient returned to normal activity one year after the surgery. Eighteen months postoperatively, she was seen because her wrist pain had returned. Roentgenographic examination revealed a recurrence or, more accurately, a persistence of the original lesion. The surgical wound and the distal one-half of the radius were resected en bloc. Pathologic examination revealed a nodule of chondrosarcoma Grade II in the intrameduallary canal of the radius. The patient will likely require an upper extremity amputation. This is an example of inadequate treatment resulting in a recurrence.

The third case is a 70-year old woman who had left thigh pain for two years. Although she had been treated for arthritis, a belated roentgenographic examination revealed a radiolucent lesion within the medullary canal that had enlarged the femur. Within the lesion were flecks of calcification. A hip disarticulation was done, and the pathologic diagnosis was chondrosarcoma Grade II. She died of metastasis two years later. This case shows a histologic low-grade chondrosarcoma occurring toward the axial skeleton which followed a highly malignant course.

The last primary bone tumor I wish to discuss is giant cell tumor. Although most of these tumors are considered benign, there is a growing tendency for surgeons to treat them aggressively for this tumor has a disturbingly high recurrence rate. About 10% metastasize to the lungs. Jaffe and Lichtenstein (1940) developed a grading system for this type of tumor. A small number, about 10%, are obviously malignant, histologically Grade III, and should be radically removed. About 10% are Grade I and have roentgenographic evidence of a reactive border. These tumors respond well to curettage. About 80% are Grade II and follow a rather unpredictable course.

Summary

Bone tumors can be graded with some degree of accuracy, but several factors must be considered to obtain a successful grading system:

1) The clinical behavior of the tumor gives a rough but helpful indication of the degree of malignancy. Obviously, rapidly-growing tumors are more malignant than slow-growing tumors.

2) Tumors within or near the axial skeleton are more malignant than distal tumors.

3) Tumors within the medullary canal tend to be more malignant than extramedullary ones.

4) Histologically, most osteosarcomas are Grade III or IV, and the majority of chondrosarcomas are Grade I and II.

References

1. Broders, A.C., Hargrave, R., and Meyerding, H.W.: Pathological features of soft tissue fibrosarcoma - With special reference to the grading of its malignancy. Surg. Gynec. Obstet. 69, 267, 1939.

2. Jaffe, Henry L., Lichenstein, Louis and Portis, Robert B.: Giant cell tumor of bone. Arch. Pathol. 30, 993, 1940.

3. Lichtenstein, Louis and Jaffe, Henry L.: Chondrosarcoma of bone. Am. J. Pathol. 19, 553, 1943.

4. Phemister, Dallas B.: Chondrosarcoma of bone. Surg. Gynec. Obstet. 50, 216, 1930.

5. Unni, Krishnah K., Dahlin, David C., and Beabout, John W.: Periosteal osteogenic sarcoma. Cancer 37, 2476, 1976a.

6. Unni, Krishnah K., Dahlin, David C., Beabout, John W., and Ivins, John C.: Parosteal osteogenic sarcoma. Cancer 37, 2466, 1976b.

Biopsy of Neoplasms of Bone and Soft Tissues: A Rational Approach

H. J. MANKIN

Introduction

All protocols for the staging of primary malignant bone and soft tissue tumors require an accurate histologic diagnosis and grading of the lesion (American Joint Committee, 1977; Enneking et al, 1980). This latter information is always provided by a segment of tissue obtained by biopsy. Although this essential procedure has been demonstrated to be prone to certain hazards, discussion of biopsy procedures in standard textbooks and in the literature generally provides only limited information and directives (Brostrom et al, 1979; Jaffe, 1958; Moore et al, 1979; Schajowicz, 1981; Schajowicz and Derqui, 1968; Spjut et al, 1971; Wood and Binder, 1978).

Surgeons who frequently treat patients with malignant tumors of bone and soft tissues quickly recognize the possible complications which may result from a biopsy. Experience has taught them that such complications are more frequently encountered than is commonly known. In a recent article published by the Musculoskeletal Tumor Society (MSTS) (Mankin et al, 1983), a study was reported assessing the hazards associated with the biopsy procedure. A summary of the results of that study may offer some suggestions on how to avoid the hazards inherent in biopsy procedures.

The MSTS Study

Materials and methods

Each member of the Musculoskeletal Tumor Society was asked to complete a questionnaire (Table I) on 20 sequentially unselected new patients treated before January 1, 1979 with a definitive diagnosis of malignant primary bone or soft tissue tumor. Moreover, the patients to be included were those in whom a biopsy and subsequent definitive surgery were performed, and who were then followed long enough to determine early outcome (Mankin et al, 1983). Pathology reports for both the original biopsy and subsequent definitive surgery were appended and the surgeon was also asked to explain how the outcome was affected by problems related to the biopsy.

A total of 329 patients fulfilled the criteria for the study of whom 173 were men (52.6%) and 156 women (47.4%). The mean age was 36.5 years (range 2 weeks to 83 years). The series comprised 222 primary bone tumors and 107 soft tissue tumors.

Current Concepts of Diagnosis and Treatment of Bone and Soft Tissue Tumors.

Table I. Musculoskeletal Tumor Society questionnaire

IDENTIFYING DATA:

Surgeon: ____________________

Case No.: ____________________ Hospital Unit No. ____________________

Name: ______________________________ Age: ________ Sex: ________

Append to this sheet copies of the pathology report for the original biopsy (whether performed in your institution or elsewhere) and for the definitive procedure (done in your institution).

Answer the following questions:

			Yes	No
A. 1. Was the biopsy representative of the tumor?			____	____
2. Was the technical preparation of the biopsy slide adequate?			____	____
3. Did either the inaccuracy or technical aspects of the biopsy alter the way in which you treated this patient?			____	____
4. If so, was the problem one of:		skin	____	____
		soft tissues	____	____
(Answer as many as applicable)		bone	____	____
		other (specify)	____	____
5. If you performed a local procedure, did the problem related to the biopsy cause you to change your local operation from: (Answer only one)				
Marginal (II)	to	Wide (III)	____	____
Wide (III)	to	Marginal (II)	____	____
Marginal or Wide (II or III)	to	Radical (IV)	____	____
Radical (IV)	to	Marginal or Wide (II or III)	____	____
6. Did the problem cause you to change your operation from a local procedure to an amputation?			____	____
B. 1. Did the (in)accuracy or technical errors of the biopsy materially alter the patient's course or ultimate outlook?			____	____
2. If so, explain:				

Table II. Distribution of tumor types

Bone tumors (n=222)		Soft tissue tumors (n=107)	
Osteosarcoma	110	Synovial sarcoma	22
Chondrosarcoma	49	Malignant fibrous histiocytoma	22
Ewing's sarcoma	15	Liposarcoma	18
Parosteal osteosarcoma	11	Fibrosarcoma	12
Fibrosarcoma	9	Rhabdomyosarcoma	5
Lymphoma	6	Unclassified sarcoma	5
Malignant fibrous histiocytoma	5	Neurofibrosarcoma	4
Chordoma	5	Soft tissue osteosarcoma	3
Hemangiosarcoma	3	Epithelioid sarcoma	3
Paget's sarcoma	2	Lymphangiosarcoma	2
Squamous cell CA in osteomyelitis	2	Clear cell sarcoma	2
Mixed sarcoma of bone	2	Hemangiosarcoma	2
Plasmacytoma	1	Hemangiopericytoma	2
Leukemia	1	Melanoma	2
Adamantinoma	1	Mesenchymal chondrosarcoma	1
		Leiomyosarcoma	1
		Mesenchymoma	1

The distribution of tumor types as shown in Table II generally conformed to the incidence of these tumors in reported series (Dahlin, 1978; Hajdu, 1979). Almost 50% of the bone tumors were osteosarcomas, 22% were chondrosarcomas and 6.7% were Ewing's tumors. The relative infrequency of the other tumors is reflected by the few examples which were reported. The data for the soft tissue tumors similarly reflected the frequency with which surgeons encounter synovial sarcoma (22%) and malignant fibrous histiocytoma (22%), liposarcoma (17%) and the now less commonly made diagnosis of fibrosarcoma (11%) (Hajdu, 1979). Some of the other soft tissue tumors were less common with only one or two seen in the series.

In 143 patients (43.5%), the initial biopsy was done in a hospital (the referring institution). These patients were subsequently transferred to a second hospital (the treating center) for definitive care. In 186 patients (56.5%), the initial biopsy procedure was performed in the treating center.

Results

Diagnostic accuracy

The diagnoses of the initial biopsies and those of the tissue obtained at subsequent surgical procedures differed in 82 of the 329 patients (25%) (Table III). The differences in the diagnoses were reviewed and arbitrarily divided into major or minor variances, depending on the likelihood that the variance would materially affect the outcome. A diagnosis was rated a major variance when it was likely to mislead the surgeon, alter the treatment plan, change the indication for adjunctive therapy or, in some way,

Table III. Diagnostic variation by tumor*

	Referring institution		Treating center	
	Variation		Variation	
	Major	Minor	Major	Minor
Osteosarcoma (110)	8	0	3	0
Parosteal osteosarcoma (11)	3	0	3	0
Soft tissue osteosarcoma (3)	3	0	0	0
Malignant fibrous histiocytoma (27)	6	2	0	3
Fibrosarcoma (21)	2	4	0	0
Synovial sarcoma (22)	2	4	0	2
Rhabdomyosarcoma (5)	0	0	1	0
Ewing's sarcoma (15)	1	0	0	0
Lymphoma (6)	1	0	1	0
Leukemia (1)	0	0	1	0
Chondrosarcoma (49)	9	0	4	1
Chordoma (5)	1	0	1	0
Mesenchymal chondrosarcoma (1)	1	0	0	0
Epithelioid sarcoma (3)	2	0	0	0
Clear cell sarcoma (2)	1	0	0	0
Mixed sarcoma of bone (2)	0	0	0	2
Liposarcoma (18)	0	1	1	0
Unclassified sarcoma (5)	1	0	0	1
Malignant mesenchymoma (1)	0	1	0	0
Melanoma (2)	1	0	0	0
Malignant vascular tumors (7)	1	1	2	0
Totals	43	13	17	9
	56		26	
	82			

* Major and minor variances: referring institutions and treating centers

interfere with appropriate treatment. Diagnostic differences unlikely to affect the patients' course were considered to be minor. Of the 82 diagnostic differences, 60 (73%) were major, of which 43 (71%) had been made in the referring institution and 17 (29%) in the treating center.

A special category of variance was that in which the initial biopsy material was considered nonrepresentative or technically unsatisfactory (ie, stained poorly; cells squashed; necrotic area only included). Thirty-four patients (10.3% of the entire series) were included in this group, and could account for the diagnostic problems in some of the lesions. It should be noted that only five of the 14 carried out in this series were needle or trochar biopsies. The accuracy rate was the same for needle as for open biopsy technique. However, patients are often preselected for needle biopsies because the presumed lesion is considered to be easily diagnosed based on a limited amount of tissue (Moore et al, 1979; Schajowicz, 1981; Schajowicz and Derqui, 1968). Twenty-eight of the 34 nonrepresentative or technically poor biopsies (82.4%) were performed in referring centers, while 6 (17.6%) occurred in the treating center. In seven patients (30.4%) the outcome was adversely affected.

Biopsy complications

Fifty-seven patients (17.3%) were reported to have complications of the biopsy procedure which were categorized according to tissue involved (skin, soft tissue, bone, combinations of these or other). Skin and soft tissue complications included wound breakdown, hemorrhage or infection. In complications involving bone, fracture was the most common (Table IV). In 30 patients (53% of the groups and 9% of the total) the complications adversely affected either treatment or outcome. Forty-four (77.2%) patients were initially treated in a referring institution and 13 (22.8%) in a treatment center.

Table IV. Biopsy procedure complications
(Total number of patients: 329)

Complication	Number
Skin alone	6
Skin and soft tissue	16
Soft tissue alone	7
Bone	4
Skin and bone	2
Skin, soft tissue and bone	5
Soft tissue and bone	2
Other* alone	7
Other*, skin and soft tissue	2
Other*, skin, soft tissue and bone	1
Total	57 (17.3%)

In these 57 patients, 12 were subjected to amputation who might have been otherwise treated; and in 18, the biopsy problems were considered to have an adverse effect on the patients' courses.

* Factors listed under "other" included: second biopsy required (2); delay and repeat biopsy (2); seeding of joint (2); needed roentgenotherapy (1); biopsy through abdomen (1); delay in therapy (2).

Effect of biopsy on treatment and outcome

From the entire group of 329 patients, 60 patients (18.2%) had a significant change of treatment or outcome due to some difficulty with the ini-

Table V. Biopsy altering treatments
(Total number of patients: 329)

Osteosarcoma (110)	9	(8.2%)
Chondrosarcoma (49)	8	(16.3%)
Malignant fibrous hystiocytoma (27)	8	(29.6%)
Fibrosarcoma (21)	5	(23.8%)
Synovial sarcoma (22)	5	(22.7%)
Chordoma (5)	3	(60%)
Malignant vascular tumor (7)	3	(42.9%)
Rhabdomyosarcoma (5)	3	(60%)
Parosteal osteosarcoma (11)	4	(36.4%)
Epithelioid sarcoma (3)	2	(66.7%)
Ewing's sarcoma (15)	2	(13.3%)
Liposarcoma (18)	2	(11.1%)
Unclassified sarcoma (5)	2	(60%)
Clear cell sarcoma (2)	1	(50%)
Lymphangiosarcoma (2)	1	(50%)
Mesenchymal chondrosarcoma (1)	1	(100%)
Neurofibrosarcoma (4)	1	(25%)
Total (329)	60	(8.2%)

45/60 (75%) were performed in referring hospitals.
15/60 (25%) were performed in treating centers.
In 24 patients (40%) the eventual outcome was adversely affected.

tial biopsy (Table V). These data are skewed because for biopsies performed in the referring institution the complication rate was 31.5% (45/143) but only 8.1% (15/186) for biopsies performed in the treating center. Altered treatment plans varied from having to perform a radical resection instead of a wide or marginal one to amputation in situations where local resection would have been possible. In fact, 15 of the 329 patients (4.5%) had unnecessary amputations. When the treating surgeons were asked whether the biopsy adversely affected the patients, they provided sufficient information for 28 of the patients (8.5% of the total series) to permit the conclusion that the prognoses and results were adversely affected by some problem directly associated with the biopsy procedure.

Comparative results between biopsies performed in the referring institution and in the treating center.

Patients whose staging biopsy was performed in the referring institution fared less well than those biopsied in the treating center (Table VI). Patients from the referring institution had more than a threefold risk of a major diagnostic error; a sixfold increase in nonrepresentative or poorly prepared biopsies; a more than fourfold frequency of problems with skin, soft tissue or bone; an almost fourfold risk of being treated differently than if the biopsy had been correct; and an almost three times greater risk of an adverse outcome. It should be clearly noted, however, that although these data are well defined in this study they do not take into account the possibility that some patients may have been transferred to the treating center from the referring institution because of biopsy problems. The possible bias introduced by this factor should be considered in interpreting these results.

Table VI. Comparison of biopsy problems between referring institutions and treating centers.

	Total series	Biopsy performed at referring institution	Biopsy performed at treating center
Total number	329	143 (43.5%)	186 (56.5%)
Major error in diagnosis	60 (18.2%)	43 (30.1%)	17 (9.1%)
Nonrepresentative or technically poor biopsy	34 (10.3%)	28 (19.6%)	6 (3.2%)
Problems of skin, soft tissue, bone or other	57 (17.3%)	44 (30.8%)	13 (7.0%)
Alteration in treatment as a result of biopsy	60 (18.2%)	45 (31.5%)	15 (8.1%)
Change in prognosis as a result of biopsy	28 (8.5%)	19 (13.3%)	9 (4.8%)

The summary data from this study show that in this group of 329 sequentially unselected patients with malignant primary bone and soft tissue tumors treated by experienced surgeons in major centers, errors in diagnoses, non-representative or poorly stained biopsy materials, and wound problems at the biopsy sites occurred in a startlingly high percentage, frequently affecting treatment and prognoses. Almost 20% of these patients had to be managed by a less than optimal treatment and at least 8% of the patients had a distinctly adverse prognosis because of biopsy problems. Fully 4.5% of those patients who might have had a limb-sparing procedure required an amputation as a result of the biopsy. Furthermore, analysis of the data strongly suggests that the risk of all these problems increases markedly when the initial biopsy is performed in a referring institution rather than in a treating center (Table VI).

Table VII. Comparison between distribution of cases biopsied in referring institutions and treating centers

	Bone			Soft tissue		
	Total	Common	Rare	Total	Common	Rare
Referring institution	88	73 (83%)	15 (17%)	55	39 (71%)	16 (29%)
Treating center	134	117 (87%)	17 (13%)	52	37 (71%)	15 (29%)

Suggested Rational Biopsy Program

On the basis of this study it is possible to formulate some rules regarding biopsy procedures which may largely reduce the biopsy hazards to a minimum. Complications and difficulties are still possible, particularly in cases of rare or unusual lesions, when the tumor site is inaccessible, or when tissue trauma results in hemorrhage, infection or fracture. With some attention to the details outlined below, however, such complications can be reduced considerably or even prevented by skillful management.

General concepts

It must be emphasized that in orthopedic oncology biopsy is the only and usually the final part of the staging process. As has been suggested by Enneking, Spanier and Goodman (1980) in their landmark text on staging of musculoskeletal tumors, a series of investigations must be carried out to define the anatomic extent of the lesions (both bone and soft tissue) and to determine whether metastases are present. Some of these procedures are such that performing them after the biopsy will decrease their value in determining either the "T" or the "M", necessary for the staging equation. Thus, angiography, bone scan, CT or lymphangiography should be performed before biopsy whether by needle or by open incision.

The purpose of the biopsy is to obtain tissue for establishing the stage of the lesion in two ways. The first is to determine a diagnosis (benign vs. malignant; tissue type, etc.) and the second to determine the grade (G_0, G_1, or G_2). To accomplish this, sufficient tissue must be obtained and, after discussion with the surgical team, the pathologist will be responsible for the final determination. Since the lesion may be nontumorous, appropriate culture and smear tests should be carried out.

The biopsy is a critical part of the surgical management of the patient and should be planned carefully. The individual who performs the biopsy should be a senior surgeon since a biopsy is not a training ground for junior house officers or interns. It should be carried out by someone with experience in the complications of such surgery.

Planning the biopsy

In planning the biopsy, the first requisite is to review all the material obtained in the staging process with the surgical team and the radiologist and pathologist who are in part responsible for establishing the tentative diagnosis. The pathologist, moreover, has to establish the definitive diagnosis and grade the lesion. This is often performed best by a "mini-conference" with these individuals as well as the oncologist and radiation therapist. They can all help to decide what sort of biopsy approach is indicated and what sort of primary and adjuvant treatment might be best for various possible diagnoses.

In planning the procedure, one should give serious consideration to the anatomic site and the consequences of violating anatomic planes. The smallest incision possible should be made, attempting to take the most direct route to the lesional area. The least amount of tissue should be contaminated by the biopsy. This can be achieved by going through a muscle rather than along fascial or intermuscular planes. If possible, the incision should be no more than a few centimeters long and should bypass sites which may cause hemorrhage, interfere with or expose major nerve trunks or vascular structures.

If an amputation is a likely outcome, then the biopsy incision should be placed where it will not imperil the flaps of the amputation or make it difficult to perform conventional amputation techniques. If a resection is likely, the biopsy incision should be placed so that it can be totally excised along with the specimen. The surgeon who performs the biopsy must avoid transverse excisions of the extremities. They are difficult to extend and limit the possibility of performing a wide resection as a subsequent surgical procedure; in fact, they may make it difficult to perform a competent amputation. With these rules in mind, one should make a small incision along the long axis of the extremity which can be fully excised and subsequently extended without undue stress on the skin and underlying soft tissues.

Biopsy technique

One of the issues currently unresolved is whether a tourniquet is desirable for the biopsy. Although definitive proof is lacking, surgeons at several centers argue against the use of the tourniquet and thus do not use this device. Their reasoning is based on its damming effect on the venous circulation. Since manipulation of any tumor has been shown to break off clumps of cells, during the course of a biopsy procedure masses of cells released by the tumor enter the venous channels and may be widely disseminated to the lung fields. The number of cells may be high but the cross-sectional diameter of the cell masses may be sufficiently small that they can pass through the lungs and be readily attacked by the immune mechanisms and/or cells of the reticuloendothelial system and destroyed. If, however, the masses are larger, they may become wedged in narrow channels within the lung field and form a site of metastatic growth. The size of the cell clumps then may contribute to the development of pulmonary metastases following manipulation of the malignant tumor. Thus, the reason not to use a tourniquet is that the adherent cells can increase massively at the level of the tourniquet. With release, these enormous masses of cells then travel to the lungs and become foci of micrometastatic growth. It should be emphasized that this postulate is anecdotal. Numerous surgeons in orthopedic oncology, although recognizing the possible validity of the concept, feel that the hemostasis and better visualization associated with the use of the tourniquet far outweigh its possible consequences.

Whether or not a tourniquet is used, the incision should be small, the tissues carefully handled, and meticulous hemostasis achieved. The dissection should be carried through muscles rather than through the fascial planes or the intermuscular septa. Bleeding should be well controlled by coagulation and if necessary ligation of vessels. Although it is sometimes necessary to retract a small wound rather excessively to arrive at the deep plane where most malignant tumors are located, it should not be done with sharp retractors but with smooth right-angle devices. Furthermore, it should be done firmly but without forces that rip or tear soft tissues.

Wherever possible, a bone biopsy should be avoided. If the lesion is confined within a bone, it is obviously necessary to enter the bone to obtain a proper biopsy but, if at all possible, the material should come from an extraosseous soft tissue mass. If such is encountered, the bone should not be violated. The principal reason is related to pathologic fracture which obviously complicates the biopsy and possibly adversely affects the results. Other reasons are the long delay required for decalcification and the distortion or squeeze artifacts which may occur with the decalcification procedure. One can usually get a much more representative section from the "leading edge" of the tumor which is almost always located in extraosseous soft tissue mass. Furthermore, bleeding from bone is difficult to control by standard hemostatic measures. If the bone must be entered, one should make drill holes to avoid causing a fracture through weakened bone. The window should be small and without sharp corners which could serve as stress risers. A small plug of polymethylmethacrylate may be introduced into the bone after removing the pathologic tissue to reduce the stress riser effect and also to prevent leakage of tumor cells and blood along the tissue planes.

Handling of the tissues obtained from the biopsy is a critical part of the procedure. First, sufficient tissue should be obtained for diagnosis. This will vary depending on the nature of the process and the comfort of the pathologist with the study of small tissue segments. A frozen section should be obtained not so much to determine the diagnosis but to assure that pathologic and representative tissue is obtained. It is also essential that the wound be cultured for anaerobic and aerobic organisms and some additional material be obtained for fungus cultures if fungi are suspected.

The tissue obtained should be sent to the pathology laboratory for routine histologic examination and special staining as deemed necessary. The pathologist should be responsible for the choice of the preservative so that special stains (such as PAS) can be used. Minute segments of tissues should be placed into glutaraldehyde fixation for possible electron microscopic examination. If a lymphoma is suspected, some tissue should be obtained and sent to the laboratory for surface markers. If metastatic carcinoma is a consideration, tissue should be placed into a special container for determination of estrogen receptors. If flow cytometry is to be used as a special technique, a block of fresh tissue obtained aseptically should be sent immediately to the laboratory in a sterile container. All these procedures should be planned before surgery so that the proper laboratories can be alerted and the surgeon, pathologist, and technicians are fully aware of the materials they will need.

Following the biopsy, most of the surgeons advocate a tight closure, in several layers, using continuous instead of interrupted sutures. Drains of any sort should be avoided.

Incisional versus excisional biopsy

Occasionally, one encounters a very small tumor accessibly placed which is logically treated by excisional biopsy as a primary procedure. This circumstance, however, is rare. Most tumors of bone and soft tissue should be treated by incisional biopsy. A frozen section should then be performed to obtain a tentative diagnosis. If the frozen section defines the lesion as clearly benign, one can then proceed to a definitive excisional biopsy. If, on the other hand, the frozen section clearly shows malignancy or is indeterminant (based on the inability of the pathologist to ascertain the nature of the lesion in a short period of study of a distorted frozen section preparation), the wound should be closed and the definitive procedure performed either at that time or later, awaiting complete staging information and full planning with the radiologist, pathologist, oncologist and radiation therapist. This issue is particularly true for soft tissue tumors where the common treatment pattern is to expose the lesion, and "shell it out". This usually results in a marginal excision which may leave not only local satellite tumor deposits but disseminate the microscopic foci throughout the wound and makes the subsequent resection or radiation therapy much more difficult. As pointed out above, incisional biopsies should be performed through very small wounds which should then be closed (if the lesion is not clearly benign) and the new approach made through a separate plane (to include the entire biopsy tract) and the adjacent structures.

Timing of the definitive procedure after the biopsy has been subject to considerable discussion in the past. There has never been any evidence to show that performing the definitive procedure at the time of the biopsy has any advantage over closing the wound, sending the patient back to the floor and then appraising the entire situation carefully before arriving at a final plan. Most of the time, such appraisal will improve the surgical results. When in doubt, particularly about the definitive diagnosis, no disadvantage will be incurred to the patient by waiting a few days and then performing the appropriate procedure after determining the nature of the lesion, its grade, and the plan of approach.

Needle versus open biopsy

One of the issues constantly raised is whether a needle biopsy (Tru-Kut, "skinny", Craig, Vim-Silverman) is a proper approach to a bone or soft tissue lesion. Such techniques rarely cause any difficulty in providing a diagnosis. As a semiclosed procedure, they obviously have the advantage of decreasing the risk of hemorrhage or interference with soft tissue planes. For certain vertebral body or pelvic lesions, needle biopsy, either using roentogenographic or CT localization, has the great advantage of providing ease of approach and reducing the risk of bleeding, possible tissue disruption or pain. This is particularly true where one suspects a highly vascular tumor such as hemangioma, renal cell tumor, or a giant cell tumor.

The difficulty with needle biopsies is that they provide only a small amount of tissue for study. Moreover, one may obtain a nonrepresentative segment, potentially altered by squeeze artifacts or difficult to assess due to the limited number of cells and patterns to study. Furthermore, even with multiple cores it is often difficult to obtain sufficient tissue to perform some of the special studies cited above. The decision whether to use a needle biopsy should be discussed with the pathologist who ultimately will be required to make the diagnosis and to grade the lesion. Some pathologists feel quite comfortable with small amounts of tissue while others require more tissue than obtainable even with a large trochar. In general, needle biopsies are of limited value in suspected cartilage tumors (which may vary considerably) or some soft tissue tumors, especially when large areas of necrosis are present.

Conclusions

Biopsy is not a simple procedure and should be carried out with great care and appropriate consideration. To reduce problems associated with biopsy procedures to a minimum and to provide optimal care for the patient, the following rules should be applied:

1. Plan the biopsy procedure as carefully as the definitive procedure.
2. Pay as close attention to asepsis, skin preparation, hemostasis, and wound closure as with any other surgical procedure.
3. Place the skin incision so that it will not compromise the subsequent definitive surgical procedure (avoid transverse incisions).
4. Be certain to obtain an adequate amount of representative tissue and that the pathologist prepares the slides so as to permit a definitive diagnosis.
5. Be certain to obtain sufficient tissue for special staging studies.
6. Always culture the wound.
7. If the pathologist cannot make a diagnosis because of unfamiliarity with bone and soft tissue tumors, consultation should be requested.
8. If the orthopedist or the institution is not equipped to perform accurate diagnostic studies or definitive surgery and adjunctive treatment, patients with suspected bone or soft tissue tumors should be referred to a fully equipped treating center.

Acknowledgement

Supported in part by grant #CA15614 from the National Cancer Institute.

References

1. American Joint Committee for Cancer Staging and End Result Reporting. Manual for Staging of Cancer. Chicago, Illinois, 1977, pp 153-158.

2. Brostrom, L.A., Harris, M.A., Simon, M.A., Cooperman, D.R., and Nelsonne, U.: The effect of biopsy on survival of patients with osteosarcoma. J. Bone Joint Surg. 61(B), 209, 1979.

3. Dahlin, D.C.: Bone Tumors: General Aspects and Data on 6221 Cases. 3rd Ed., C.C. Thomas, Springfield, Illinois, 1978.

4. Enneking, W.F., Spanier, S.S., and Goodman, M.A.: Current concepts review: The surgical staging of musculoskeletal sarcoma. J. Bone Joint Surg. 62(A), 1027, 1980.

5. Hajdu, S.I.: Pathology of Soft Tissue Tumors. Lea and Febiger, Philadelphia, 1979.

6. Jaffe, H.L.: Tumors and Tumerous Conditions of Bone and Joints. Lea and Febiger, Philadelphia, 1958, pp 14-17.

7. Mankin, H.J., Lange, T.A., and Spanier, S.S.: The hazards of the biopsy in patients with malignant primary bone and soft tissue tumors. J. Bone Joint Surg. 64(A), 1121, 1983.

8. Moore, T.M., Myers, M.H., Potzakis, M., Terry, R., and Harvey, J.P., Jr.: Closed biopsy of musculoskeletal lesions. J. Bone Joint Surg. 61(A), 375, 1979.

9. Schajowicz, F.: Tumors and Tumor-like Lesions of Bone and Joints. Springer-Verlag , New York, 1981, pp 5-16.

10. Schajowicz, F., and Derqui, J.C.: Puncture biopsy in lesions of the locomotor system. Review of results in 4050 cases including 941 vertebral punctures. Cancer 21, 531, 1968.

11. Spjut, H.J., Dorfman, H.D., Fechner, R.E., and Ackerman, L.V.: Tumors of Bone and Cartilage. Washington Armed Forces Institute of Pathology, 1971, pp 26-28.

12. Wood, W.D., and Binder, S.C.: Biopsy principles. In, Cancer: A Manual for Practitioners, 5th Ed. Boston American Cancer Society, Massachusetts Division, 1978, pp 18-21.

Bone Tumors - Clinical and Radiologic Investigation

F. R. EILBER and E. CAULKINS

Introduction

Malignant neoplasms of bone are rare. Their estimated annual incidence is less than one case per 100,000 population, but the incidence of these bone cancers is age related. They constitute approximately 5% of all malignant tumors in young persons under 18 years of age, and reach a peak incidence of 8% of all cancers in the age group of 15-19 years. Their incidence declines to only a fraction of 1% of all malignant neoplasms after the age of 30. Osteosarcoma constitutes 40% of these tumors. Chondrosarcoma represents 20%, Ewing's sarcoma 12%, and fibrosarcoma 5%. In the peak age group, ie, ages 12-18 years, 60% of the bone cancers are osteosarcoma and 30% are Ewing's sarcoma. In the 30-60 year age group, the most common primary malignant tumor of bone is chondrosarcoma.

Although the clinical, pathologic, and radiographic characteristics of all these lesions are distinct and relatively unique for each subtype, this discussion will focus on clinical diagnosis and evaluation of patients with primary osteosarcoma.

Osteosarcoma occurs most frequently in young individuals at a peak incidence in the 10-25 year age group, although it does occur in Paget's disease in a small proportion of older adults. Osteosarcoma occurs almost twice as frequently in men as in women, most commonly around the knee, distal femur, proximal tibia, and proximal humerus. Approximately 70-85% of the patients have the disease in these three bones (Dahlin and Coventry, 1967). Other common sites are the ilium, proximal femur, mandible, fibula, and vertebrae. The knowledge of the frequency and distribution of these tumors is important for clinical and radiologic assessment. Because most osteosarcomas occur about the knee and shoulder in teenagers, this information is important for evaluation of patients with symptoms related to these areas. One should suspect a primary malignant tumor of bone in young patients who do not have a specific history of marked trauma to the area but who complain of pain. This suspicion could be the prime factor in early diagnosis and treatment of these tumors. Although pain about the knee and shoulder are relatively common complaints in the older patient, persistent pain and discomfort are rare in the young.

Current Concepts of Diagnosis and Treatment
of Bone and Soft Tissue Tumors.

Radiographic Evaluation

Standard roentgenograms are the most common first line diagnostic tests for malignant bone tumors. Malignant tumors of bone have no pathognomic signs or symptoms. Diagnosis is arrived at by a series of clues or radiologic findings. Radiologic findings can be grouped in terms of 1) zone of transition; 2) marginal sclerosis; 3) trabeculation; 4) cortical breakthrough; 5) periosteal reaction; 6) tumor size; and 7) overall roentgenographic appearance. Table I lists the variables used by the radiologists to differentiate between a benign and malignant tumor (Osbourne, 1974). Additional clues include the patient's age, the anatomic site of origin, and the tumor substance. These clues, alone or in combination, however, are just indications and not of themselves diagnostic of a primary malignant bone tumor. Clearly, the problem of diagnosis can only be solved by the pathologic interpretation of a biopsy specimen. Furthermore, the pathologist needs the roentgenograms to carefully evaluate the histologic material. This requirement is not necessary for histologic interpretation of a tumor biopsy from other anatomic locations. If the roentgenogram is suggestive of a primary malignant bone tumor, then additional diagnostic procedures are usually undertaken.

Table I. Primary bone roentgenograms

	Malignant	Benign
Zone of transition	Wide, infilt.	Narrow, sharp
Marginal sclerosis	Thin	Thick
Trabeculation	Fine	Coarse
Cortical break	Extensive	Rare (Fx)
Periosteal reaction	Multilayer	Single layer
Tumor size	Large	Small
Overall appearance	Moth eaten	Geographic

Laboratory Evaluation

A number of laboratory tests have been investigated to determine whether serologic analysis can be used for diagnosis of primary malignant bone tumors. Elevated sedimentation rates are usually found in most patients with malignant bone tumors; however, this finding is nonspecific and has seldom proved of diagnostic value. Complete blood count and white blood count are usually within normal limits. The only finding that appears to be of some diagnostic use for osteosarcoma or Ewing's sarcoma is a marked serum elevation of alkaline phosphatase levels. This complex enzyme can be associated with both primary liver disease and bone pathology; however, it is apparently not specific for osteosarcoma. Many investigators have found that approximately 80% of the patients with primary osteosarcoma have marked serum elevations in alkaline phosphatase and that these elevations generally tend to decline when the primary tumor has been treated. Thorpe et al reported in 1979 the prognostic significance of alkaline phosphatase levels in patients with osteosarcoma. They measured alkaline phosphatase levels in 30 patients and found that of 17 patients with elevated alkaline phosphatase, 12 developed tumor recurrences postoperatively. Only four out of 13 patients with normal serum alkaline phosphatase levels developed recurrences. These findings suggested that markedly elevated levels of alkaline phosphatase were of prognostic significance. A review of the experience at the University of California at Los Angeles (UCLA) showed that many patients had elevated alkaline phosphatase levels postoperatively without evidence of tumor recurrence. Others also have suggested that there is no direct relationship between elevated alkaline phosphatase levels, tumor recurrence, and prognosis. Therefore, although the measurement of alkaline phosphatase levels may be useful clinically, it does not necessarily have prognostic or diagnostic significance.

Rosenberg and his group have found a high correlation between high tissue alkaline phosphatase levels in biopsy specimens and eventual systemic disease recurrence (Levine and Rosenberg, 1979). Again, this study concerned a relatively small group of patients (30), and the findings have not been confirmed by other investigators. Certainly, investigation of enzymes and tumor relationship should be continued for this disease.

Several investigators have found that measurements of serum lactic dehydrogenase (LDH) are of some prognostic significance for patient with Ewing's sarcoma (Glaubiger et al, 1980). It is not clear at the present time whether LDH levels can be directly correlated with hepatic involvement in this disease, or whether LDH has something to do with the primary tumor. Most physicians, however, request routine pretreatment and post-treatment determination of LDH levels in patients with Ewing's sarcoma.

Additional Radiographic Investigation

Additional diagnostic studies done for patients with suspected or proven malignant bone tumor by and large fall into two categories: systemic evaluation and local tumor evaluation.

Systemic evaluation for metastases

The evaluation for metastases in patients with suspected bone tumors is predicated on the most likely sites. For example, for most of the malignant bone tumors, the sites include the pulmonary parenchyma, additional bones, and, for Ewing's sarcoma, the liver and bone marrow.

Pulmonary investigation

With the advent of whole lung tomography, it became apparent that many patients whose chest roentgenograms were normal had abnormal whole lung tomograms. The initial work suggested that approximately 30% of the patients with normal chest roentgenograms and malignant bone tumors, such as osteosarcoma, did, in fact, have positive whole lung tomograms, presumably because the resolution variable is higher with whole lung tomography and lesions of smaller size could be detected. This information could be very important for the patient who might undergo primary amputation and then die a short time later from unsuspected metastases. This patient would be better served by nonamputative procedures. The fact that routine chest roentgenograms failed to show unsuspected metastatic disease became a clinically important phenomenon. Since 1968, many investigators have reported the usefulness of whole lung tomography. Most medical centers include whole lung tomograms for patients with primary malignant bone tumors as a portion of the routine workup. Almost all investigators agree that tomograms are clinically useful, both for the initial diagnosis of malignant bone tumors and for follow-up examinations. In 1978, however, Parker et al, from the Department of Radiology at Stanford University, reported the results of a comparison study of whole lung tomograms and radiographs of 45 children with osteosarcoma and 17 children with Ewing's sarcoma from 1972 to 1976 (Parker and Marglin, 1978). Although the tomograms were positive for metastatic disease in 17 patients, they stated that the preliminary chest radiograph was positive in 15 or equivocal in two. Therefore, they questioned the routine use of whole lung tomograms as a diagnostic test.

Several other investigators have clearly shown that approximately 15% of patients with normal chest roentgenograms will have positive tomograms. These patients are treated as having a different stage of the disease. Clearly, this issue is not completely resolved but the consensus favors whole lung tomograms in the workup of patients suspected of having malignant osteosarcoma. In retrospective studies, chest roentgenograms that were at first read as normal often appear to be abnormal at a later reading. Tomograms can resolve this issue to some extent.

CT scanning of chest

With the advent of computerized tomography, an additional dimension was added to the radiographic evaluation of the pulmonary parenchyma. The CT scanner has a much higher resolution and can detect small lesions in the subpleura. Because osteosarcomas and chondrosarcomas commonly metastasize to the subpleura, CT scanning appeared to add an additional dimension to this evaluation. Investigators at the Mayo Clinic compared CT scanning, chest roentgenography, and whole lung tomography. They agreed that CT scanning permitted higher diagnostic accuracy. They reported a study of 50 patients with osteosarcoma, in which 22% of the patients had positive CT scans and negative chest roentgenograms, and 15% had positive CT scans, negative chest roentgenograms and negative whole lung tomograms (Pritchard, 1982). Other investigators, however, have found many false positive results with the CT scan. At the present time, there is no consensus that CT scanning is of routine value in the initial staging of patients with malignant bone tumors or for follow-up examinations.

The technetium-99 bone scan is an additional systemic test for patients with malignant bone tumors. Autopsy examinations have shown that at least 60% of the patients who die from osteosarcoma have disease in multiple bones. Historically, these multiple metastases were not a problem because most patients died from uncontrolled pulmonary disease. Chemotherapy and more aggressive throacotomies have changed this dire prognosis, however, and the clinical problem of multiple bony metastases has become much more apparent. In a series of patients at UCLA, 50% who developed recurrent disease did so in the lungs; in an additional 50%, bone was the first site or was present simultaneously with pulmonary disease. It therefore seems reasonable to do routine whole body bone scans on patients with osteosarcoma to find either sites of synchronous tumor development or occult metastatic disease to other bones.

Finally, patients with Ewing's sarcoma may require liver/spleen scans, especially those patients with elevated alkaline phosphatase levels. Routine liver scans for patients with primary osteosarcoma or chondrosarcoma do not appear to be of value.

Evaluation of primary tumor extent

Additional diagnostic tests for local anatomic extent of the primary tumor fall into the category of the bone scan. Most patients with malignant bone tumors have positive technetium-99 bone scans. The increased uptake in the area of the tumor appears to represent reactive bone formation rather than size of the primary tumor itself. Thus, bone scans tend to overestimate the extent of bony or marrow involvement rather than predict the anatomic extent of the tumor. A recent report by Lange from Wisconsin suggests that at an early phase, almost immediately after its injection, the technetium-99 bone scan will display the tumor vascularity. Thus, it may be useful to consider doing both early and late phases of the bone scan to determine anatomic extent (Lange, 1983). The early, or vascular, phase is clearly positive for highly malignant bone tumors but may be negative for benign tumors such as giant cell tumors or the enchondromas that have little or no neovascularity. In these cases, the delayed film is positive. Additional confirmation of these early encouraging results is needed.

Other investigators are testing isotopes for scanning purposes. Rikitake, from Japan, has compared sodium-18-F bone centigrams and positron camera scans with technetium-99 pyrophosphate scans (Rikitake et al, 1979). He reports that there was some value in the sodium-18-F bone scans over the routine bone scanning for patients with osteosarcoma in that osteosarcomas accumulated sodium-18-F at much higher levels than the technetium. Additional reports from Rosen's group in New York have suggested that galium scans may be useful in evaluating primary tumor response to preoperative therapy.

These latter two methods are still experimental at this time and are not available in most medical centers.

Angiography

Several investigators have looked at the role of routine angiography for malignant tumors. Enneking and his group studied 47 consecutive patients with osteosarcoma: six with parosteal osteosarcomas, nine with giant cell tumors of bone, four with malignant fibrous histiocytomas, 11 with Ewing's sarcomas, and 23 with benign bone tumors (Hudson et al, 1975). Their purpose was to evaluate angiography in terms of its contribution to the management of the patient. The value of the angiograms was assessed in conjunction with the physical examination, standard roentgenograms, and bone scans. In general, they found that the angiogram was useful for only a small percentage of the patients. Although tumor neovascularity, vessel invasion, and pathologic displacement of vessels were visualized, in many cases compressed normal tissue rather than the tumor itself was the cause. As a diagnostic test, angiography was not accurate in their view in distinguishing benign from malignant tumors and was not helpful for detecting residual disease or recurrent tumor. The test has potential value, however, for staging tumors as to intraosseous or extraosseous extent.

Most investigators do not use routine angiograms to determine local anatomic extent of tumor, but in selected cases, particularly with pelvic tumors, arteriography has been very useful (Chuang et al, 1982).

CT scanning of tumor site

Computerized tomography of local lesions has significantly affected the evaluation of the anatomic extent of the local tumor. In 1979, Mueller et al reported the results of CT scan evaluation of malignant bone tumors, including 36 primary bone tumors. They found that the CT scans displayed the soft tissue involvement, its axial extension, and the tumor's relationship to contiguous visceral and vascular structures. At UCLA we have found CT scans extremely useful for accurately predicting the lateral anatomic extent of the tumor, although the scans have been relatively inaccurate in determination of the intraosseous or marrow extent of the tumor. Often the CT scan, in our experience, has overestimated the length of marrow involvement by at least 25%. The scans have been extremely useful, however, for determining the extraosseous extent of tumor and are valuable for planning the limits of the surgical resection.

Experimental Diagnostic Tests

The clonogenic assay recently described by Salmon is being tested in clinical practice. The basic premise of the assay is to test cloned stem cells against various drugs. Tumor specimens are minced and placed in soft agar culture. The growing tumor colonies are separated and then tested against various antitumor antibiotics or chemotherapeutic agents. The reduction in number of tumor cells in relation to nontreated, nondrug-exposed clones is then compared. Many investigators see promise in this procedure for its potential usefulness in testing the drug sensitivity of primary malignant bone tumors. Our experience at UCLA has shown us that only approximately 30% of the primary bone tumors will grow in soft agar (Bertelsen et al, 1983). Of this 30%, only 50% are sensitive to one of the drugs added to the culture. Thus, for every 100 patients with osteosarcoma, the assay would be potentially valuable for only 15. Our results and those of Salmon indicate that the test may be of more value for determining drug-resistant cells rather than drug-sensitive cells. The general correlation of *in vivo* with clonogenic assay resistance and *in vitro* patient resistance is extremely high indeed. The *in vitro* drug response to patient response ratio is only approximately 50%. Additional prospective investigations will be necessary to determine the eventual clinical benefit of this assay.

Huth et al (1983) have described a urinary antigen in patients with malignant bone tumors. This antigen, found in concentrated 24-hour urine samples of patients with osteosarcoma, may represent an antigenic moiety related to the destruction of tumor cells. Urinary antigen excretion is elevated in patients with actively growing tumors and increases when these tumors are destroyed *in vivo* by drugs and radiation but decreases following surgical resection. Postoperative follow-up examinations of 50 patients showed that a correlation between tumor recurrence and reappearance of this urinary antigen is approximately 90%. In patients who remain free of disease, however, urinary antigen titers have been continuously negative.

These studies are still experimental, but as more is learned about the biochemical characteristics of this urinary antigen, the assay can be standardized for application to large groups of patients in a randomized, prospective fashion.

Summary

Clinical evaluation of patients with malignant bone tumors involves many medical disciplines. A sense of suspicion is of value when a young person, 10-25 years of age, complains of persistent pain in a joint. Routine roentgenograms together with the clinical symptoms can assist in the diagnosis. Biopsy remains the only unequivocal diagnostic test.

Bone scans, arteriograms, and CT scans have been useful for determining the local anatomic extent of disease. Systemic evaluations for metastic spread include chest roentgenograms, whole lung tomography, and chest CT scanning. Serologic evaluation of serum alkaline phosphatase levels may be a valuable tumor marker in follow-up examinations.

Acknowledgement

Supported by Grants CA29605 and CA12582, awarded by the National Cancer Institute, DHHS.

References

1. Bertelsen, C.A., Sondak, V.K., Mann, B.D., Korn, E.L., Morton, D.L., and Kern, D.H.: Chemosensitivity testing of human solid tumors - A review of 1582 assays with 263 clinical correlations. Cancer 1983, in press.

2. Chuang, V.P., Charnsangavey, C., Soo, C.S., Wallace, S., Benjamin, R.S., and Jaffe, N.: Radiographic and angiographic changes of osteosarcoma following intraarterial chemotherapy. Proc. Roent. Rad. A251, 1982.

3. Dahlin, D.C., and Coventry, M.B.: Osteosarcoma: a study of six hundred cases. J. Bone Joint Surg. 49, 101, 1967.

4. Glaubiger, D.L., Makuch, R., Schwarz, J., Levine, A., and Johnson, R.E.: Determination of prognostic factors and their influence on therapeutic results in patients with Ewing's sarcoma. Cancer 45, 2213, 1980.

5. Hudson, T.M., Haas, G., Enneking, W.F., and Hawkins, L.: Angiography in the management of musculoskeletal tumors. Surg. Gynecol. Obstet. 141, 11, 1975.

6. Huth, J.F., Gupta, R.K., Eilber, F.R., and Morton, D.L.: A prospective postoperative evaluation of urinary tumor-associated antigens in sarcoma patients: correlation with disease recurrence. Cancer 1983, in press.

7. Huth, J.F., Gupta, R.K., and Morton, D.L.: Sequential analysis of urinary antigen(s) in sarcoma patients. Surg. Forum 30, 150, 1979.

8. Lange, T.: Personal communication, 1983.

9. Levine, A.M., and Rosenberg, S.A.: Alkaline phosphatase levels in osteosarcoma tissue are related to prognosis. Cancer 44, 2291, 1979.

10. Mueller, P.R., Wittenberg, J., Ferrucci, J., Murray, W.T., Mankin, H., and Shiller, A.: CT scanning in bone and soft-tissue tumors. J. Comput. Asst. Tomogr. 3, 570, 1979.

11. Osbourne, R.: Differential radiologic diagnosis of bone tumor. CA 24, 194, 1974.

12. Parker, B.R., and Marglin, S.I.: Evaluation of full-lung tomography in children with newly diagnosed bone and soft-tissue sarcoma. Clin. Res. 26, 196A, 1978.

13. Pritchard, D.: Personal communication, 1982.

14. Rikitake, T., Tateno, Y., and Arimizu, N.: Osteosarcomas in 18F bone positron scintigram - Comparative study with 99MTC-pyrophosphate. Nippon Igaku Hoshasen Gakkai Zoshi 39, 620, 1979.

15. Thorpe, W.P., Reilly, J.J., and Rosenberg, S.A.: Prognostic significance of alkaline phosphatase measurements in patients with osteogenic sarcoma receiving chemotherapy. Cancer 43, 2178, 1979.

Scintigraphy of Bone Tumors

R. KLOIBER

Bone imaging with technitium 99m diphosphonates has brought unsurpassed sensitivity to detection of occult skeletal lesions. Localization of radiophosphates depends on blood flow for delivery as well as local binding. Most skeletal lesions produce hyperemia and induce reactive new bone formation. This immature bone has a large exchangeable surface area of hydroxyapatite crystals and nonmineralized matrix which bind the tracers. Detectability, therefore, depends on metabolic activity and not on the substantial change in local mineral content required to make the lesions visible radiographically. Conditions which do not incite bone response, such as multiple myeloma, plasmacytoma, metastatic neuroblastoma, or aggressive lesions that completely destroy bone, will produce photopenia and may be missed scintigraphically. A diagnostic approach comprising initial plain radiography of symptomatic sites followed by scintigraphy in undiagnosed cases will give the greatest sensitivity. Scintigraphically-detected sites must be further characterized by laminography or transmission computerized tomography (TCT) for specific diagnosis. In complex areas such as the spine, carpals, tarsals, pelvis and proximal femora, lesions may not be demonstrable even by these specialized radiographic techniques (Fig. 1) (Lisbona and Rosenthall, 1979; Smith and Gilday, 1980). High resolution, high countrate scintigrams in multiple projections are mandatory for precise anatomic localization. Optimum technique is also critical in assessing bone involvement by contiguous soft tissue sarcomas (Chew et al, 1981).

Increased radiophosphate deposition has been described with bone sarcomas, metastatic deposits, and benign conditions including osteoid osteoma, osteoblastoma, chondroblastoma, fibrous dysplasia, and histiocytosis X (Gilday and Ash, 1976; Murray, 1980; Rosenthall and Lisbona, 1980; Theodore Harke, 1978). Even with lesions of inherent low metabolic activity, such as unicameral bone cysts, prominent uptake can be seen after pathologic fracture or when the size and location is such as to place abnormal stress on adjacent normal bone or joints. Bone islands have also been demonstrated scintigraphically during certain phases of their development (Hall et al, 1980) and with continued improvements in imaging devices, radiopharmaceuticals and techniques, visualization of benign lesions with very low grade metabolic activity can be anticipated. The scan can thus determine vascularity and osteogenic activity of a process but cannot provide histologic information.

Current Concepts of Diagnosis and Treatment
of Bone and Soft Tissue Tumors.

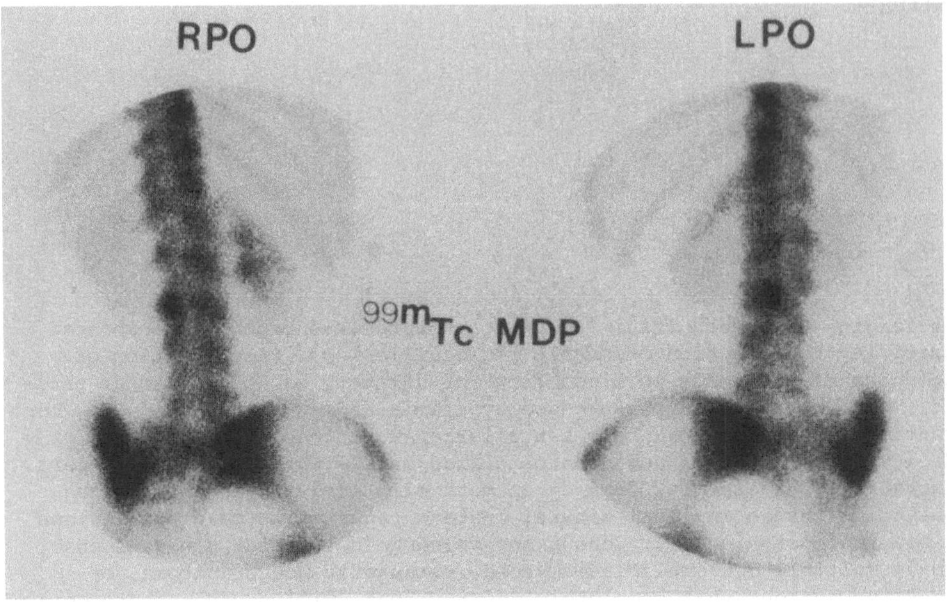

Fig. 1. Osteoid osteoma in the superior facet of L3 on the left, not demonstrable by plain radiography, laminography or T.C.T.

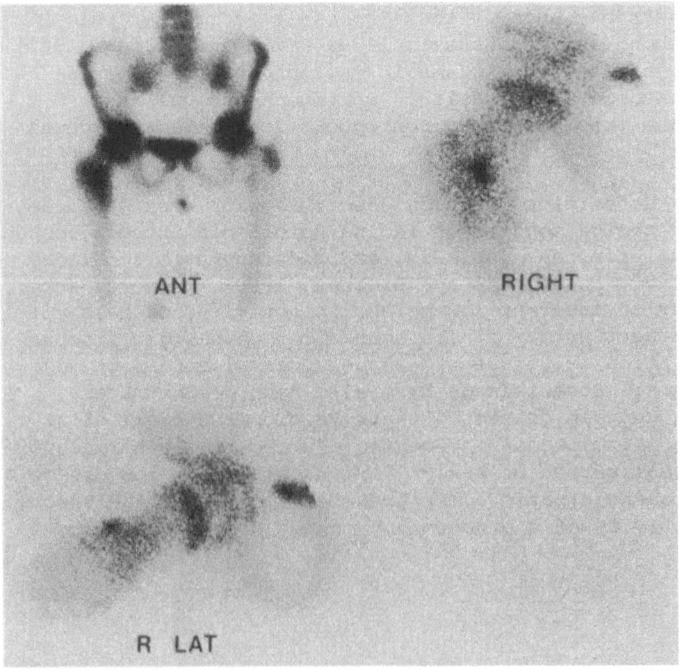

Fig. 2. Osteoid osteoma in the right proximal femur. Routine anterior view shows extensive bone reaction but pinhole views with appropriate exposure localize the nidus.

Lesions which appear sclerotic radiographically and have very low grade radiophosphate uptake may be considered benign, but even high grade activity does not imply malignancy. If serial scans of known lesions, such as central osteochondromas, are available a progressive increase in activity may signify malignant degeneration. Lytic lesions or those not well defined radiographically cannot be reliably separated into benign and malignant categories based on scan appearance alone.

Accurate preoperative determination of the extent of a lesion is a prerequisite to planning appropriate management. Because of the extensive change in mineral content required to visualize a lesion on plain radiographs, these studies often significantly underestimate extent. Scintigraphy gives a better indication of size but may overestimate involvement because of adjacent increased activity due to hyperemia or acute disuse osteoporosis. Exposure at multiple intensity settings or gray scale variation of computerized images is required to demonstrate internal structure of lesions and can often assist in separating the lesion from surrounding reaction (Fig. 2). Most malignant tumors and some benign types such as osteoid osteoma are highly vascular; therefore, images shortly after injection, when radiotracer is predominantly in the blood and interstitial fluid (blood pool image), can assist in selecting an appropriate site for biopsy. T.C.T. can add information regarding soft tissue extension and intramedullary spread, particularly with peripheral lesions in older patients with predominantly fatty marrow. A combined approach gives the most accurate depiction.

Recently, intraoperative scintigraphy has been used to guide resection of lesions not visible radiographically. Modern mobile gamma cameras can be used in the operating suite as easily as image intensifying systems and afford the best definition of complex areas (Sty and Simons, 1982). Alternatively, if the active site is well defined and not in close proximity to growth plates or other normal structures of high tracer uptake, a hand-held probe system is adequate and more maneuverable (Harvey and Lancaster, 1981). Scintigraphy of the specimen can corroborate removal of the lesion and autoradiography may aid in pathologic assessment (Ghelman and Vigorita, 1983).

Total body scintigraphy adds no additional radiation exposure and should be performed at the time of discovery or evaluation of an undiagnosed lesion to screen for metastases or multiplicity of sites of involvement. Though the incidence of multifocal lesions or metastases at the time of presentation is low for most sarcomas, a total body survey is of value because of the potentially profound effect on therapy. It also serves as a base line study, particularly in older individuals with degenerative or coexistent disease. Chemotherapy appears to be more effective in controlling non-osseous lesions and the number of patients presenting with bone metastases before demonstrable pulmonary metastases is increasing, particularly in Ewing's and osteogenic sarcoma (Goldstein et al, 1980). Regular follow-up is therefore warranted.

Postoperative evaluation of the original site of the lesion is complicated by bone reaction induced by the surgery itself. Increased uptake can persist for variable periods but should show a progressive decrease during follow-up. Any subsequent reactivation would signify a complication.

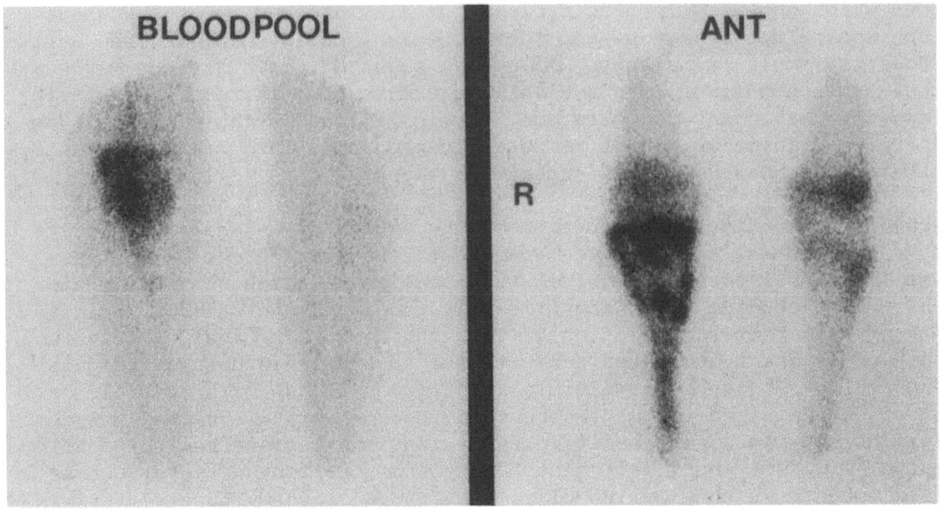

Fig. 3. Blood pool image shows recurrence of giant cell tumor in the tibial metaphysis. Delayed image (right) shows extensive reaction lower in the shaft from previous surgery.

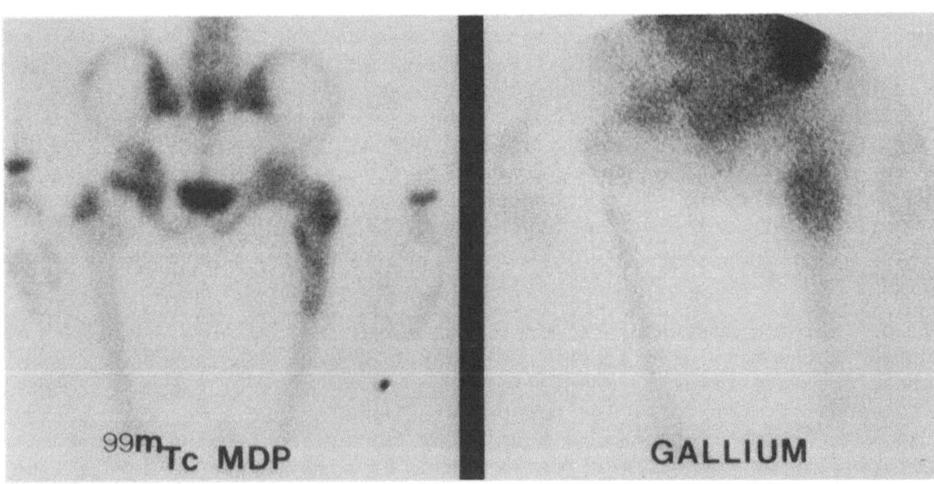

Fig. 4. Recurrence of Ewing's sarcoma demonstrated by gallium imaging.

Without the benefit of serial studies, blood pool images may pinpoint a focal disproportionate hypervascular area (Fig. 3) and should always be included. Supplementary gallium imaging can detect tumor recurrence (Fig. 4) or infection but cannot differentiate the two. Because sterile bone reaction will also accumulate gallium, scans must not be interpreted in isolation and are only considered positive if gallium uptake is disproportionate or different in distribution compared with radiophosphate uptake. Radiolabelled leucocytes have shown high sensitivity in detecting abdominal abscesses (Goodwin, 1983). Accumulation at neoplastic sites has not been a problem, although indium-leucocyte uptake has been shown in a bone metastasis from adenocarcinoma (Sfakianakis et al, 1982). The orthopedic applications of this technique merit further investigation.

More frequent attempts at limb salvage and extensive reconstructive surgery have resulted in greater use of free vascularized bone grafts and osteocartilagenous allograft transplantation. In revascularized grafts, radiophosphate imaging can be used to assess patency of vascular anastomosis and bone viability when plain radiographs afford little information beyond structural integrity of the bone (Lisbona et al, 1980). Initial studies should be carried out within one week of surgery when loss of blood supply will result in a photon-deficient zone. Subsequently, regional increased phosphate uptake may result from appositional new bone formation on the surface of an avascular graft (Berggren et al, 1982). Soft tissue deposition in surrounding necrotic muscle should not be confused with normal uptake. Serial imaging will demonstrate incorporation of conventional autogenous bone grafts and allografts, although, particularly with the latter, this is a protracted process spanning many years (Mankin et al, 1976). The sensitivity of radionuclide imaging for detection of nonunion, loss of fixation, stress fracture or infection in partially incorporated grafts remains to be established.

In summary, the main contributions of radionuclide imaging is detection of occult disease before more specific studies or surgery. It can aid in demonstrating the extent of the process but the contribution to differential diagnosis is limited. At the present time, this is the most sensitive technique for detecting metastatic spread of most forms of malignancy and can also disclose other complications during the follow-up period.

References

1. Berggren, A., Weiland, A. and Ostrup, L.: Bone scintigraphy in evaluating the viability of composite bone grafts revascularized by microvascular anastomosis, conventional autogenous bone grafts, and free non-revascularized periosteal grafts. J. Bone Joint Surg. (Am.) 64A, 799, 1982.

2. Chew, F., Hudson, T. and Enneking, W.: Radionuclide imaging of soft tissue neoplasms. Sem. Nucl. Med. 11, 266, 1981.

3. Ghelman, B. and Vigorita, V.: Postoperative radionuclide evaluation of osteoid osteomas. Radiology 146, 509, 1983.

4. Gilday, D. and Ash, J.: Benign bone tumors. Sem. Nucl. Med. 6, 33, 1976.

5. Goldstein, H., McNeil, B., Zufall, E. and Treves, S.: Is there still a place for bone scanning in Ewing's sarcoma? J. Nucl. Med. 21, 10, 1980.

6. Goldstein, H., McNeil, B., Zufall, E., Jaffe, N. and Treves, S.: Changing indications for bone scintigraphy in patients with osteosarcoma. Radiology 135, 177, 1980.

7. Goodwin, D.: Clinical use of In-111 leukocyte imaging. Clin. Nucl. Med. 8, 36, 1983.

8. Hall, F., Goldberg, R., Davies, J. and Fainsinger, M.: Scintigraphic assessment of bone islands. Radiology 135, 737, 1980.

9. Harvey, W. and Lancaster, J.: Technical and clinical characteristics of a surgical biopsy probe. J. Nucl. Med. 22, 184, 1981.

10. Lisbona, R. and Rosenthall, L.: Role of radionuclide imaging in osteoid osteoma. AJR 132, 77, 1979.

11. Lisbona, R., Rennie, W. and Daniel, R.: Radionuclide evaluation of free vascularized bone graft viability. AJR 134, 387, 1980.

12. Mankin, H., Fogelson, F., Thrasher, A. and Jaffer, F.: Massive resection and allograft transplantation in the treatment of malignant bone tumors. N. Engl. J. Med. 294, 1247, 1976.

13. Murray, L.: Bone scanning in the child and young adult. Skel. Radiol. 5, 1, 1980.

14. Rosenthall, L. and Lisbona, R.: Role of radionuclide imaging in benign bone and joint diseases of orthopedic interest. Nuclear Medicine Annual, p. 267. Raven Press, New York, 1980.

15. Sfakianakis, G., Mnagmneh, L., Ghandur-Mnaymneh, L., Al-Sheikh, W., Hourani, M. and Iteal, A.: Positive Indium-111 leukocyte scintigraphy in a skeletal metastasis. AJR 139, 601, 1982.

16. Smith, W. and Gilday, D.: Scintigraphic appearances of osteoid osteoma. Radiology 137, 191, 1980.

17. Sty, J. and Simons, G.: Intraoperative 99m technetium bone imaging in the treatment of benign osteoblastic tumors. Clin. Orthop. 165, 223, 1982.

18. Theodore Harke, H.: Bone imaging in infants and children. J. Nucl. Med. 19, 324, 1978.

Basic Concepts of the Resistance of Cartilage to Tumor Invasion

K. E. KUETTNER and B. U. PAULI

Invasion is a characteristic of many physiologic and pathologic conditions. It may be broadly defined as the ability of a cell to enter another tissue. Non-neoplastic invasive growth is an important event in the fertilization of the ovum, the growth of the trophoblast, embryonic morphogenesis, the development of the mammary gland during pregnancy, osteoclastic bone remodelling and neovascularization (Kuettner and Pauli, 1982). A common but often overlooked feature of non-neoplastic invasion is that the invading cell or tissue induces a response in its host. For example, trophoblast implantation stimulates uterine collagen synthesis and osteoclastic bone resorption triggers osteogenesis. Connective tissue cells and matrices do not necessarily accept an invader; rather, they actively interact with it. In many instances, this cooperation is highly regulated and may even be obligatory for the invasive process to succeed.

Progressive proliferation of neoplastic cells yield an increasing bulk of tissue which must somehow be accommodated within the tissues of the host. Benign tumors grow expansively and are often circumscribed by a distinct capsule: a neoplastic tissue which does not intermingle but compresses or displaces neighboring tissues. In contrast, malignant tumors penetrate the surrounding tissues which they may even destroy. Invasiveness is considered to be a hallmark of malignancy in solid tumors but is not by itself sufficient. Tumor cell invasion is a local process in which tissue barriers are crossed by migrating and/or proliferating cells, and, in the case of metastasis, is probably associated with decreased cellular adhesion that favors primary tumor disaggregation.

Connective tissues consist of a meshwork of collagen fibers which, together with other structural macromolecular complexes (ie, proteoglycans, glycoproteins, hyaluronic acid, and elastin), provide physical barriers within which most tissue cells are retained (Pauli et al, 1983). Malignant cells have the capacity to penetrate these mechanical connective tissue barriers including organ stromata. Movement of cells through the extracellular connective tissue matrix is probably dependent on a variety of factors, including cell motility, chemotaxis, and the physicochemical nature of each matrix. Certain cells penetrate collagenous structures such as basement membranes or bone. There is now considerable support for the concept that enzymatic degradation of the extracellular matrix facilitates such transitional movements of cells within the invaded tissues (Strauli et al, 1980). Certain tissues and structures, however, act as local barriers to tumor spread. Mammalian cartilage is the best example of a tissue highly

Current Concepts of Diagnosis and Treatment
of Bone and Soft Tissue Tumors.

resistant to tumor invasion (Kuettner and Pauli,1981; Pauli and Kuettner, 1982), although in the rat Yoshida sarcoma cells have been shown to invade cartilage matrix (Machianami, 1972). With this one exception, cartilages generally appear to resist destruction by tumors. Based on available data and extensive studies in our laboratory, a confirmable primary hypothesis has been proposed: certain normal tissues such as cartilage, which are rich in protease inhibitors (or protease regulators), restrict the degradation of their extracellular matrix components (Kuettner and Pauli, 1981; Pauli and Kuettner, 1982). Therefore, the pattern of tumor cell invasion may be dependent on the relative abundance of these protease regulators within tissues.

In our original *in vitro* system, we studied the resistance of cartilage to invasion by osteosarcoma, mammary carcinoma, or urinary bladder carcinoma cells and found that these tumor cells were unable to penetrate this tissue. The same cells, however, had no difficulty in invading other connective tissues, such as bone and tendon (Kuettner and Pauli, 1981). In our original organ-cell culture system, tumor cells were unable to adhere or attach to the cartilaginous explants unless the tissue had been altered by extraction (Kuettner and Pauli, 1981; Kuettner et al, 1978). It has been suggested that tumor cell attachment may be a prerequisite for tumor cell invasion. We, therefore, devised a new organ culture system in which the above-mentioned cells were grown in direct contact with viable cartilage (Pauli et al, 1981; Sorgente et al, 1975). This invasion-chamber is composed of a hollow, stainless-steel cylinder with a sharpened edge at one end so that a disc can be cut from a 1 mm thick slice of bovine articular cartilage. Cells are plated onto the cartilage disc in the bottom of the cylinder and the medium is changed daily. In this invasion chamber, tumor cells were unable to invade the cartilage. Large numbers of cells, in intimate contact with the cartilage, covered the surface irregularities of the tissue without showing any signs of matrix destruction or penetration as determined by electron microscopy of random samples (Pauli et al, 1981). Comparative studies with tendon discs within the chamber showed invasion by the malignant tumor cells but not by control fibroblasts.

Treatment of fresh articular cartilage with 1 M GuHCl or NaCl depleted the tissue of a part of the proteoglycan pool and the factor which inhibited neovascularization on the chorioallantoic membrane (Sorgente et al, 1975). No detectable amounts of collagen were solubilized, and the collagen fibers were not denatured, as shown ultrastructurally, by the preservation of the normal cross-banding pattern. Several layers of tumor or control cells were plated in the invasion chamber onto the modified cartilage discs. Osteosarcoma cells, but not fibroblasts, growing in contact with the devitalized, extracted cartilage began to penetrate the matrix with numerous microvilli extending into the matrix between intact collagen fibers (Pauli et al, 1981). After a culture period of 14 days, deep invasion was observed in most of the discs. Single tumor cells or tumor cell clusters penetrated as far as 300 µm into the 1 mm thick disc. Penetration was often associated with rarefaction of the matrix and widening of interfibrillar spaces (Pauli et al, 1981). Adjacent to tumor cell membranes, collagen fibers were fragmented and had lost the characteristic cross-banding pattern. Entire fibers were depolymerized into thin fibrils, and amorphous, electron-dense material occasionally accumulated. These morphologic observations suggest that tumor cells use collagenolytic enzymes (eg, of the collagenase, elastase crosslinkase or cathepsin type). In a critical experiment, we added the cartilage derived anti-invasion factor (AIF, detailed below) at low concentrations and observed that invasion by osteosarcoma cells was prevented (Pauli et al, 1981). This observation supports our postulate that hyaline cartilage contains extractable matrix components that inhibit invasion (Kuettner and Pauli, 1981; Pauli and Kuettner, 1982).

In our laboratory, we have been able to show that component(s) within AIF can inhibit the proteases trypsin and chymotrypsin (Pauli and Kuettner, 1982), a subcomponent recently identified as Trasylol (aprotinin) by Rifkin et al (Rifkin and Crowe, 1977). We further showed that AIF could inhibit collagenases from mammalian skin (Kuettner et al, 1976), from *in vitro* stimulated osteosarcoma cells as well as collagenolytic activity from mammary carcinoma cells (Kuettner et al, 1977). In addition, AIF inhibits a cartilage-derived metalloproteinase and the general proteolytic enzyme spectrum of the buffy coat. Recently, we were able to show that AIF can also inhibit elastase activity derived from human leukocytes. Moreover, in collaboration with Dr. L.A. Liotta, NCI-NIH*, we have shown that AIF also expresses strong inhibitory activities against neutral metalloproteinases derived from metastatic murine PMT sarcomas and from metastatic murine M-5076 reticulum cell sarcomas that cleave collagens of Types IV and V (Pauli and Kuettner, 1982).

The cartilage-derived anti-invasion factor (AIF) also expresses growth inhibitory activity selectively directed against actively dividing endothelial cells (Pauli and Kuettner, 1982). We have provided evidence in favor of the hypothesis that this endothelial cell growth inhibitory activity (EGIA) may well be responsible for the avascularity of cartilage, thus making it a poor site for tumor growth. It may be identical to the cartilage-derived antiangiogenesis activity which was originally described by us and has received great attention through the work of Folkman and his collaborators (Folkman and Cotran, 1976). These investigators designed experiments to show that intact cartilage as well as a preparation equivalent to AIF can inhibit the effect of the tumor angiogenesis factor which induces endothelial cell proliferation and migration, both *in vivo* and *in vitro* (Brem and Folkman, 1975; Folkman and Cotran, 1976). Recently, we were able to fractionate AIF with the aid of HPLC and to separate EGIA from the proteolytic enzyme inhibitory activity.

With regard to the mode of action of EGIA, many unanswered questions remain. Experiments show that the unfractionated AIF is cytotoxic for endothelial cells (Pauli and Kuettner, 1982). However, once purified, it simply increases the endothelial cell doubling time, arresting cells in G1 phase without killing them (Sorgente and Dorey, 1980). This discrepancy between unfractionated AIF and purified endothelial cell growth inhibitor may be explained by a simultaneous action of proteinase inhibitors and EGIA, which may in concert express a cytotoxic effect after the growth arrest of endothelial cells has occurred.

---

* National Cancer Institute-National Institute of Health

In this brief review, the following basic assumptions have been made:

1) tumor cells use a variety of matrix-degrading enzymes in the process of invasion into the surrounding tissue (Pauli et al, 1983; Strauli et al, 1980); and

2) the resistance of cartilage to invasion is probably due to specific molecules which inhibit (regulate) matrix degrading enzymes.

Investigation of the local inhibitors of invasion may well yield information that is basic to an understanding of tumor cell behavior and potentially useful in future cancer therapy.

Acknowledgement

This work was supported by NIH grants CA-21566, CA-25034 and, in part, by AM-09132, and the Council for Tobacco Research-USA, Inc., grants R-1206 and 1394.

References

1. Brem, H., and Folkman, J.: Inhibition of tumor angiogenesis mediated by cartilage. J. Exp. Med. 141, 427, 1975.

2. Folkman, J. and Cotran, R.: Relation of vascular proliferation to tumor growth. Int. Rev. Exp. Pathol. 16, 207, 1976.

3. Kuettner, K.E., Hiti, J., Eisenstein, R. and Harper, E.: Collagenase inhibition by cationic proteins derived from cartilage and aorta. Biochem. Biophys. Res. Commun. 72, 40, 1976.

4. Kuettner, D.K. and Pauli, B.U.: Resistance of cartilage to invasion. In, Bone Metastasis, Weiss, L. and Gilbert, H.A. (eds). G.K. Hall, Boston, 1981, pp 131-165.

5. Kuettner, K.E., and Pauli, B.U.: Vascularity of cartilage. In, Cartilage, Vol. 1. Hall, B.K. (ed), Academic Press, New York, 1982. pp 281-312.

6. Kuettner, K.D., Pauli, B.U. and Soble, L.: Morphological studies on the resistance of cartilage to invasion by osteosarcoma cells *in vitro* and *in vivo*. Cancer Res. 38, 277, 1978.

7. Kuettner, K.E., Soble, L., Croxen, R.L., Marczynska, B., Hiti, T., and Harper, E.: Tumor cell collagenase and its inhibition by a cartilage-derived protease inhibitor. Science 196, 653, 1977.

8. Machianami, R.: A study of the invasive growth of malignant tumors. I. An electron microscopic study of the invasive growth of Yoshida sarcoma cells in rat articular cartilage. Acta Pathol. Jap. 22, 19, 1972.

9. Pauli, B.U., and Kuettner, K.E.: The regulation of invasion by a cartilage-derived anti-invasion factor. In: Tumor Invasion and Metastasis, Liotta, L.A., and Hart, I.R. (eds). Martinus Nijhoff Publishers, The Hague, 1982. pp 267-290.

10. Pauli, B.U., Memoli, V.A., and Kuettner, K.E.: *In vitro* determination of tumor invasiveness using extracted hyaline cartilage. Cancer Res. 41, 2084, 1981.

11. Pauli, B.U., Memoli, V.A., and Kuettner, K.E.: Regulation of tumor invasion by cartilage-derived anti-invasion factor *in vitro*. J. Natl. Cancer Inst. 67, 65, 1981

12. Pauli, B.U., Schwartz, D.E., Thonar, E.J.-M., and Kuettner, K.E.: Tumor invasion and host intracellular matrix. Cancer Metastasis Reviews (in press) 1983.

13. Rifkin, D.B. and Crowe, R.M.: Isolation of a protease inhibitor from tissues resistant to tumor invasion. Hoppe-Seyler's Z. Physiol. Chem. 358, 1525, 1977.

14. Sorgente, N. and Dorey, L.K.: Inhibition of endothelial cell growth by a factor isolated from cartilage. Exp. Cell Res. 129, 63, 1980.

15. Sorgente, N., Kuettner, K.E., Soble, L.W. and Eisenstein, R.: The resistance of certain tissues to invasion. II. Evidence for extractable factors in cartilage which inhibit invasion by vascularized mesenchyme. Lab. Invest. 32, 217, 1975.

16. Strauli, P., Barrett, A.J. and Baici, A.: Proteinase and Tumor Invasion. Raven Press, New York, 1980. p. 1

Chemotherapy for Advanced Osteogenic Sarcoma and Ewing's Sarcoma

B. HERNANDEZ and J. E. HARRIS

Osteogenic Sarcoma

Malignant bone and cartilage tumors account for less than 1% of all malignant neoplasms. Osteogenic sarcoma is the most common primary malignant neoplasm of bone, most frequently occurring in the diaphysis, especially in the femur. The incidence of osteogenic sarcoma peaks between ten and 25 years of age. Although many of the patients with osteogenic sarcoma present with clinically localized disease, the five-year survival rate for osteogenic sarcoma achieved with local treatment alone (surgical excision or amputation, radiation therapy, or both) is only about 20%. Preoperative radiation and surgery does not improve survival over surgery alone. In 1972, Friedman and Carter reviewed 17 reported series of patients with osteosarcoma and found a five-year survival ranging from 16-23% (Friedman and Carter, 1972). Some have suggested that the five-year disease-free survival after surgery alone has been raised to 50% (Carter and Friedman, 1978; Taylor et al, 1978). This has, however, been disputed and the figure of 20% remains generally accepted (Gehan et al, 1978; Mike and Marcove, 1978). The 10-year survival after surgery alone is 16% so that most patients who survive to five years may be regarded as cured. The natural history of osteogenic sarcoma does not appear to have changed over the past several decades (Mike and Marcove, 1978; Jaffe et al, 1974; Harvei and Solheim, 1981). The mean time to the appearance of pulmonary metastases -- the most common metastases -- after surgery is 10 months (Friedman and Carter, 1972). Before the 1970's, the average time between the recognition of pulmonary metastases and death was six months (Friedman and Carter, 1972). These survival and natural history variables provide consistent clean end points which may be used in judging the effectiveness of new therapies for osteogenic sarcoma. The rapid clinical appearance of distant metastatic disease has led many clinical investigators to conclude that most patients present with distant micrometastases which are not clinically evident at the time of their initial diagnosis. It is therefore expected that these patients will fail when given local therapy only.

Attempts to improve survival by treating metastatic disease when it appears had in the past been largely unsuccessful. Before the 1970's, therapy of metastatic disease with radiation and a few alkylating agents had met with only limited success. Over the past 13 years, however, a number of new agents have been developed which individually may produce significant responses in metastatic osteogenic sarcoma. Their use in combination may prove to be even more effective. The change in prognosis for metastatic

Current Concepts of Diagnosis and Treatment
of Bone and Soft Tissue Tumors.

osteosarcoma comparing patients diagnosed in 1971 or later with those diagnosed in 1970 or earlier has been reviewed by Perez et al (1978). They found a postmetastatic survival time advantage for the former compared with the latter. They attributed this to the results of the more intensive and more effective chemotherapy used from 1971 on. Their conclusion was that chemotherapy for metastatic disease before 1971 could not be considered truly effective.

In the 1960's, the use of cyclophosphamide, phenylalanine mustard, DTIC ** and mitomycin C as single agents was reported to give transient and usually incomplete responses in metastatic osteosarcoma (Evans, 1961; Friedman and Carter, 1972; Pratt, 1980; Sullivan and Sutow, 1963). These agents had little impact on patient survival. In the 1970's, however, agents such as cisplatin and doxorubicin* became available and the use of high-dose methotrexate was introduced. These new treatments began to affect significantly patient survival. The use of these new treatment approaches in metastatic osteosarcoma will be considered in detail below. Selected papers will be highlighted to illustrate progress in disseminated disease made with these drugs used alone or in combination. Their use in the adjuvant and neoadjuvant therapy of osteosarcoma is considered elsewhere in this symposium's proceedings.

Doxorubicin, a cytotoxic antibiotic derived from *Streptomyces peucentius*, was shown in the early 1970's to produce objective tumor regressions in a variety of neoplasms. In one of the first major investigations of doxorubicin in bone tumors in North America, Cortes et al from the Roswell Park Memorial Institute in Buffalo treated 13 patients with metastatic osteogenic sarcoma beginning in July 1970 (Cortes et al, 1972). Doxorubicin was given at a dosage of 17.5-35 mg/M^2 per day for three to four days. This course was repeated monthly. They reported one complete response lasting over 22 months. Additionally, there were three partial responders, and a fourth patient had a measurable decrease of 25% in tumor size. All patients showing a response received more than 20 mg/M^2 of doxorubicin for three consecutive days, monthly, and responses were observed beginning within the first month of treatment. One of the partial responders had failed before on 5-FU, methotrexate and cyclophosphamide therapy. This patient's partial response lasted ten months. The patient died of cardiomyopathy with evidence of tumor progression. The second patient in this study, a partial responder for 12 months, died of progressive disease. The third partial responder obtained a complete response by pneumonectomy following therapy with doxorubicin. This patient did well for 12.5 months and then died of chemotherapy-induced cardiomyopathy without evidence of tumor at autopsy. This early study included two deaths secondary to chemotherapy-induced cardiomyopathy, but in both incidents the patients had received more than 900 to 1000 mg/M^2 cumulative doxorubicin doses.

From November 1971 to August 1974, the Acute Leukemia Group B conducted a Phase I followed by a Phase II trial of doxorubicin in patients with

----

* Adriamycin

** Dacarbazine

metastatic osteogenic sarcoma (Cortes et al, 1975). In the Phase I portion of the study, doxorubicin was administered at 17.5 to 35 mg/M^2 per day for three to four days. The course was repeated every three weeks. Of 17 patients studied, all had measurable pulmonary metastases. One attained a complete response, five a partial response, and one patient showed a 25% measurable decrease of tumor size. The complete responder was alive and disease free at 54 months. Twenty-six patients were admitted to the Phase II portion of the study and received similar therapy. Of this group, only two partial responses were observed, in two of the five patients who had received a dose of 30 mg/M^2 per day for three days. These studies showed that responsiveness to doxorubicin was dependent on adequate dosage. The need to limit the total cumulative doxorubicin dose to avoid cardiac toxicity was also confirmed. Doxorubicin combined with DTIC has produced responses in metastatic osteosarcoma (Gottlieb et al, 1972; 1975). Doxorubicin with high-dose cyclic methotrexate has also yielded responses and symptom-free survival in patients with advanced osteogenic sarcoma (Rosen, 1974).

In the late 1960's, investigators had begun to use high-dose methotrexate with citrovorum rescue (HD-MTX-CR) with some success in the treatment of primary carcinoma of the lung. The major early side effects with this therapy were stomatitis, myelosuppression, and nephrotoxicity. By decreasing the time of methotrexate infusion to six hours or less and with vigorous hydration and the use of sodium bicarbonate for urinary alkalinization, the incidence of the severe forms of these toxicities was successfully reduced. Since responses to HD-MTX-CR were seen in patients who had been refractory to low-dose methotrexate therapy, it was recognized that tumors that had previously been resistant to methotrexate might indeed respond to methotrexate in high doses. Disseminated osteogenic sarcoma was known to be resistant to low-dose methotrexate or methotrexate given in conventional doses of 30 mg/M^2 twice weekly or 3 mg/M^2 daily.

In June 1973, Jaffe et al reported their experience using HD-MTX-CR in ten patients with metastatic osteogenic sarcoma. Their regimen consisted of high-dose methotrexate, 100 to 200 mg/M^2 as a six-hour infusion, followed by citrovorum rescue and repeated every two to three weeks. All but one patient had easily measurable pulmonary metastases; the remaining patient had measurable soft tissue and bone metastases. In the ten patients treated, two had complete disappearance of their pulmonary metastases, with two others attaining a partial remission. The complete responses lasted five to seven months, but the partial responders relapsed after only one to two months. In five of the patients, radiation therapy was given in addition to high-dose methotrexate. Although all five of these patients were partial responders, the responses remained short, the longest lasting two months. Jaffe et al (1977) have also demonstrated that HD-MTX-CR given with vincristine on a weekly basis may be more effective than the same regimen given every three weeks and will even produce some responses in patients with metastatic disease who have failed on the triweekly schedule. Rosen has stressed the importance of HD-MTX-CR therapy as a precise pharmacokinetic procedure (Rosen and Nirenberg, 1982). He has cautioned that to achieve effective therapy, careful attention must be given to an adequate methotrexate dosage, fluid intake, serum methotrexate measurements, alkalinization of urine, and leucovorin rescue.

With the appreciation that several drugs had demonstrable effectiveness in metastatic osteogenic sarcoma, St. Jude Children's Hospital embarked on a nonrandomized study conducted between March, 1972 and July, 1979 (Pratt et al, 1980), testing the efficacy of HD-MTX-CR alone or combined with doxorubicin and cyclophosphamide in patients with metastatic osteogenic sarcoma. Methotrexate was given at a dosage of 2-15 g/M^2, followed by citrovorum rescue. Twenty-seven patients were included in the study, with 13 receiving doxorubicin and cyclophosphamide in addition to the HD-MTX-CR. Six of the patients had received prior adjuvant chemotherapy with doxorubicin, cyclophosphamide and HD-MTX-CR. None of these patients responded. Three of the 27 patients achieved a complete response (CR). The one treated with HD-MTX-CR alone had a short response lasting three weeks. The remaining two patients on CR lasted seven and 40 months respectively and were in the combination chemotherapy group. Additionally, two partial responses (PR) were seen lasting five and seven months respectively. When compared with historical controls, the improved survival in patients responding to HD-MTX-CR, with or without doxorubicin or other drugs, was very encouraging. However, it remained unclear whether improved survival could be achieved with doxorubicin and cyclophosphamide alone without the addition of HD-MTX-CR. In 1978, A.S. Levine, from the National Cancer Institute (Levine et al, 1978), published a nonrandomized study of the effectiveness of cyclophosphamide, 45 mg/kg per day for one to four days plus doxorubicin 35 mg/M^2 per day for two consecutive days with gradual increase of the cyclophosphamide dosage, followed by two cycles of HD-MTX-CR in patients who had not previously received any methotrexate therapy. The HD-MTX was omitted in patients who were known to have been methotrexate failures. All responders were in the former group of patients, with one CR and one short PR. None of the patients responded to an increase of the cyclophosphamide dosages.

HD-MTX-CR has been used with radiation and aggressive resection of pulmonary metastases in the management of metastatic osteogenic sarcoma. From 1973 to 1976, a group of investigators at the Sidney Farber Institute in Boston (Weichselbau et al, 1977) treated ten patients with HD-MTX-CR plus low-dose radiation therapy to both lungs with additional boosting to areas of known gross disease. This was followed, whenever possible, by surgical resection of pulmonary metastases. Eight of the ten patients had failed on adjuvant chemotherapy, three with HD-MTX-CR plus doxorubicin, three with high-dose methotrexate alone, one with doxorubicin alone, and one with transfer factor. In this aggressively treated group, three patients attained a complete remission with disease-free survival continuing at 12, 18, and 24 months. In selected patients, such an aggressive multimodality approach may drastically change the survival outlook.

In 1965, cisplatin, a chelated heavy metal, was developed as an antitumor agent. It was shown to be active against testicular carcinomas and epidermal carcinomas. In 1978, Ochs et al from the Roswell Park Memorial Institute published a report on the use of cisplatin in eight patients with metastatic osteogenic sarcoma. All patients had been previously treated with adjuvant doxorubicin and six patients had also received HD-MTX-CR. One patient with CR survived more than 18 months, and four patients with PR lasting two to five months. In 1979, the Children's Cancer Study Group (Baum et al, 1979) published the results of a Phase II trial of cisplatin in metastatic osteogenic sarcoma. Three to 4.5 mg/kg were given every three weeks. Vigorous hydration and diuresis was maintained throughout the administration of cisplatin. Of the 17 patients studied, all had metastatic osteogenic sarcoma resistant to doxorubicin and HD-MTX-CR. One of these patients obtained a complete response. This patient refused further therapy after three cycles. Despite this, the patient was disease free two months after stopping chemotherapy. Partial responses lasting seven and five months were seen in two additional patients. Cisplatin appeared to be an active agent in osteogenic sarcoma, with responses and prolonged survival even in patients previously resistant to doxorubicin and HD-MTX-CR. In a Phase II trial conducted by the EORTC in patients with osteogenic sarcoma previously treated with one or more of doxorubicin, methotrexate, cyclophosphamide and vincristine, cisplatin gave a 20% response rate (Bonadonna et al, 1983). Cisplatin has been given by the intra-arterial route to obtain regressions in primary osteogenic sarcoma (Jaffe et al, 1983).

With advances in treating osteogenic sarcoma with adjuvant chemotherapy as well as with advances in treating metastatic disease, a growing number of patients are now presenting with new or recurrent metastatic disease which is resistant to many of the chemotherapeutics, including HD-MTX-CR, doxorubicin, and cisplatin. In 1977, Rosen et al, from Memorial Sloan Kettering Cancer Center in New York, reported the results of a trial with bleomycin, cyclophosphamide, and dactinomycin in patients failing on adjuvant therapy (Mosende et al, 1977). Therapy was given on the first two days of a biweekly cycle with bleomycin, 12 mg/M^2, cyclophosphamide 600 mg/M^2, and dactinomycin 450 mg/M^2. Eight of the 13 patients had received previous treatment with high-dose methotrexate with citrovorum rescue, cyclophosphamide and doxorubicin. Three of these eight patients obtained a partial remission lasting 2+ to 3+ months. Of the five previously untreated patients, all had evidence of tumor regression. The overall response rate for both groups of patients was 61.5%. Unfortunately, no complete remissions were obtained. The major side effects of this regimen was severe nausea and vomiting in 100% of the patients, which could be controlled with antiemetics. Mucositis was rare; alopecia was virtually universal. Only mild to moderate myelosuppression was observed.

No one would dispute that currently available cytotoxic drugs offer relatively effective therapy for metastatic osteosarcoma. However, certain considerations qualifying this conclusion need emphasis (Friedman and Carter, 1972; Pratt, 1980; Rosen and Nirenberg, 1982). Osteosarcoma is an uncommon disease and the various reported studies dealing with its chemotherapy involve small numbers of patients. Randomized clinical trials have been virtually impossible because of the small numbers of patients available for study. Only rarely are the various stratification factors affecting clinical outcome considered when the results of therapy are evaluated. These are:

(i) *Location of the primary tumor:* patients with primary lesions of the femur have a worse prognosis than those with lesions in the humerus or tibia.

(ii) *Age of the patient:* patients older than 21 years have a more favorable prognosis, although this is not a uniformly held position.

(iii) *Histopathologic subgroup:* The osteoblastic, telangiectatic, fibrosarcomatous and fibrous histiocytomatous variants of osteosarcoma have a more favorable prognosis than the small cell variant. Parosteal (or juxtacortical) subtypes of osteogenic sarcoma have a lower malignant potential and are more readily resected than other subtypes of osteogenic sarcoma. Some osteogenic sarcomas may have chondrosarcomatous elements and ought more properly be classified as osteochondrosarcomas. The results obtained with therapy for studies, especially if they are small, which comprise patients with more favorable histopathologic findings will be difficult to evaluate in comparison with other series with less favorable histopathologic patterns.

(iv) *Thoroughness of initial workup:* for purposes of interstudy comparisons, all patients should have bone scans and either tomographic roentgenograms of the chest or computerized axial tomographic examinations of the chest.

(v) *Time from diagnosis to metastases:* The longer the time from initial diagnosis to the development of metastases the better the patient prognosis.

(vi) *Tumor size:* patients with tumors larger than 10 cm have 7-fold less likelihood of surviving five years than patients with smaller tumors.

These factors may all influence the response to treatment, patient prognosis and survival. This is seldom discussed in clinical reports. Some drugs used in combination therapies, such as vincristine and bleomycin, have not been clearly shown to be effective by themselves in osteogenic sarcoma. Yet, they are included in therapeutic regimens not only for disseminated disease but also for adjuvant treatment. The drugs used in the 1970's and the 1980's are certainly superior to those used in the 1960's although this could not be demonstrated in Phase III studies. Indeed, hardly any Phase III studies have been done in disseminated osteosarcomas. There are no Phase III studies to establish definitively the value of any combination of drugs over single agent therapies in that condition; nor are there any Phase III trials to show, statistically convincingly, that a particular high-dose methotrexate regimen is superior to another. The literature is filled with the competing claims of numerous Phase II clinical investigations. Their significance depends on questionable comparisons with historical controls. They provide shaky justification for the complicated and difficult adjuvant and neoadjuvant multiple agent chemotherapy studies currently being advocated. Agents such as cisplatin and doxorubicin are combined for adjuvant therapy without testing the combination first to observe whether it is effective in metastatic disease (Ettinger et al, 1979). The increasing use of pulmonary metastasis resection with and without radiation to the lungs combined with cytotoxic chemotherapy for metastatic disease makes it difficult to evaluate the relative contribution to patient survival of any one modality. What seems obviously necessary based on all the above data, is to concentrate in future on the care of osteogenic sarcoma in a small number of specialized centers. To conduct properly controlled and adequately stratified future studies will require that even these centers must pool their clinical resources for national and intergroup cooperative clinical trials. Those types of studies will form the basis for future advances in the management of disseminated osteogenic sarcoma. Without them, the field will become confused as is currently the case in studies of adjuvant therapy for osteosarcoma (Lange et al, 1982).

In the meantime, what should be done with the patient who suffers from advanced metastatic disease? That patient should, whenever possible, be entered on a cooperative group or single institution clinical trial protocol. A practical approach where that is not possible would be to use the T-10 protocol popularized by Rosen (1981) or a variation of that aggressive multiple agent regimen. Single agent therapy used sequentially would be wholly justifiable. The treatment of metastatic disease resulting after a surgical resection not combined with adjuvant or neoadjuvant chemotherapy would depend on the disease-free interval and the pattern of metastatic development. If metastases develop within six to nine months after resection, then sequential therapy with high-dose methotrexate to be followed (should responses not be obtained) by doxorubicin and then cisplatin is a justified approach. Pulmonary metastases should be resected whenever possible. A patient who has pulmonary metastases six to nine months or later following a course of adjuvant therapy should have the metastases surgically resected and then be continued on the original chemotherapeutic regimen, perhaps with the addition of another drug. An early relapse while on adjuvant chemotherapy or shortly after its cessation is an indication for more aggressive chemotherapy with multiple agents.

Ewing's Sarcoma

Ewing's sarcoma is another malignant tumor of bone, although it occasionally arises in soft tissue. It is a relatively uncommon tumor accounting for only about 10% of all primary bone tumors. It is, however, an unusually aggressive tumor. Radical surgery has been of little benefit for the primary tumor since pulmonary and bone metastases invariably develop within several months. Seventy-five percent of patients die of metastatic disease, even when local disease is adequately controlled with surgery and/or radiation therapy. Progress was not made in the treatment of this disease until the advent of effective chemotherapeutic agents. The remaining portion of this review will focus on a sampling of those significant papers which over the past 10-14 years have contributed to the development of effective chemotherapy for Ewing's sarcoma.

In 1970, the National Cancer Institute reported on using actinomycin-D for five to seven days in patients with metastatic Ewing's sarcoma (Senyszy et al, 1970). All five patients had been refractory to previous therapy with cyclophosphamide and vincristine. Despite this, three patients obtained a partial remission. Although all responses were brief, the longest lasting about two months, these results suggested a lack of cross-resistance with actinomycin-D and the combination cyclophosphamide/vincristine. In 1971, the Southwest Cancer Chemotherapy Study Group (Sutow, et al, 1971) reported their Phase I trial of multiple chemotherapeutic drugs in metastatic Ewing's sarcoma. They reported one partial remission in 12 patients treated with uracil mustard. Using cyclophosphamide, they had five partial remissions in 13 patients. Five of these patients had been previously treated with regimens including cyclophosphamide. Vincristine gave one partial remission in seven patients. Daunorubicin gave four responses in nine patients. All but one of these had been previously treated with vincristine and cyclophosphamide. Six patients had also received actinomycin-D. In 1972, Oldham reported on his experience at the National Cancer Institute using doxorubicin, 60-65 mg/M^2, given every three weeks. All 11 patients had widely metastatic disease. Nine of these were refractory to previous therapy with cyclophosphamide, vincristine, and actinomycin-D. Four patients had also received prior therapy with BCNU. Despite this, six patients achieved a partial remission.

These early encouraging results with single agent chemotherapy in advanced metastatic Ewing's sarcoma led to the investigation of combination chemotherapy given concurrently with radiation to the area of primary disease to improve cure rate and prolong disease-free survival. In 1979, Chan et al, from the M.D. Anderson Hospital, reported results with vincristine plus cyclophosphamide with concurrent radiation therapy in 21 patients with Ewing's sarcoma. Fourteen patients died of distant metastatic disease with a mean survival of 29 months. No local recurrences were encountered. Seven patients remained disease free at 63+ to 99+ months. They simultaneously reported on 15 patients treated with cyclophosphamide, doxorubicin, vincristine, and actinomycin-D plus local radiation therapy. One patient died with local recurrence and diffuse metastatic disease at 16 months. Five additional patients died of distant metastases with a mean survival of 29 months. Nine had no recurrence from 26+ to 60+ months. From 1964 to 1972, the group at St. Jude Children's Hospital (Hustu et al, 1972) treated 13 patients with the combination vincristine, cyclophosphamide, and radiation and achieved complete remission in all 13 patients. Eleven were still disease free at four to 91 months following completion of their chemotherapy. Chemotherapy in the initial group was continued for one year, but two of the earliest patients treated relapsed one week and seven months after completing their one-year chemotherapy. Because of these early failures all other patients were continued on chemotherapy for two years with no subsequent relapses. Continued follow-up of these patients would be necessary in assessing ultimate cure.

In 1976, the group from the Sidney Farber Cancer Institute in Boston (Jaffe et al, 1976) reported the use of vincristine, actinomycin-D and cyclophosphamide with simultaneous radiation to the areas of primary disease. Nine of their 14 patients had no evidence of metastatic disease at the onset of therapy. Two of the nine died of pulmonary metastases. Permanent local control was achieved in all patients. All seven patients suffering from localized disease obtained a complete remission with disease-free survival ranging from four months to four and one-half years following completion of two years of chemotherapy. Three of the five patients with distant metastases achieved a complete remission lasting from 9 to 24+ months. Efforts to consolidate these complete remissions were made with radiation therapy to the lungs in two patients. Despite this, both died of metastatic disease within six months following completion of radiation therapy. All patients who failed treatment with this chemotherapeutic regimen were subsequently treated with doxorubicin, 75 to 90 mg/M^2, at three week intervals. A minimum of four courses were given to each patient. There were no responders in this group, pointing out the ineffectiveness of doxorubicin in patients failing on primary therapy with combination vincristine, actinomycin-D, and cyclophosphamide.

In the T-2 protocol, the Memorial Sloan-Kettering Cancer Center Group (Rosen et al, 1978) attempted to improve results by the addition of doxorubicin to actinomycin-D, vincristine, and cyclophosphamide. This combination was tried in 20 patients previously untreated and with localized disease. All obtained a complete remission but five patients relapsed, all at the end of the second year of chemotherapy or shortly after the end of chemotherapy. Of the 15 patients remaining in complete remission, the mean survival at the time of publication was 46+ months with actuarial survival estimated at 75%. Additionally, eight patients with metastatic disease were treated in a similar fashion. All obtained a complete remission with seven relapses also occurring at the end of their chemotherapy. This led the investigators to propose the hypothesis that aggressive initial chemotherapy as well as intensive consolidation chemotherapy might be necessary to improve long-term survival and cure. Their current T-11 protocol involves intensive chemotherapy with cycles of cyclophosphamide, doxorubicin, and high-dose methotrexate alternating with cycles of bleomycin, cyclophosphamide and actinomycin-D; vincristine is given on a weekly basis throughout both cycles (Rosen, 1981).

The results of the Cancer and Leukemia Group B Study (Pilepich et al, 1981; Vietti et al, 1981), using cyclophosphamide, doxorubicin, and vincristine, followed by actinomycin-D, vincristine, cyclophosphamide, and followed again with a second course of doxorubicin, resulted in 74% complete remissions in 44 patients. An additional 18% achieved a partial remission. Once again, all relapses occurred within 24 months. These results were similar to those previously reported from Sloan-Kettering.

Summary

New and progressively intensive chemotherapy regimens have been devised using agents with known effectiveness in Ewing's sarcoma. These regimens have included vincristine, doxorubicin, cyclophosphamide, and DTIC. Protocols in some high-risk patients treated at the National Cancer Institute (Tepper et al, 1980) have included all these agents combined with total body radiation and autologous bone marrow rescue. The early results of such aggressive therapy have been encouraging but further follow-up is needed.

The success of combination chemotherapy including vincristine, cyclophosphamide, doxorubicin and actinomycin-D together with radiotherapy for the management of metastatic disease (Pilepich et al, 1981; Vietti, et al, 1981) have allowed the Advanced Ewing's Sarcoma Intergroup Study to conclude "that intensive multimodal management in metastatic Ewing's sarcoma can produce prolonged disease-free survival and possibly cure in a significant percentage of patients." That accurately summarizes the present state of therapy for disseminated forms of Ewing's sarcoma. Up to 40% and more patients may obtain long-term disease-free survival or cure with currently available multimodality therapies. The chemotherapy portion of these multimodality programs will now uniformly include combinations of vincristine, actinomycin-D and cyclophosphamide. Doxorubicin is sometimes added or substituted for actinomycin-D although without evidence that it is more active than actinomycin-D in Ewing's sarcoma.

Patients with pulmonary metastases who relapse following completion of perioperative and adjuvant chemotherapy may have their disease controlled with pulmonary radiation (+ actinomycin-D) and pulmonary metastatic resection. They may be further benefited by continuing their previous adjuvant chemotherapy.

References

1. Baum, E.S., Gagnon, P., Greenberg, L., Krivit, W., and Hammond, D.: Phase II study of cis-Dichlorodiammineplatinum (II) in childhood osteosarcoma: Children's Cancer Study Group Report. Cancer Treat. Rep. 63, 1621, 1979.

2. Bonadonna, G., Gasparini, M., Rouesse, J., Van Oosterom, A., Wagener, T., Voute, T., Thomas, P. and Sylvester, R.: Phase II study of CDDP in advanced osteogenic sarcoma (OS). Proc. Am. Soc. Clin. Oncol. 2, 240, 1983.

3. Carter, S.K., and Friedman, M.: Osteogenic sarcoma treatment overview and some comments on interpretations of clinical trial data. Cancer Treat. Rep. 62, 199, 1978.

4. Chan, R.C., Sutow, W.W., Lindberg, R.D., Samuels, M.L., Murray, J.A., and Johnston, D.A.: Management and results of localized Ewing's sarcoma. Cancer 43, 1001, 1979.

5. Cortes, E.P., Holland, J.F., Wang, J.J. and Sinks, L.F.: Doxorubicin in disseminated osteosarcoma. JAMA 221, 1132, 1972.

6. Cortes, E.P., Holland, J.F., Wang, J.J., and Glidwell, O.: Adriamycin (NSC-123127) in 87 patients with osteosarcoma. Cancer Chem. Rep. 6, 305, 1975.

7. Ettinger, L.J., Douglass, H.O. Jr., Higby, D.J., et al: Adriamycin (Adria) and cis-diamminedichloroplatinum (CDDP) as adjuvant therapy in primary osteosarcoma (OS). Proc. Am. Assoc. Cancer Res. 20, 438, 1979.

8. Evans, A.E.: Mitomycin C. Cancer Chem. Rep. 14, 1, 1961.

9. Friedman, M.A., and Carter, S.K.: The therapy of osteogenic sarcoma: current status and thoughts for the future. J. Surg. Oncol. 4, 482, 1972.

10. Gehan, E.A., Sutow, W.W., and Uribe-Botero, G.: Osteosarcoma: the M.D. Anderson experience, 1950-1974. Prog. Cancer Res. Ther. 6, 271, 1978.

11. Gottlieb, J.A., Baker, L.H., Quagliana, J.M., Luce, J.K., Whitecar, J.P. Jr., et al: Chemotherapy of sarcomas with a combination of adriamycin and dimethyl triazeno imidazole carboxamide. Cancer 30, 1632, 1972.

12. Gottlieb, J.A., Baker, L.H., O'Bryan, R.M., Sinkovics, J.G., Hoogstraten, B., et al: Adriamycin (NSC-123127) used alone or in combination for soft tissue and bony sarcomas. Cancer Chem. Rep. 6, 271, 1975.

13. Harvei, S., and Solheim, O.: The prognosis in osteosarcoma. Norwegian national data. Cancer 48, 1719, 1981.

14. Hustu, H.O., Pinkel, D., and Pratt, C.B.: Treatment of clinically localized Ewing's sarcoma with radiotherapy and combination chemotherapy. Cancer 6, 1533, 1972.

15. Jaffe, N., Farber, S., Traggis, D., Geiser, C., Kim, B.S., Dass, L., Frauenberger, G., Djerassi, I., and Cassady, J.R.: Favorable response of metastatic osteogenic sarcoma to high-dose methatrexate with citrovorum rescue and radiation therapy. Cancer 31, 1367, 1973.

16. Jaffe, N., Frei, E. III, Traggis, D., and Bishop, Y.: Adjuvant methotrexate and citrovorum-factor treatment of osteogenic sarcoma. N. Engl. J. Med. 291, 994, 1974.

17. Jaffe, N.J., Traggis, D., Salian, S., and Cassady, J.R.: Improved outlook for Ewing's sarcoma with combination chemotherapy (vincristine, actinomycin-D and cyclophosphamide) and radiation therapy. Cancer 38, 1925, 1976.

18. Jaffe, N., Frei, E.III, Traggis, D., and Watts, H.: Weekly high-dose methotrexate-citrovorum factor in osteogenic sarcoma. Pre-surgical treatment of primary tumor and of overt pulmonary metastases. Cancer 39, 45, 1977.

19. Jaffe, N., Knapp, J., Chuang, V.P., Wallace, S., Ayala, A., Murray, J., Cangir, A., Wang, A., and Benjamin, R.S.: Osteosarcoma: intra-arterial treatment of the primary tumor with Cis-Diammine-Dichloroplatinum (CDP). Cancer 51, 402, 1983.

20. Lange, B., Kramer, S., Gregg, J.R., Toledano, S., Wimmer, R., and Evans, A.E.: High-dose methotrexate and adriamycin in osteogenic sarcoma. The Children's Hospital of Philadelphia Study. Am. J. Clin. Oncol. 5, 3, 1982.

21. Levine, A.S., Appelbaum, R.G., Graw, R.G., Magrath, I.T., Pizzo, P.A., Poplack D.G., and Ziegler, J.L.: Sequential combination chemotherapy (containing high-dose cyclophosphamide) for metastatic osteogenic sarcoma. Cancer Treat. Rep. 62, 247, 1978.

22. Mike, V., and Marcove, R.C.: Osteogenic sarcoma under the age of 21: experience at Memorial Sloan-Kettering Cancer Center. Prog. Cancer Res. Ther. 6, 283, 1978.

23. Mosende, C., Gutierrez, M., Caparros, B., and Rosen, G.: Combination chemotherapy with bleomycin, cyclophosphamide and dactinomycin for the treatment of osteogenic sarcoma. Cancer 40, 2779, 1977.

24. Ochs, J.J., Freeman, A.I., Douglass, H.O., Higby, D.S., Mindell, E.R., and Sinks, L.F.: cis-Dichlorodiammineplatinum (II) in advanced osteogenic sarcoma. Cancer Treat. Rep. 62, 239, 1978.

25. Oldham, R.K., and Pomeroy, T.C.: Treatment of Ewing's sarcoma with adriamycin. Cancer Chem. Rep. 56, 635, 1972.

26. Perez, C., Herson, J., Kimbal, J.C., and Sutow, W.W.: Prognosis after metastases in osteosarcoma. Cancer Clin. Trials 1, 315, 1978.

27. Pilepich, M.V., Vietti, T.J., Nesbit, M.E., Tefft, M., Kissane, J., Burgert, E.O., and Pritchard, D.: Radiotherapy and combination chemotherapy in advanced Ewing's sarcoma. Cancer 47, 1930, 1981.

28. Pratt, C.B.: Chemotherapy of osteosarcoma - an overview. In, Therapeutic Progress in Ovarian Cancer, Testicular Cancer and the Sarcomas. Eds., A.T. Van Oosterom et al. Martinus Nijhoff Publishers. The Hague, 1980.

29. Pratt, C.B., Howarth, C., Ransom, J.L., Bowles, D., Green, A.A., Kumar, A.P.M., Rivera, G., and Evans, W.E.: High-dose methotrexate used alone and in combination for measurable primary or metastatic osteosarcoma. Cancer Treat. Rep. 64, 11, 1980.

30. Rosen, G.: Current management of malignant bone sarcomas. In, Cancer. Achievements, Challenges and Prospects for the 1980s. Eds., J.H. Burchenal and H.F. Oettgen. Gruen and Stratton, New York, 1981.

31. Rosen, G., Suwansirikal, S., Kwon, C., Tan, C., Wu, S.J., Beattie, E.J., Jr., and Murphy, M.L.: High-dose methotrexate and citrovorum factor rescue and adriamycin in childhood osteogenic sarcoma. Cancer 33, 1151, 1974.

32. Rosen, G., Caparros, B., Mosende, C., McCormick, B., Huvos, A.G., and Marcove, R.C.: Curability of Ewing's sarcoma and considerations for future therapeutic trials. Cancer 41, 888, 1978.

33. Rosen, G., and Nirenberg, A.: Chemotherapy for osteogenic sarcoma: an investigative method, not a recipe. Cancer Treat. Rep. 66, 1687, 1982.

34. Senyszy, J.J., Johnson, R.E., and Curran, R.E.: Treatment of metastatic Ewing's sarcoma with actinomycin-D (NSC-3053). Cancer Chemo. Rep. 54, 103, 1970.

35. Sullivan, M.P., and Sutow, W.W.: Phenylalanine mustard as a treatment for metastatic osteogenic sarcoma in children. J. Ped. 63, 227, 1963.

36. Sutow, W.W., Vietti, T.J., Fernback, D.J., Lane, D.M., and Lonsdale, D.: Evaluation of chemotherapy in children with metastatic Ewing's sarcoma and osteogenic sarcoma. Cancer Chem. Rep. 55, 67, 1971.

37. Taylor, W.E., Ivins, J.C., and Dahlin, D.C.: Trends and variability in survival from osteosarcoma. Mayo Clin. Proc. 53, 695, 1978.

38. Tepper, J., Glaubiger, D., Lichter, A., Wackenhut, J., and Glatstein, E.: Local control of Ewing's sarcoma of bone with radiotherapy and combination chemotherapy. Cancer 9, 1969, 1980.

39. Vietti, T.J., Gehan, E.A., Nesbit, M.E. Jr., Burgert, E.O. Jr., Pilepich, M., Tefft, M., Kissane, J., and Pritchard, D.J.: Multimodal therapy in metastatic Ewing's sarcoma: An intergroup study. Nat. Cancer Inst. Monograph 56, 279, 1981.

40. Weichselbau, R.R., Cassady, J.R., Jaffe, N., and Filler, R.M.: Preliminary results of aggressive multimodality therapy for metastatic osteosarcoma. Cancer 40, 78, 1977.

Malignant Transformation of Bone Tumors

T. VIZKELETY, L. KÉRY, and Z. CSATÓ

Introduction

The malignant transformation of benign tumors and tumor-like conditions has always attracted the attention of clinicians and pathologists. Moreover, it has also been a serious problem both as to diagnosis and to therapy. In this paper the difficulties we had to face in daily practice will be discussed, based on examples of a few cases. Osteoblastoma, a classic semimalignant tumor, is widely known and the generally accepted principles for its therapeutic management can be clearly outlined. We diagnose this tumor and treat it according to the guidelines published in the literature, and our results are statistically similar to those reported by authoritative investigators. In this presentation, we will not deal with the osteoblastoma but with other rarely-occurring tumors with malignant transformations. The desmoid tumor of the extremities is generally treated by the orthopedic surgeon but, although it is not a bone tumor, we believe it is worthwhile to report one of our cases.

Clinical Results

We have treated five patients with a desmoid tumor of the lower extremities, one of which underwent a malignant transformation.

On December 19, 1977 a 12-year old girl was admitted with a flexion contracture of the right knee without previous trauma. No other pathologic alteration was found. The contracture was corrected with a plaster cast. She had no complaints for some time, but later the flexion contracture recurred and a 10° equinus contracture had developed as well.

During surgery on June 11, 1978 we found adjacent to the medial gastrocnemius a dense tumor with a broad base which, however, was not attached to the surrounding soft tissue. The popliteal artery, found laterally, was surrounded and obliterated by a similar tumor. Only three vascular lumina, as thin as needles, could be observed. The tumor tissue was excised as much as possible. The tibial nerve, showing a finger-thick swelling proximal to the tumor, was dissected. Histologic examination showed a desmoid (Fig. 1).

Current Concepts of Diagnosis and Treatment
of Bone and Soft Tissue Tumors.

a) H.E. 25x

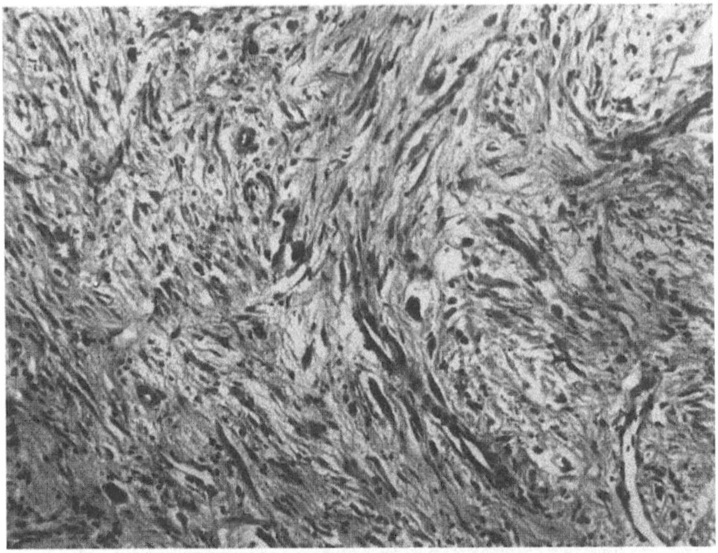

b) H.E. 80x

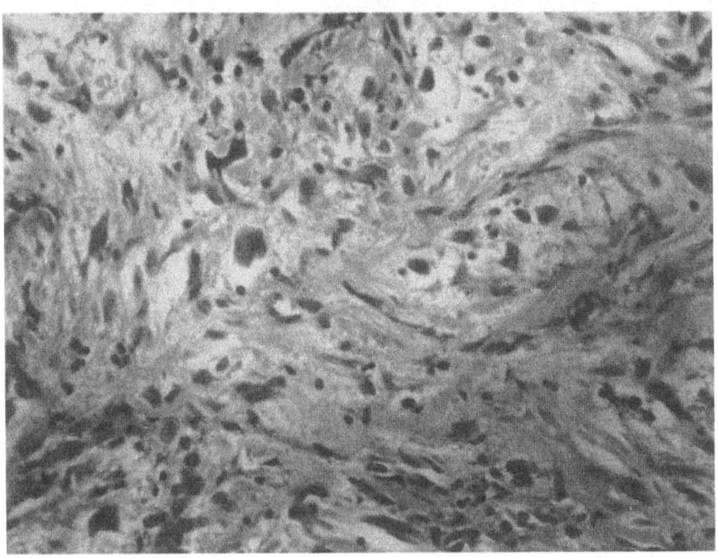

Fig. 1. Between the interwoven bundles of intercellular collagen, large numbers of fibroblast- and fibrocyte-like cells and bands of striated muscle are present.

Postoperatively, the circulation of the limb was satisfactory despite total absence of the dorsalis pedis pulse. The foot was somewhat paler and colder than the contralateral foot.

In January, 1979 we found a widespread recurrence, with an extension of the tumor to the thigh and lower leg. Since the circulation of the distal parts of the limb was secured by the collateral vessels only, a new angiogram was done preoperatively and a radical excision of the tumor was decided, followed by a simultaneous substitution of the popliteal artery. The tumor in the gastrocnemius and the thigh was removed and we made a bypass with the venae saphena parva. The histologic pattern was the same as before. Due to the rapid invasive growth of the tumor, irradiation was recommended.

The next recurrence was seen in June, 1980 when a daily increase in the size of the tumor was observed. At reoperation, the borders of the tumor were cleared. During the operation the remaining part of the popliteal artery was freed together with the transplanted venae saphena parva which could be isolated easily.

The tumor which was inseparable from the sciatic nerve was excised with it as much as possible in healthy tissue. Wound healing was uneventful and the lack of the sciatic nerve did not disturb the patient's gait. Histologic findings showed a recurrence with signs of malignant transformation (Fig. 2).

In February, 1981 a large tumor was observed, again on the lower leg, apparently localized on the peroneal muscles. Exposure revealed that the tumor had invaded a large part of the triceps surae, the peroneal muscles and the extensors. In the upper third of the lower leg intact muscle could hardly be seen.

One week earlier a brownish tumor of 15 mm diameter developed in the right elbow at the site of a former vena dissection. This, too, was removed. Over the left iliac crest a sensitive region could be observed and a tumor palpated. Whereas previous histologic examinations consistently showed the characteristic picture of a desmoid, the last histologic examination confirmed at both sites the presence of a typical fibrosarcoma (Fig. 3).

The case is interesting from two aspects:

The tumor that proved to be a desmoid on repeated histologic examinations transformed after 2-1/2 years to a fibrosarcoma, appearing also on the elbow, on the skin, and in the lumbosacral area.

The transplanted vessel behaved in the tumor tissue in some respects like a foreign body. In contrast to the original popliteal artery and nerves, it was not invaded and could be isolated from the tumor tissue.

Until recently, only a few papers on the malignant transformation of cysts of aneurysmal bone have been published. Three cases were described by Hirst et al (1970) and one case by Cremer and Münzenberg (1980). Schajowicz (1981) has reported that malignant transformation occurred after irradiation. We have observed two cases.

A 13-year old boy was admitted because of pain in the right shoulder region and a slight swelling. The roentgenograph revealed in the proximal end of the right humerus a rarification of the whole metaphysis with ill-defined borders. The lesion was exposed and after an incision

a) H.E. 25x

b) H.E. 80x

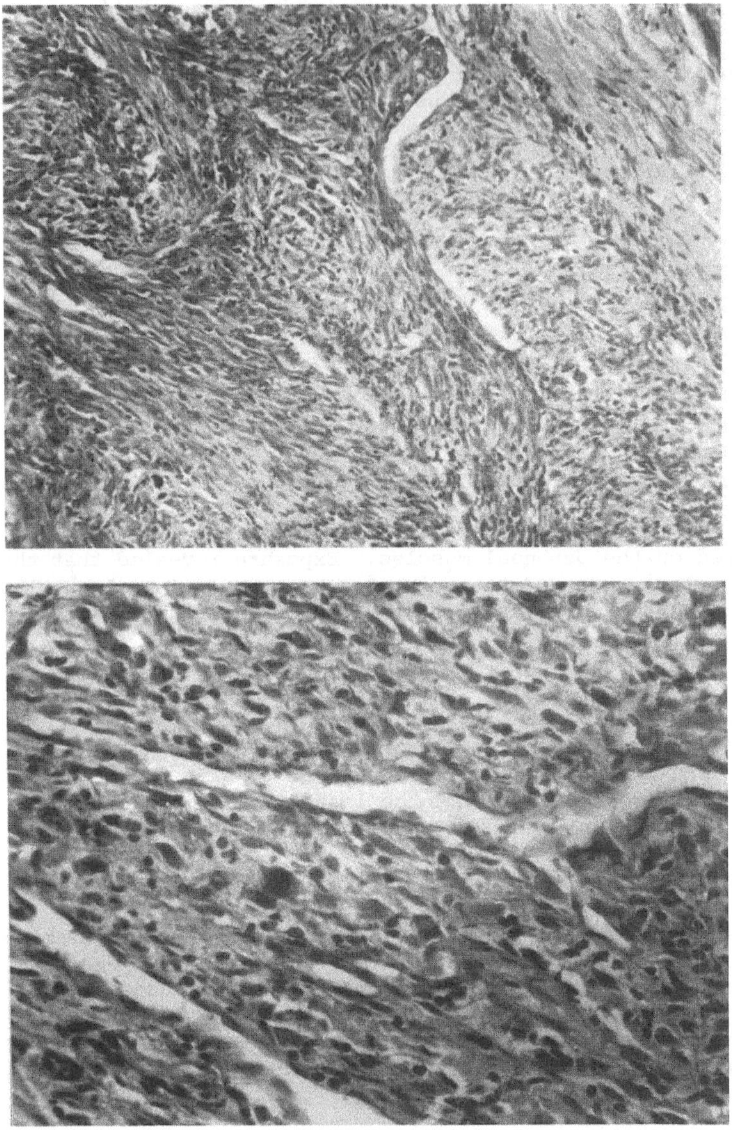

Fig. 2. Similar to previous stage, but numerous vascular spaces with layer of endothelial cells are present between the intercellular collagen bundles. Wide variety of cellularity can be observed around the vascular spaces varying in size, shape and cell density. Atypical forms of mitoses are frequently observed.

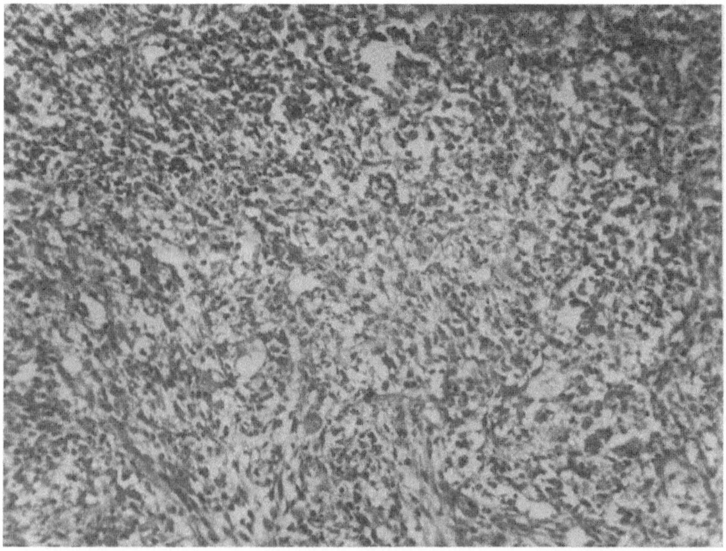

a) H.E. 25x

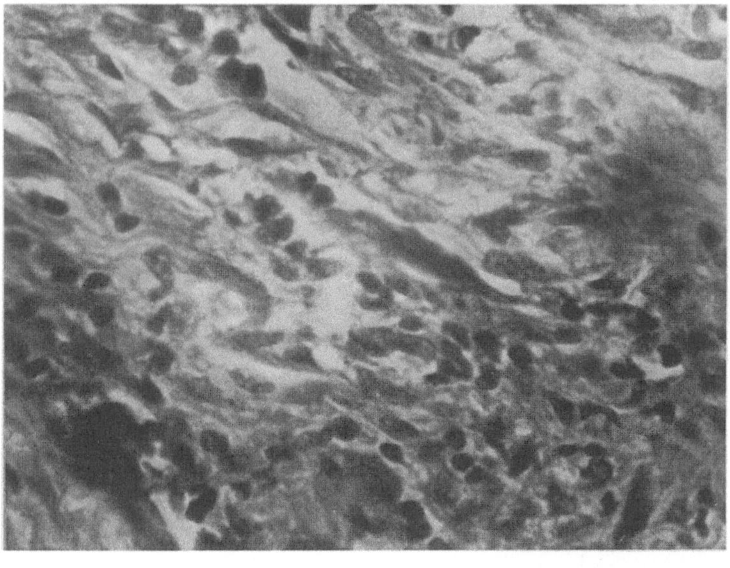

b) H.E. 160x.

Fig. 3. The cell-rich tumor consists of polymorphic cells with hyperchromic, bizarre nuclei. Atypical forms of mitosis are numerous. Intercellular collagen bundles are rare.

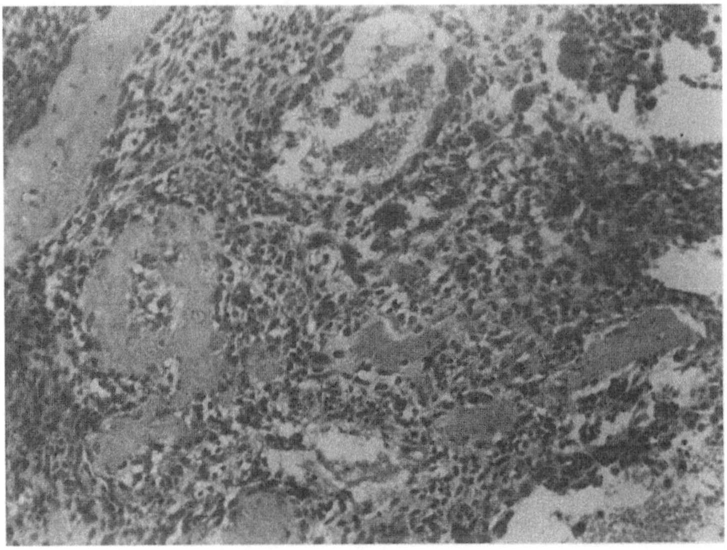

Fig. 4. H.E. 40x. Irregular vascular spaces not lined by endothelium, separated by fibrous septae of different widths, contain osteoclast giant cells and osteoid or immature bone trabeculae. The stroma shows hemosiderin and histiocytic elements. The mitotic forms of cells are not rare.

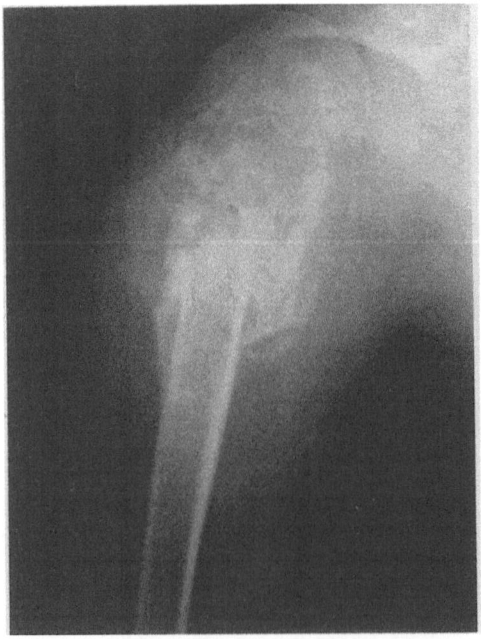

Fig. 5. Aneurysmal bone cyst with transverse fracture on the proximal end of the right humerus. The bone is surrounded by a nonhomogeneous shadow of calcific density.

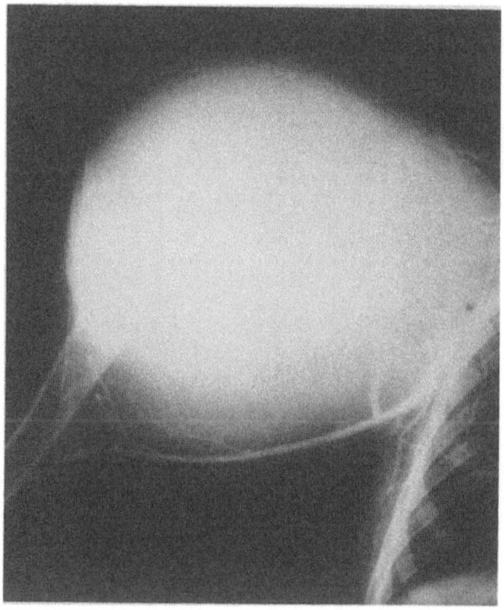

Fig. 6. Large swelling of the proximal part of the upper arm, 3 weeks after the fracture. The angiograph showed filling of the tumor and the displacement of the brachial artery.

of the periosteum on the anterior surface, a blood-filled cavity was opened and excochleated. An aneurysmal bone cyst was presumed and the wound closed after a biopsy was taken from the cyst wall for histologic examination. Histologic examination proved the presence of an aneurysmal bone cyst (Fig. 4).

After wound healing, roentgenotherapy was suggested and the patient received irradiation of 5000 r. In another institution, during gymnastic exercises for his shoulder, the patient's humerus fractured at the site of the cyst (Fig. 5). Following this fracture, swelling increased rapidly. The shoulder was immobilized in a plaster cast without, however, preventing progression of the swelling. It was decided to operate again. The preoperative angiograph showed only the displacement caused by the tumor and no other alterations (Fig. 6). The subclavian artery was exposed to stop possible perioperative bleeding through direct compression. The cavity which was now of the magnitude of two fists was exposed and the cyst's wall, now boneless, removed. The epiphysis remained in the glenohumeral joint. Despite hemostatic procedures, severe postoperative bleeding occurred. Histologic examination of the material obtained at the second operation was the same as observed previously; nevertheless, the possibility of a malignant transformation was considered.

The depressed area, after the resection, filled again during the postoperative days, and on the 10th day reached its original form and a tumor the size of an infant's head had developed. The circulation of the extremity was normal. Due to rapid deterioration, however, and the histo-

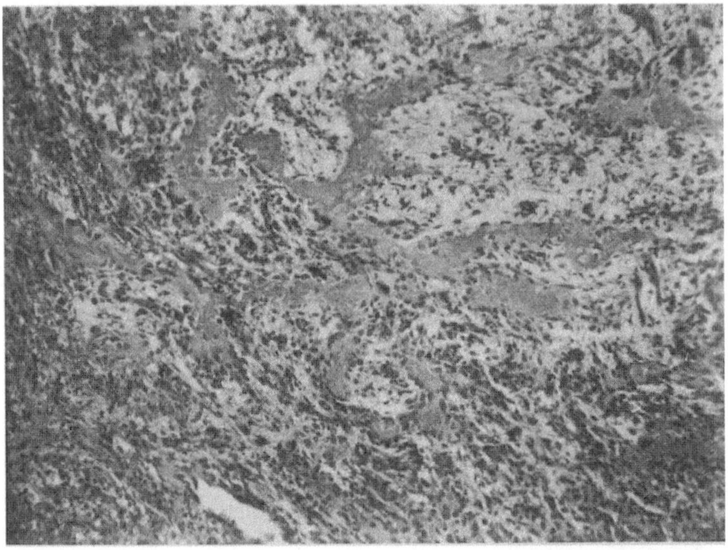

a) H.E. 25x

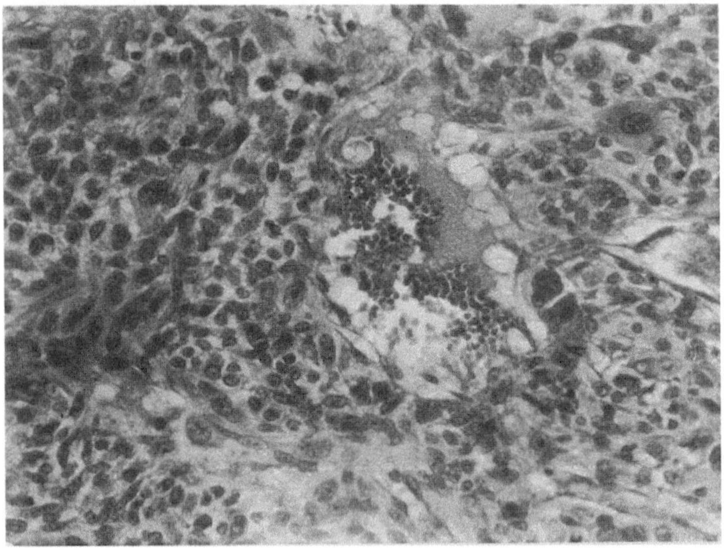

b) H.E. 80x

Fig. 7. The histologic picture is similar to the previous stage, but the stroma cells vary widely in size, shape and density. Mitotic forms of cells are numerous.

logic findings, it was decided to amputate. The remnant epiphysis and the extremity were ablated. Histologic diagnosis was an aneurysmal bone cyst with signs of malignant transformation (Fig. 7).

Six weeks later, bluish-purple nodules appeared under the skin of the stump and a lesion on the neck of the scapula was observed on the roentgenograph. Because of this recurrence, the scapula, together with its muscles, was removed and the clavicle in its middle-third resected. The histologic finding was not essentially different from the former.

A prosthesis was fitted and the child went back to school. The loss of the extremity was well tolerated. Fifteen months after the initial symptoms, and three months after the last operation, the bluish-purple nodules reappeared and metastasis to the chest was observed. The child remained at home and lung function worsened due to a gradually diminishing respiratory capacity. The patient died with the symptoms of respiratory insufficiency 18 months after the beginning of the disease. No autopsy was performed.

We also treated another patient with an aneurysmal bone cyst with malignant transformation. This, however, was a malignancy when seen the first time.

In the first case described, the question must be raised whether or not radiation had played a role in the malignant transformation. No definite answer can be given. We can, however, definitely state that when a surgical procedure is possible it should be carried out instead of irradiation.

The incidence of malignant transformation of osteochondroma, according to some investigators, is 1% (Dahlin, 1957; Glauber et al, 1979). The two case histories presented below will demonstrate, first of all, the therapeutic management which has to be considered.

A man, 30 years of age, had a hard tumor in his left supraclavicular fossa. Three months had elapsed before the patient could be operated. During that time, the tumor had protruded from the supraclavicular fossa and could be palpated under the clavicle in the fossa axillaris. The roentgenograph revealed a calcifying lobulated tumor, originating from the ventral surface of the scapula. Angiography and lymphography confirmed displacement of the blood and lymphatic vessels (Fig. 8 and 9). Since the roentgenographs suggested an osteochondroma or a chondrosarcoma, a biopsy was performed. The histologic finding revealed a chondrosarcoma (Fig. 10).

Following confirmation of this finding, a limb-conserving radical operation was planned. According to this plan, the functionally important parts of the scapula and the muscle origins and insertions were to be preserved. After excision of the acromion and the spina scapulae, we succeeded in completely everting the scapula on its medial edge, together with the tumor, which could be adequately separated from surrounding tissues.

We observed that the stem of the tumor had originated in the middle of the ventral surface of the scapula. The margo axillaris and the margo vertebralis were osteotomized, and the tumor with the medial part of the bone excised. The margo axillaris and vertebralis were fixed on the distal pole with a wire loop. The medial end of the spina scapulae was also fixed with a wire loop. The osteotomized acromion was attached to the

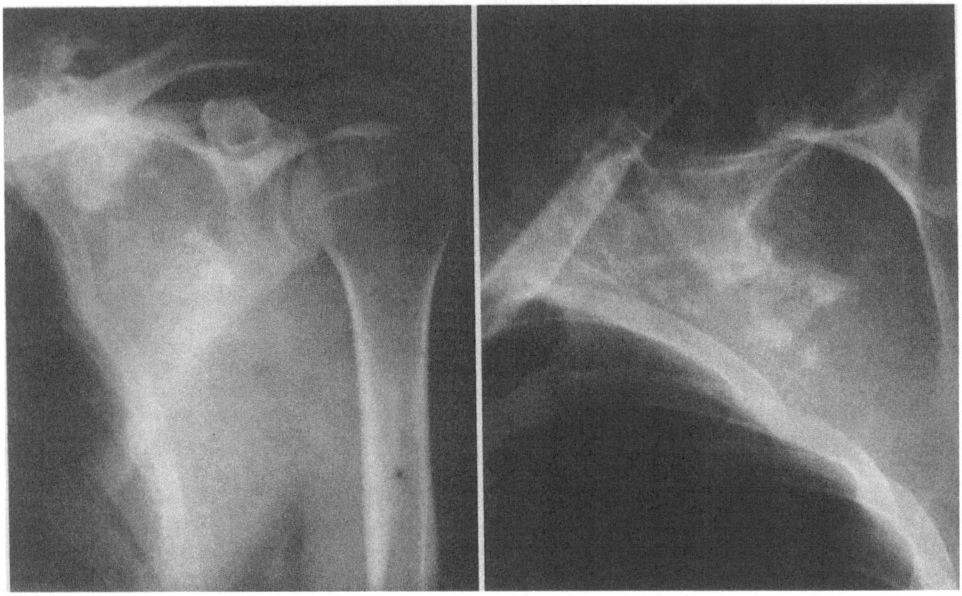

a) b)

Fig. 8. Roentgenographs: a) anteroposterior view of left shoulder; b) projected view of the left scapula. A tumor with nonhomogeneous calcification seen on the ventral surface of the scapula.

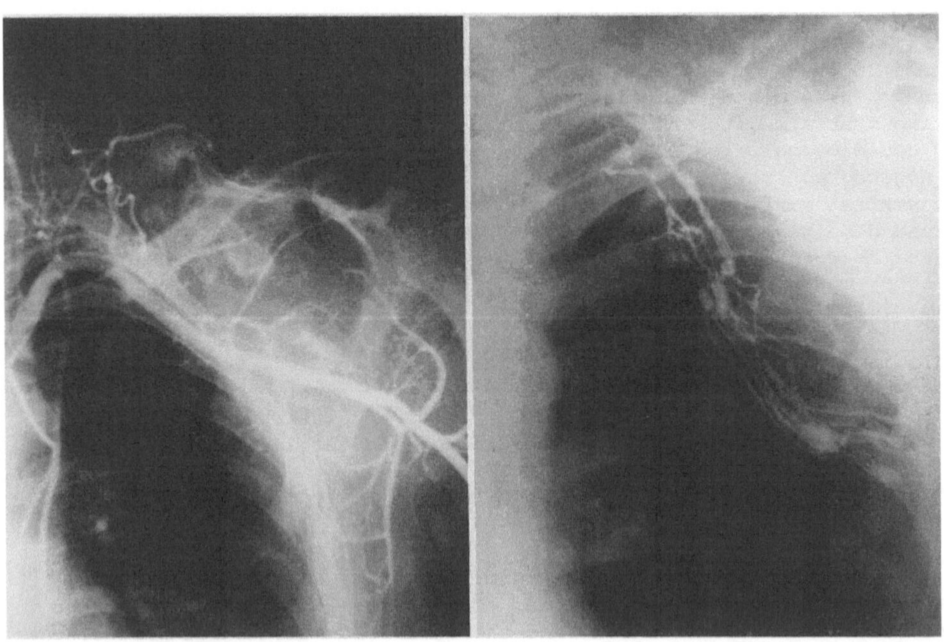

a) b)

Fig. 9. Angiograph and lymphograph show the displaced subclavian artery and lymphatic vessels.

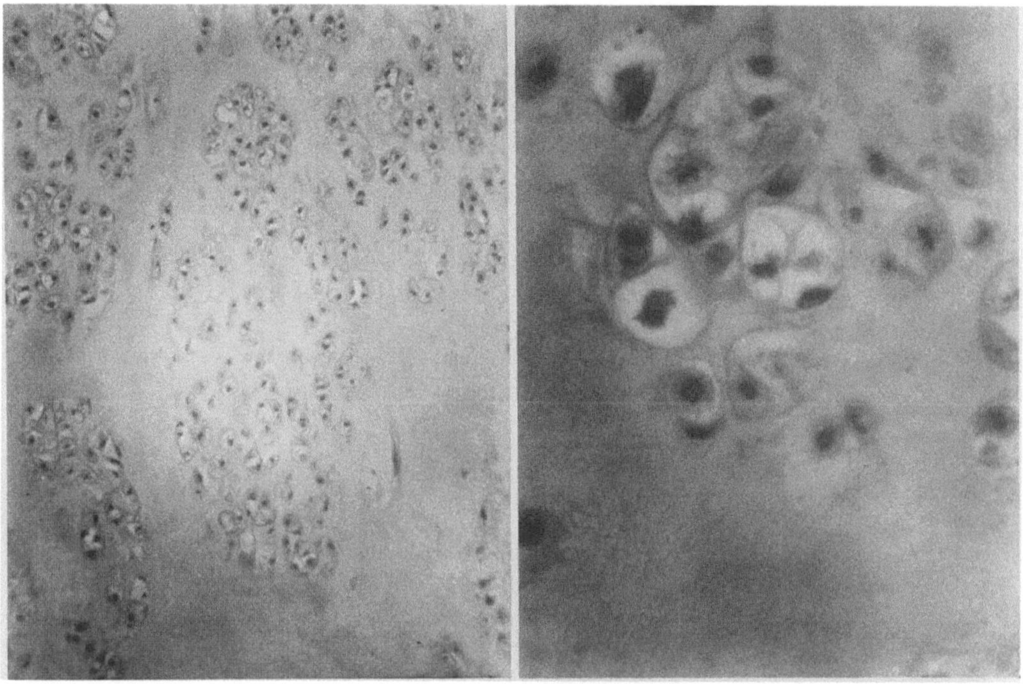

a) H.E. 25x　　　　b) H.E. 160x

Fig. 10. Lobulated, cartilaginous tumor. The tumor cells vary widely in size and cell density. The cells are plump and have double, bizarre, hyperchromatic nuclei. There are numerous atypical forms of mitoses.

lateral side and the end of the spina scapulae with a screw to the upper end of the margo axillaris. Thus, we produced a triangular frame, conserving the origin and insertion of the most important muscles. A plaster cast was applied for four weeks (Fig. 11).

One year later, a nut-sized tumor was removed from the neck. Most of the tumor was necrotic, but showing the histologic characteristics of a chondrosarcoma. Shoulder function was completely restored, and the patient can work, take part in sports, run, and ski without difficulty.

The other patient with an osteochondroma, who is now 43 years old, has been followed up for 10 years. He had previously refused an operation. During the last three months, he has noted a rapid growth of the tumor.

Based on the roentgenographs made during the last 10 years, the growth of the tumor could be easily followed (Fig. 12). Due to the extension of the tumor and uncertain vascular supply, it was questionable whether or not a limb-preserving operation could be performed. We decided to try it and were also prepared for a substitution of the vessel. The well demarcated, cauliflower-like tumor had its broad origin on the dorsal surface

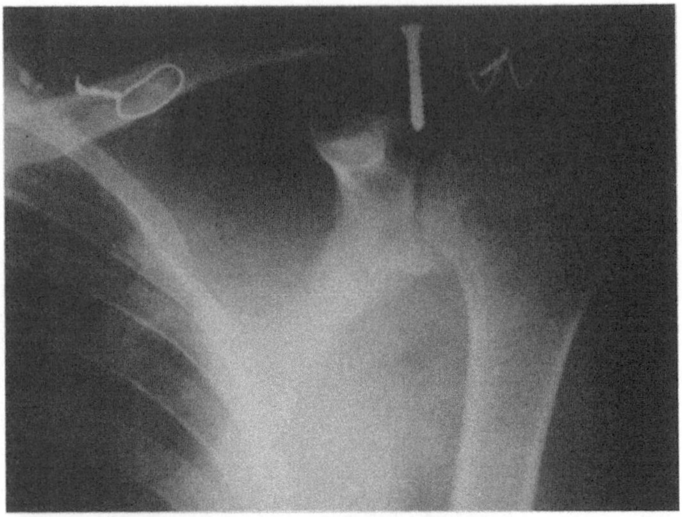

a)

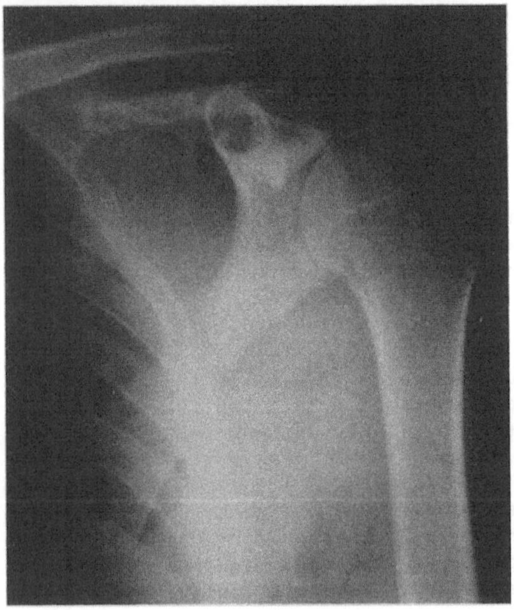

b)

Fig. 11. Roentgenographs of the scapula: a) 3 months; b) 1 year postoperative. Triangular, frame-like form of the bone.

of the tibia. We succeeded in removing the tumor while preserving the vessels and nerves. Circulation remained normal and the patient was soon able to walk. Histologic examination revealed an osteochondroma.

Based on the cases presented above, we would like to call attention to the fact that malignant transformation of benign tumors or tumor-like conditions is possible and that radiation may play a role in this process. Therefore, if possible, surgical methods are preferred over radiotherapy. The possibility of a limb-preserving operation exists even in

a) b)

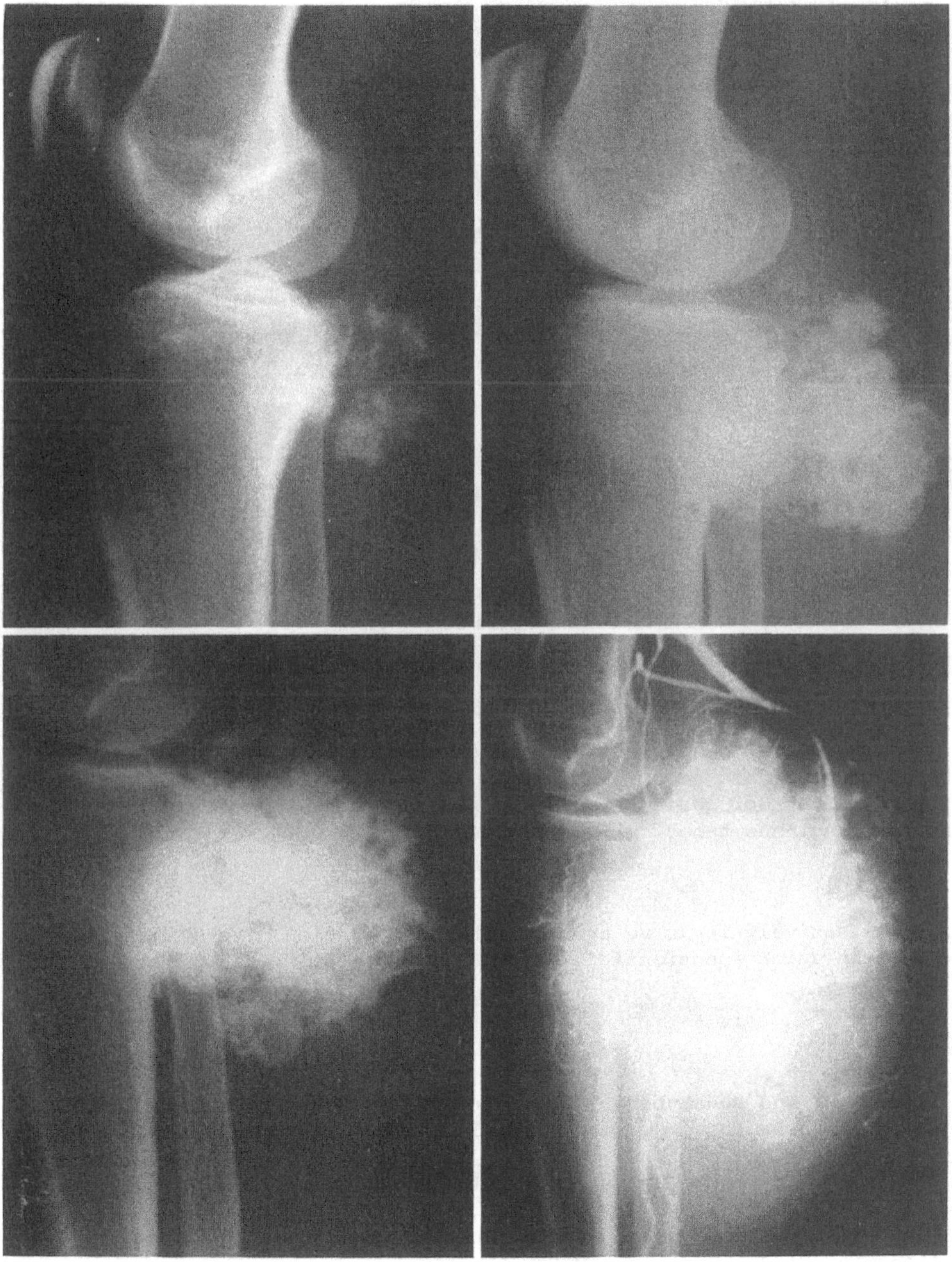

c) d)

Fig. 12. Roentgenographs. The proximal end of the left tibia: a) in 1972; b) in 1976; c) in 1979; d) angiogram in 1982. Further growth of the tumor. The popliteal artery narrows at the proximal margin of the tumor.

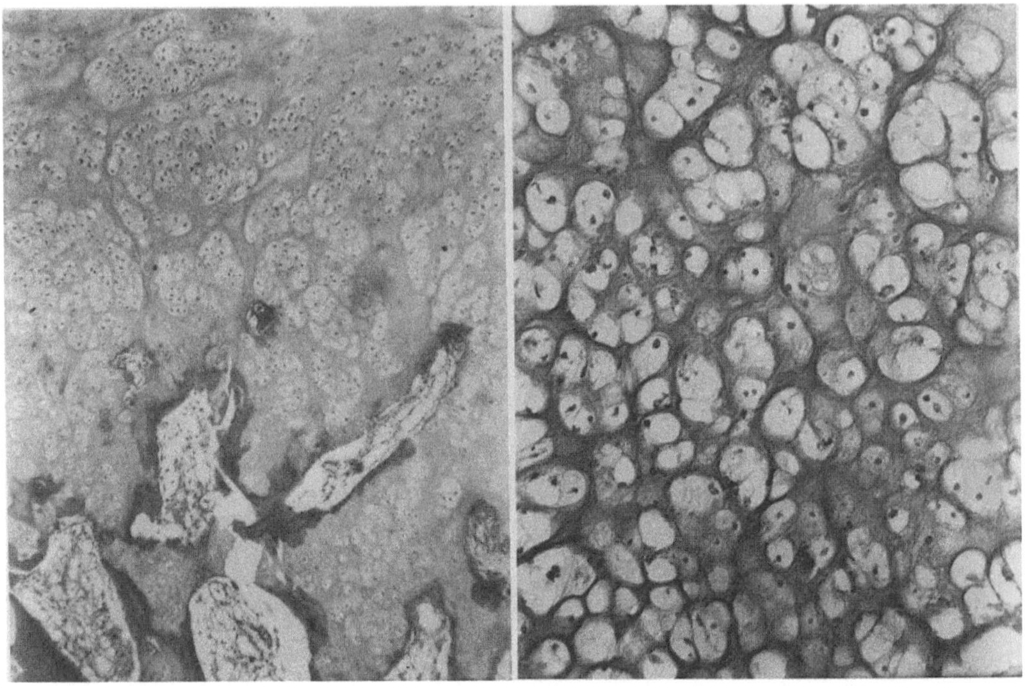

a) H.E. 25x b) H.E. 80x

Fig. 13. The normal cancellous bone is covered by lobulated cartilaginous tissue of varying thickness. The border of bone and cartilage is regular; no sign of infiltration is present. In some views the pycnosis and necrosis of the cartilaginous tumor cells are present.

cases of extensively invasive tumors. The preservation of the important parts assures quick functional restoration.

References

1. Cremer, H. and Münzenberg, K.J.: Aneurysmatische Knochenzyste mit malignem Verlauf. Z. Orthop. 118, 147, 1980.

2. Dahlin, D.C.: Bone Tumors. Springfield, Illinois, 1957.

3. Glauber, A., Zsuzsa Csató, and J. Juhász: Knochengeschwülste Akadémiai Kiadó, Budapest 1979.

4. Hirst, E., Kellar, C.C., Ellis, M. and Smith, V.: Malignant aneurysmal bone cyst. J. Bone Joint Surg. 52B, 791, 1970.

5. Schajowicz, F.: Tumors and Tumorlike Lesions of Bone and Joints. Springer-Verlag, New York, 1981.

Radiation Treatment of Primary Bone Tumors

R. D. T. Jenkin

Introduction

Primary malignant bone tumors may be divided into two distinct groups relative to their response to radiation treatment. Those which respond rapidly and with a good chance for permanent local control are Ewing's sarcoma, solitary osseous plasmacytoma, and non-Hodgkin's lymphoma. Those with a slow and usually incomplete response to high radiation doses are osteosarcoma, malignant fibrous histiocytoma, fibrosarcoma and, to a lesser extent, chondrosarcoma.

Thus, the usual appropriate treatment for the first group is irradiation and for the second group resection. Effective adjuvant treatment is available for Ewing's sarcoma and non-Hodgkin's lymphoma and is obligatory to obtain high cure rates. No effective adjuvant treatment is available for malignant fibrous histiocytoma, fibrosarcoma or chondrosarcoma. In osteosarcoma, adjuvant chemotherapy may be of value but has little, if any, effect on the primary tumor cure rate by irradiation. In chondrosarcoma, radiation treatment offers unresectable patients some chance for cure.

Osteosarcoma

Radiation is no longer indicated as initial treatment for primary osteosarcoma of the extremities. The response rate and duration to tolerable radiation doses is inadequate. Thus, in 27 patients less than 30 years of age, a complete response was achieved in three patients, a partial response in 17, and no response in seven patients. All patients suffered a relapse at the primary site (Jenkin et al, 1972).

Some quantitative radiation dose-response data have been reported for 18 patients who underwent amputation six months after irradiation. No residual tumor was found at doses above the range of 3200-5000 rad in 10 days to 8000-10000 rad in 60-70 days. Below this range, however, all tumors recurred with a variable response (Gaitan-Yangas, 1981).

Massive irradiation of the primary tumor before amputation (8000-12000 rad in 8 to 10 fractions in eight to 12 days) has been evaluated since, theoretically, fewer viable malignant cells should be liberated into the circulation and since such treatment might be a beneficial tumor-specific

Current Concepts of Diagnosis and Treatment
of Bone and Soft Tissue Tumors.

immunologic stimulus. Absolutely no advantage could be demonstrated (Caceres and Zaharia, 1972).

Treatment of the primary tumor is therefore by resection. Only in the uncommon unresectable tumors should irradiation be considered. In a recent trial of adjuvant treatment of osteosarcoma conducted by the National Cancer Institute of Canada and open to patients of any age and with any primary site, only three of 113 patients were judged to be unresectable. All had medial pelvic primary sites with sacral involvement.

Current combination chemotherapy, while not curative for the primary tumor, when given before resection may produce both gross and histologic response in most patients (Rosen et al, 1982). The radiation response of primary osteosarcoma, when effective combination chemotherapy is given concurrently and sequentially, remains to be evaluated. Unresectable tumors are rare; consequently, few reports of response are available as to the combination of not less than 5000 rad in 25 fractions with at least doxorubicin* and high-dose methotrexate. In five such patients known to me, a primary tumor response was obtained in all patients for 10, 10, 16, 60 and 54+ months. Thus, radiation is indicated for gross unresected osteosarcoma and with probability of control for a number of years. When only microscopic residue remains after resection of, for example, a pelvic tumor, postoperative irradiation is indicated and may be expected to obtain superior results compared with the use of radiation alone.

There may be a role for elective whole lung irradiation as an adjuvant in osteosarcoma. The data are conflicting. At the Mayo Clinic, a randomized trial (n=53) showed no benefit from 1500 rad of Cobalt-60 irradiation in 14 days (Rab et al, 1976), whereas a multi-institutional trial conducted by EORTC (n=86) which used 2000 rad in 10 fractions in two weeks resulted in a 43% five-year disease free rate for the irradiated patients compared with 28% for those not irradiated (P=0.059)(Breur et al, 1978). In Europe, an on-going study is comparing adjuvant radiation with adjuvant chemotherapy and a combination of both modalities (Van der Schuren, 1979).

In summary, high-dose radiation is indicated in primary osteosarcoma when resection is not feasible or is refused. When radiation and effective systemic agents are combined short-term response is the rule, but only occasionally does the response last longer than one year. Adjuvant lung irradiation remains experimental. Radiation remains the chief palliative treatment for metastatic bone disease and occasionally in the lung for localized unresectable disease.

Chondrosarcoma

Chondrosarcomas are resected whenever possible. Overall, a rough ten-year survival rate of about 65% is obtained by surgical management of tumors arising in the extremities, limb girdles and spine (Pritchard, 1980). Local recurrence remains a major problem, however. In Grade I tumors,

----

* Adriamycin

23 of 74 patients relapsed locally and no patients had any distant metastases at first relapse. In patients with Grade II tumors, 16 of 44 relapsed locally with five additional patients having distant metastases at first relapse; in Grade III, four of 12 patients relapsed with four additional patients having metastases (Pritchard, 1980). Of the 43 patients who relapsed locally at least 26 have died because of local relapse. Thus, local relapse is the chief obstacle to cure in patients treated initially by resection alone. Others have noted that local relapse is commoner at trunk and head and neck tumor sites (85%) compared with extremity sites (15%) (McKenna et al, 1966).

In general, in bone and soft tissue sarcomas radiation treatment is valuable in preventing local recurrence when there is a high probability of microscopic local residue following resection. In chondrosarcoma little published data exist to show whether radiation is of practical value, perhaps because clasically this tumor has been labelled "radioresistant".

Useful radiation data have recently been published. In Toronto, radical irradiation of 12 patients with gross, mainly recurrent, Grade I and II chondrosarcomas gave a complete response in six patients (3+, 5+, 5+, 15+, 15+, 16+ years); relapse at eight years occurred in one patient following a complete remission, and relapse between three and five years in three patients following a partial response. Apparently, stationary disease occurred for one to two years in the remaining two patients before relapse. Thus, response of greater or lesser degree in all patients was evident and three of the 12 patients remained in remission for over 10 years. These results compare favorably with those obtainable with radiation in many soft tissue sarcomas of similar size. The radiation dose was for the most part 5000 rad for four to five weeks. Poor local control was obtained in dedifferentiated and mesenchymal chondrosarcomas (Harwood et al, 1980).

Similar results have since been reported from the M.D. Anderson Hospital (McNaney, 1982). Primary irradiation of 11 patients with 6100-7000 rad in six and one-half to seven weeks prevented local relapse in five patients at 12+, 26+, 30+, 30+ and 87+ months; four patients relapsed locally after 26, 42, 132 and 156 months of local control and two patients did not respond. In addition, two of three patients treated by local resection and elective postoperative irradiation remain relapse free at 44 and 52 months.

Following the radiation treatment of gross chondrosarcoma, response is slow, taking many months, and is difficult to measure because the radiologic appearances of tumor in bone may not change. Absence of progression is perhaps the best measure of radiation effect.

It therefore appears that postoperative irradiation should be given to all patients in whom adequacy of resection is doubtful and to all patients in whom the primary site is not resectable.

Solitary Osseous Plasmacytoma

Approximately 3% of patients with plasma cell neoplasms have a solitary osseous plasmacytoma. The axial skeleton (spine, skull, sacrum) is the common primary site, but the ribs, limb girdles and extremities may occasionally be involved. The mean age at onset, 50 years, is a decade younger than for multiple myeloma, but the sites of skeletal involvement are similar. Solitary osseous plasmacytoma appears to be an early presentation of multiple myeloma.

Local irradiation (3500 rad in 15 treatments to 4250 rad in 20 treatments) is effective. In Toronto, of 23 irradiated patients 11 remain in a first complete remission from five to 197 months; 11 progressed to multiple myeloma and eventually succumbed (eight to 84 months), and one patient developed a second solitary osseous plasmacytoma at 96 months and was successfully retreated (22+ months) (Knowling et al, 1983).

Radiation is the treatment of choice for most patients with solitary osseous plasmacytoma. Small lesions at easily resected sites, rib for example, may be both diagnosed and treated by local excision.

Ewing's Sarcoma

Radiation has been the classic treatment of the primary tumor in Ewing's sarcoma because it gives rapid pain relief and resolution of the tumor mass with subsequent slow remodelling and healing of the involved bone. Permanent tumor control often ensues. The prospect for a local cure depends on whether or not occult metastatic disease is present. For example, in 35 patients with a primary Ewing's sarcoma of a limb bone and without overt metastases at diagnosis, the primary tumor was controlled in nine of ten (90%) patients who never subsequently developed metastatic disease compared with 17 of 25 (68%) who subsequently had metastases (Jenkin, 1977). Primary tumor relapse in this comparison is defined as isolated relapse or relapse concurrent with the onset of distant metastases; thus, the true primary tumor failure rate is underestimated in patients who subsequently develop metastases. The absence of occult metastases at diagnosis may only define a subset of patients, about 20%, with small, mainly distal, tumors but the observation is important since such patients are often considered particularly suitable for resection or amputation.

Overall, the relapse rate at the primary site following radiation treatment, in the absence of effective adjuvant treatment, is 10-25% (Jenkin et al, 1977; Tepper et al, 1980; Zucker et al, 1977). The relapse rate depends on tumor site: distal -- 7%; proximal limb -- 25%; central -- 33% (Tepper et al, 1980). Postradiation relapse at the primary site remains a problem even when effective adjuvant treatment is given: 11% (mean follow-up 2+ years)(Perez et al, 1981) and 20% (Rosen et al, 1981). In recent years, moderately high radiation doses have been used, but within the range 4000-6000 rad given in 20-30 fractions it has not yet been possible to define the shape of the dose response curve for any given site (Perez et al, 1981). There is no advantage to doses greater than 6000 rad and the obvious disadvantage of marked radiation morbidity. It seems that the dose response curve is rather flat in these dose ranges. Currently, doses in the range of 5000-6000 rad are employed.

Relapse at the primary site is uniformly central. The volume to be irradiated, classically the whole bone with a very generous soft tissue margin, may be modified when effective adjuvant treatment is given to a wide local field, allowing in cases of tumors located at one end of the long bone the opposite epiphysis to be spared. This is of advantage for young patients since it decreases the risk of growth defects (Tefft et al, 1981). Long-term results of such a treatment regimen are not yet available.

Local relapse can be prevented by resecting the tumor and, when the completeness of the procedure is in doubt, by giving postoperative irradiation (Rosen et al, 1981). It is not clear whether survival is improved after initial surgical treatment. The decision must therefore depend on the relative morbidity of irradiation and resection. Radiation doses above 6000 rad are excessively morbid (Lewis et al, 1977) whereas at lower doses in distal tumors the morbidity is minimal (Kliman et al, 1982). I therefore believe that Ewing's sarcoma in the long bones is still best treated by irradiation, but with a dose limit of 5500 rad. At the least favorable site, the pelvis, the barrier to successful treatment is uncontrolled systemic disease. Until better systemic treatment is available, little can be gained by a higher control rate at the primary site as might be achieved by postradiation resection of the tumor site, unless this can be achieved with minimal surgical morbidity. At the present time, surgical treatment is carried out at a number of centres and the results should soon be available, allowing a more objective comparison between resection and irradiation.

Systemic irradiation as a component of successful adjuvant treatment remains anecdotal (Jenkin et al, 1977; 1983; Perez et al, 1981).

Non-Hodgkin's Lymphoma

Before the introduction in 1956 of the Rappaport system of the classification of non-Hodgkin's lymphoma, primary lymphomas of bone were generally classified as reticulum cell sarcoma. It has been recognized for many years that, even with cursory systemic investigation, the adult patient who appeared to have a solitary bone lesion alone could be cured by local treatment, usually irradiation; for example, the 44% five-year survival rate in a large series of 98 patients (Boston et al, 1974). More recently, detailed evaluation of these patients has revealed that most patients with an apparently solitary lesion have, in fact, widespread disease. For example, detailed evaluation of eight clinical stage I patients showed only one patient with pathologic stage I disease (Reimer et al, 1977). Using the Rappaport classification, most of these patients have diffuse histiocytic disease (12 of 14) (Bacci et al, 1982), with diffuse undifferentiated disease accounting for most of the remainder.

Local irradiation alone in pathologic stage I disease produces good results, with cure rates likely 80% or higher, in keeping with pathologic stage I presentations of non-Hodgkin's lymphoma in other tissues (Sweet et al, 1981). When bimodal therapy is used, a combination chemotherapy regimen of proven value for histiocytic lymphoma with local irradiation, even in clinical stage I disease, shows impressive results, 13 of 15 patients remaining disease free for 42-104 months (Bacci et al, 1982). It seems reasonable at this time not to undertake a staging laparotomy in clinical stage I disease and to select a bimodal treatment.

In children, bimodal treatment is now routine, obtaining in clinical stage I disease equally high cure rates; for example, 10 of 10 patients disease free one to five years (Weinstein and Link, 1979).

Thus, at all ages effective systemic treatment combined with irradiation has resulted in a major improvement in survival for the patients with apparently localized bone disease.

Traditionally, radiation doses in the range 3500-5000 rad given at approximately 1000 rad/week have been used for the treatment of primary lymphoma of bone. When bimodal treatment is used, the low end of this dose range should not be exceeded while an optimal dose, which could well be lower, is being determined.

In most of these patients the surgical procedure will be a diagnostic biopsy, and the treatment will be the responsibility of other members of the treatment team, the radiotherapist and the medical oncologist.

References

1. Bacci, G., Picci, P., Bertoni, F., Gherlinzoni, F., Calderoni, P., and Campanacci, M.: Primary non-Hodgkin's lymphoma of bone. Results in 15 patients treated by radiotherapy combined with systemic chemotherapy. Cancer Treat. Rep. 66, 1859, 1982.

2. Boston, H.C. Jr., Dahlin, D.C., Ivans, J.C. and Cupps, R.E.: Malignant lymphoma (so called reticulum cell sarcoma) with bone. Cancer 34, 1131, 1974.

3. Breur, K., Cohen, P., Schweisguth, O., and Hart, A.M.M.: Irradiation of the lungs as an adjuvant therapy in the treatment of osteosarcoma of the limbs - An EORTC randomized study. Europ. J. Cancer 14, 461, 1978.

4. Caceres, E., and Zaharia, M.: Massive preoperative radiation therapy in the treatment of osteogenic sarcoma. Cancer 30, 634, 1972.

5. Gaitan-Yangas, M.: A study of the response of osteogenic sarcoma and adjacent normal tissue to radiation. Int. J. Radiat. Oncol. Biol. Phys. 7, 593, 1981.

6. Harwood, A.R., Krajbich, J.I., and Fornasier, B.L.: Radiotherapy of chondrosarcoma of bone. Cancer 45, 2769, 1980.

7. Jenkin, R.D.T.: Radiation treatment of Ewing's sarcoma and osteogenic sarcoma. Canad. J. Surg. 20, 530, 1977.

8. Jenkin, R.D.T.: Sarcomas of the distal extremities. Int. J. Radiat. Oncol. Biol. Phys. 1983 (in press).

9. Jenkin, R.D.T., Allt, W.E.C., and Fitzpatrick, P.J.: Osteosarcoma: an assessment of management with particular reference to primary irradiation and selective delayed amputation. Cancer 30, 393, 1972.

10. Kliman, M., Harwood, A.R., Jenkin, R.D., Cummings, B.J., Langer, F., Quirt, E. and Fornasier, V.L.: Radical radiotherapy as primary treatment for Ewing's sarcomas distal to the elbow and knee. Clin. Orthop. Rel. Res. 165, 233, 1982.

11. Knowling, M.A., Harwood, A.R. and Bergsagel, D.E.: Comparison of extramedullary plasmacytomas with solitary and multiple plasma cell tumors of bone. J. Clin. Oncol. 1, 255, 1983.

12. Lewis, R.J., Marcove, R.C. and Rosen, G.: Ewing's sarcoma - functional effects of radiation therapy. J. Bone Joint Surg. 59(A), 325, 1977.

13. McKenna, R.J., Schwin, C.P., Soong, K.Y. and Higinbotham, N.L.: Sarcomata of the osteogenic series (osteosarcoma, fibrosarcoma, chondrosarcoma, parostealosteogenic sarcoma and sarcomata arising in abnormal bone). An analysis of 552 cases. J. Bone Joint Surg. 48(A), 1, 1966.

14. McNaney, D., Lindberg, R.D., Ayala, A.G., Barkley, H.T. and Hussey, D.H.: Fifteen year radiotherapy experience with chondrosarcoma of bone. Int. J. Radiat. Oncol. Biol. Phys. 8, 187, 1982.

15. Perez, C.A., Tefft, M., Nesbit, M., Burgert, E.O., Vietti, T., Kissane, J., Pritchard, D.J. and Gehan, E.A.: The role of radiation therapy in the management of non-metastatic Ewing's sarcoma of bone. Report of the Intergroup Ewing's Sarcoma Study. Int. J. Radiat. Oncol. Biol. Phys. 7, 141, 1981.

16. Pritchard, D.J., Lunke, R.J., Taylor, W.F., Dahlin, D.C. and Medley, B.E.: Chondrosarcoma: a clinical pathologic and statistical analysis. Cancer 45, 149, 1980.

17. Rab, G.T., Ivans, J.C., Childs, D.S., Cupps, R.E. and Pritchard, D.J.: Elective whole lung irradiation in the treatment of osteogenic sarcoma. Cancer 38, 939, 1976.

18. Reimer, R.R., Chabner, B.A., Young, R.C., Reddick, R. and Johnson, R.E.: Lymphoma presenting in bone. Results of histopathology, staging, and therapy. Ann. Intern. Med. 87, 50, 1977.

19. Rosen, G., Caparros, B., Nairenberg, A., Marcove, R.C., Huvos, A.G., Kosloff, C., Lane, J., and Murphy, M.L.: Ewing's sarcoma: Ten year experience with adjuvant chemotherapy. Cancer 47, 2204, 1981.

20. Rosen, G., Caparros, B., Huvos, A.G., Kosloff, C., Nairenberg, G.A., Cacario, A., Marcove, R.C., Lane, J.M., Mehta, B. and Urban, C.: Preoperative chemotherapy for osteogenic sarcoma. Cancer 49, 1221, 1982.

21. Sweet, D.L., Mass, D.T., Simon, M.A. and Shapiro, C.M.: Histiocytic lymphoma (reticulum-cell sarcoma) of bone. J. Bone Joint Surg. 63(A), 79, 1981.

22. Tefft, M.: Treatment of Ewing's sarcoma with radiation therapy. Int. J. Radiat. Oncol. Biol. Phys. 7, 277, 1981.

23. Tepper, J., Glaubiger, D., Lichter, A., Wackenhut, J. and Glatstein, E.: Local control of Ewing's sarcoma of bone with radiotherapy and combination chemotherapy. Cancer 46, 1969, 1980.

24. Van der Schueren, E., Breur, K., Schweisguth, O., Voute, P.A., Cohen, P. and Machin, D.: Assessment of the role of chemotherapy and radiotherapy as adjuvants in the treatment of osteosarcomas of the limbs - a trial of the EORTC and of the SIOP. Europ. J. Cancer 15, 1001, 1979.

25. Weinstein, H.J. and Link, M.P.: Non-Hodgkin's lymphoma in childhood. Clin. Haematol. 8, 699, 1979.

26. Zucker, J.M. and Henry-Amar, M.: Therapeutic control trial in Ewing's sarcoma. Europ. J. Cancer 13, 1019, 1977.

Adjuvant Chemotherapy of Osteogenic Sarcoma and Surgical Resection of Pulmonary Metastases

S. K. CARTER

Introduction

Adjuvant chemotherapy of osteogenic sarcoma can be viewed as a test case for the validity of the matched historical control as a clinical research tool in oncology. The positive case that can be made for the value of adjuvant chemotherapy rests entirely on historically-controlled data and as such its biologic significance is still open to serious question. While it is true that a prospective randomized trial was not needed to establish the value of penicillin for the treatment of pneumococcal pneumonia, adjuvant chemotherapy is not penicillin. The value of improving the two-year relapse-free survival (RFS) from 20% to 50-60% in patients with osteogenic sarcoma would be great but such differences, to be meaningful, require the discipline of a randomized trial with adequate numbers. Such data do not exist at this time. Historical controls do exist, however, which show a 20% RFS at two years and small single arms series which show a two-year RFS of 50-60%. The only prospective randomized trial which has been carried out shows that RFS after surgery and after surgery plus adjuvant chemotherapy is not significantly different and that surgery alone is significantly superior to earlier historical controls. Therefore, the value of adjuvant chemotherapy must be viewed as unproven and still open for clinical investigation.

For a long time it was felt that the results with local control regimens alone were so poor and consistent that osteosarcoma was a disease in which historical controls could be safely used. This, combined with the relatively low incidence of the disease and the exciting potential of adjuvant chemotherapy, led major research centers to eschew collaborative studies in favor of individualized single institution series using historical controls. With the power of hindsight, this was unfortunate since today the entire question whether or not adjuvant chemotherapy is of value remains controversial and unresolved.

Local Control Background

Friedman and Carter in 1972 reviewed surgical results from the literature, compiling data on some 1,337 patients. Despite differences in pretherapeutic work-up, surgical techniques, and demography, there was a striking consistency, over 25 years of investigation, in five-year survival rates.

Current Concepts of Diagnosis and Treatment of Bone and Soft Tissue Tumors.

The overall five-year survival rate in this review was 19.7% (253/1286). The range was fairly small, 16-23%, in the series reviewed. Of 438 patients with ten-year follow-up, the survival rate was 16% suggesting that the majority of patients who were living five years apparently free of disease were in fact cured.

Other institutions subsequently reviewed their experience and found similar results. Marcove in 1978 reviewed 210 patients as an update from the 145 patients he had earlier reported who had been diagnosed between 1949 and 1965. The additional 65 patients were collected between 1966 to 1974. Nineteen patients received no adjuvant therapy while the other 46 were given one of two types of autogenous tumor vaccines. The updated analysis showed no difference from the previous results, the updated estimate of two-year disease-free survival rate being 21%.

Gehan (1978) reported an historical control of patients with osteosarcoma, treated at the M.D. Anderson Hospital by surgery alone between 1948 and 1974. The control group totalled 89 patients, all 20 years of age or under. All had disease in the extremities, amputation, disarticulation, or hemipelvectomy for treatment of their disease and were nonmetastatic after surgery. The disease-free survival rates for three periods were reported as follows:

1948-1962	38 patients	11% disease-free at 2 years
1963-1968	31 patients	18% disease-free at 2 years
1969-1974	20 patients	15% disease-free at 2 years

Although the results for disease-free survival were consistent by time period, survival showed an improvement with chronologic time. The percentages of patients surviving at two years for the same time periods were 23%, 40% and 41%. The improvement in survival is most probably due to the fact that intensive combined therapy for metastases began in 1963.

The picture becomes clouded, however, by data from the Mayo Clinic (Taylor et al, 1978) showing superior historical control data for surgery alone. In their retrospective analysis, disease-free survival at two years was 13% for 1963-1968, 42% for 1969-1971, and 30% for 1972-1974. Overall two-year survival was 50% for 43 patients receiving surgery in 1969-1971 and 62% in 41 patients receiving surgery in 1972-1974. This percentage of 62% is, statistically, insignificantly different from the results with many adjuvant regimens. The two-year relapse-free survival of 42% is also not significantly different from the range of most chemotherapy studies. Given the small numbers, patient selection alone could explain the small differences.

Historical Controls - The Problems

Historical controls cause multiple problems associated with their interpretation. One of the most important is the inability to control unknown prognostic variables. While it is possible to match historical controls for known important variables, it cannot be done for factors which are either unknown or for which data do not exist. An example of a recently uncovered prognostic variable in breast cancer is the estrogen receptor

(ER) status. In the original trials of the NSABP and the Milan Cancer Institute, data on the ER status do not exist since the importance of ER was not appreciated when the studies were designed. It would therefore be quite difficult to use these studies for historical controls today since one could not match these studies for the ER status. Even with known variables data are not always available for all these factors in the historical population. The tremendous value of prospective randomization is that it offers the best chance that all important prognostic factors will be evenly distributed among the treated populations to be compared. The chance for an imbalance in an important factor is greatest when the numbers of patients under study are relatively small as it is in osteosarcoma. While randomization does not insure comparability, the lack of randomization in no way can be viewed as improving the situation. Therefore, in osteosarcoma the heavy use of historical controls is particularly unfortunate.

While osteosarcoma is a relatively rare form of malignancy, it encompasses within its diagnosis a great deal of heterogeneity. The site is a critically important aspect of this heterogeneity. More distal lesions have a better prognosis. Tumor size is another element of prognostic importance. The prognosis with radical amputation alone has always been better when the primary lesion was small. It is uncommon in adjuvant studies, with historical controls, to see matching of primary tumor sizes. If tumor size were smaller in adjuvant-treated patients they could explain a better result compared with an historical control with a larger primary tumor size. An example of this disparity was observed in the NCI study (Rosenberg et al, 1979). In this study the mean diameters of tumors on anteroposterior radiographs was 113 cm^2 in historical controls and 65.9 cm^2 in the adjuvant chemotherapy group. Earlier diagnosis possibly accounts for this discrepancy.

Another prognostic factor is the duration of symptoms before diagnosis. In the past, a long time-lag between symptoms and diagnosis was considered indicative of an indolent lesion and therefore a good prognostic sign. This was reflective of a generally pessimistic view that local control alone could achieve cure. In today's climate of greater optimism, an early diagnosis might mean a smaller primary lesion and a better chance to achieve both local and metastatic control.

The histologic profile is still another important variable. Some variants of osteosarcoma are considered less malignant and carry with them a better chance for cure with effective local control. These include patients with parosteal sarcoma, periosteal osteosarcoma, low-grade intraosseous osteosarcoma, and chondrosarcoma. The tumor grade is also important with high-grade lesions having a poorer prognosis. It is difficult to match histologic findings among historical controls pathologically evaluated years earlier and a newly treated group with prospective histologic studies.

Histologic controls suffer from the additional difficulty that the critically important aspect of staging does not always remain stable. Staging tends to become more aggressive and sophisticated with the use of newer and more sensitive techniques. The impact of this increased sophistication is to stage some patients in the advanced disease category who in the past would have been early-stage patients. Shifting some early stage patients into more advanced staging categories tends to improve the prognostic mix within all stages. In osteosarcoma, this has occurred in the search for pulmonary metastases at the time of initial diagnosis. Increased thoroughness in the search for metastatic disease at diagnosis by whole lung tomography and computerized axial tomography will eliminate many more patients from the localized disease category than had been eliminated historically. This results in a group of patients being selected for adjuvant chemotherapy today who have a better prognosis than those chosen for curative surgery years earlier.

Still another difficulty with historical controls relates to comparability of surgical management. In the last 15 years, changes in surgical technique have evolved as well as changes in the selection of the operative procedures used to attempt cure. The assumption in all historically-controlled studies is that all patients (adjuvant treated and control) have had comparable surgical techniques executed. This aspect, however, has not been specifically discussed in any of the historically-controlled adjuvant studies in the literature. Another confounding problem regarding surgery is the newer approaches of limb salvage within a combined therapeutic framework. It is impossible for many reasons to compare this type of therapy meaningfully with an earlier group of patients who had undergone radical amputation alone, including the selection bias that goes into choosing patients for a limb salvage procedure.

Adjuvant Chemotherapy Trials

The incidence of osteosarcoma is quite low compared with other malignancies in which adjuvant chemotherapy has been evaluated. It is an interesting phenomenon that despite its low incidence the bulk of ajuvant trials has been single institution studies. This has meant that these trials have consisted of small numbers and, with one exception, the use of historical controls. In other pediatric solid tumors where adjuvant therapy has been evaluated, multi-institutional studies have predominated to obtain large numbers of patients within a reasonable time-frame. In Wilms' tumor and embryonal rhabdomyosarcoma, several cooperative groups in the USA have joined to undertake intergroup studies aiming to collect large trial numbers. Why was this not done a long time ago with studies of osteosarcoma? It appears that each institution was so sure that it would accomplish a major therapeutic advance that it did not want to share the credit. As a result, a mass of studies exists, with small numbers, each using a somewhat different mix of adjuvant regimens including HDMTX and/or doxorubicin* with or without the addition of other drugs.

When the many adjuvant chemotherapy studies are evaluated (Table I) (Rosenberg et al, 1982) it is difficult to select one regimen which stands out as clearly superior. The results with HDMTX alone and doxorubicin alone appear to be comparable to combinations of the two or to the addition of other drugs. This consistent lack of difference among the various regimens depends on interpretation of the validity of the older

----

* Adriamycin

Table I. Adjuvant chemotherapy regimens in osteogenic sarcoma*

Chemotherapy regimen	No. of patients	Percentage disease-free at 2 years	Institution
VCR + cyclophosphamide	14	21	St. Judes
VCR + cyclophosphamide + dactinomycin	11	27	SWOG
HDMTX + VCR ± BCG	39	38	NCI
Compadri III	44	42	SWOG
HDMTX + VCR	12	42	Dana-Farber
HDMTX + VCR + doxorubicin + cyclophosphamide	83	50	Stanford
Cyclophosphamide	6	50	Shero et al
HDMTX + doxorubicin + cyclophosphamide + (T-4)	54	50	MSKCC
Compadri II	60	51	SWOG
Cisplatin + doxorubicin	20	52	Roswell-Park
Conpadri I	44	55	SWOG
Doxorubicin	132	56	CALGB
CYVADIC	25	60	M.D. Anderson
HDMTX + VCR + doxorubicin	59	62	Dana-Farber
T-7	61	84	MSKCC
T-7 ± cisplatin (T-10)	57	93	MSKCC

* Derived from Rosenberg et al, 1982

VCR = Vincristine
HDMTX = High-dose methotrexate with leucovorin
Conpadri = Cyclophosphamide, phenylalanine mustard, VCR, doxorubicin
Compadri II-III = Conpadri I + HDMTX
CYVADIC = Cyclophosphamide, vincristing, doxorubicin, DTIC
T-7 = HDMTX + doxorubicin + bleomycin + cyclophosphamide + dactinomycin ± vincristine
doxorubicin = Adriamycin

SWOG = Southwest Oncology Group
MSKCC = Memorial Sloan Kettering Cancer Center
CALGB = Cancer and Leukemia Group B

historical controls which showed a two-year survival rate in the range of 20%. If one assumes that a modern day control would be in the 40-50% range, the reason for this consistency could be interpreted as meaning that none of the adjuvant regimens is meaningfully effective. This interpretation is strengthened by the observation that adjuvant trials with therapies as diverse as transfer factor, interferon, coumadin and pulmonary radiation (Table II)(Carter, 1982) have given two-year relapse-free survivals in the same range. It is much more likely that all the chemotherapy regimens and these other procedures are equally ineffective than to assume that they are all positive.

Table II. Noncytotoxic chemotherapeutic adjuvant regimens in osteogenic sarcoma (Carter, 1982)

Regimen	No. of patients	Percentage 2 year relapse-free survival	Institute
Interferon	17	41	Karolinska Institute
Coumadin	9	44	NCI
Pulmonary irradiation			EORTC
1. Paris	26	33	
2. Other	18	55	
Transfer factor	9	55	Mayo Clinic

The first group to report on the adjuvant use of HDMTX was the Sidney Farber Cancer Center (Jaffe et al, 1974), now named the Dana-Farber Cancer Center. In their initial series, 22 consecutive patients with osteosarcoma received adjuvant chemotherapy beginning three weeks after local treatment of the primary tumor with surgery or radiation. The chemotherapy consisted of vincristine 2 mg/m^2 intravenously followed after 30 minutes by methotrexate in a continuous infusion over a six-hour period. The initial dose was 1500 mg/m^2. The treatment was given at two-week intervals with a dosage increase to a maximum of 7600 mg/m^2 every six hours for 12 doses. Subsequent to this study, adequate hydration, alkalinization of the urine, and monitoring of serum methotrexate levels were shown to be of great importance to avoid a significant drug-related mortality.

Compared with a historical control group from the same institution, in which 78% developed pulmonary metastases at 12 months, only two of 20 study patients had recurrence at the time of the initial report in the New England Journal of Medicine (Jaffe et al, 1974). This created a great stir among oncologists.

In this first study, only 15 of 20 patients had classic osteogenic sarcoma; four patients had parosteal osteogenic sarcoma, and one had multifocal sclerosing osteogenic sarcoma. Three of 15 patients with the classic osteogenic sarcoma in the study group had chondrosarcoma elements in their tumor, which might be classified as osteochondrosarcoma. Of the 15 patients with classic osteogenic sarcoma, only 12 received definitive local treatment and obtained local control. Eight of these had amputation alone and four received radiotherapy followed in three to five months by amputation. The results were compared with a historical control of 71 patients with local control who were treated between 1950 and 1972. These two groups were comparable in age distribution and sex but were not matched for any other prognostic variables, such as methodology of local control, histologic pattern, tumor location, tumor size, or pretreatment work-up for evidence of metastatic disease.

In the follow-up report of this study (Jaffe et al, 1978), seven of 12 patients remained disease-free with a minimum follow-up time of two years. This was followed by a second study where doxorubicin was added to high-dose methotrexate. In 22 patients, the life table plot at the time of the first literature report showed 60% to be disease-free but with a short follow-up. It was clear from this first report that the combination would not be superior to HDMTX alone and might in fact ultimately be worse.

The first report of adjuvant doxorubicin came from Cortes and Holland (1974), reporting for Acute Leukemia Group B. They described 21 patients who were given doxorubicin 30 $mg/m^2/d$ x 3 every four to six weeks for six courses, two weeks (mean) after surgical amputation of the primary lesion. Four patients, however, were treated more than eight weeks after amputation. The data by Marcove et al (1978) was used as a historical control. Five of the 21 study patients had relapsed at the time of the initial report in the New England Journal of Medicine, again creating a stir of enthusiasm (subsequent reports showed about half of the patients free of disease at two years). Of the 21 study cases, only 15 had radical amputation. Of the six who had subradical amputation, four had recurrence at the time of the report.

The Pediatric Division of the Southwest Oncology Group has a long history of using adjuvant chemotherapy to treat osteogenic sarcoma, beginning in 1963, as reported by Sutow et al (1975). The first study involved 14 consecutive patients treated with phenylalanine mustard with only two having a long survival. Beginning in 1968, a group of 11 children with nonmetastatic osteogenic sarcoma were treated with a three-drug pulse regimen of vincristine, actinomycin D and cyclophosphamide. Of these 11 patients, three (27%) were long-term survivors.

In 1971, a study was undertaken of a four-drug adjuvant regimen called Conpadri-I for the four drugs used: C (cyclophosphamide), ON (oncovin), P (phenylalanine mustard), ADRI (Adriamycin*). Patients received chemotherapy as soon as wound healing was satisfactory. Eighteen patients were treated with this regimen, and ten (55%) remained disease-free for prolonged periods up to the present. In subsequent studies, (Compadri-II and Compadri-III), high-dose methotrexate was added to the initial four-drug regimen. The results were not significantly superior.

The Surgery Branch (Rosenberg et al, 1979) of the National Cancer Institute has reported on a study of 39 patients with histologically proven osteogenic sarcoma treated with adjuvant HDMTX. All were deemed to be free of metastatic disease after chest roentgenograms, full lung tomograms, meta-

----

* doxorubicin

static bone survey, technetium polyphosphate bone scan, and the usual blood and laboratory tests. After a curative intent amputation, the patients were randomized to receive either high-dose methotrexate with leucovorin rescue or the same plus BCG. Chemotherapy was begun within two weeks of surgery. Dosages ranged from an initial dose of either 50 or 100 mg/kg with escalations of 50 mg/kg every three weeks up to 250 mg/kg for all patients. The highest dosage was then maintained for 18 months. The drug was given as a six-hour infusion with leucovorin therapy being started (15 mg/m^2) two hours after completion. Leucovorin was repeated, intramuscularly or intravenously, every six hours for a total of eight doses. BCG (fresh frozen from Connaught Laboratories) was given on day 14 and 21 of each drug cycle.

The results were compared with an historical control of 23 patients treated at the NCI between 1953 and 1974. All the patients met the criteria for entrance into the study.

The study was first reported with a mean follow-up time of 27 months. The actuarial analysis estimated that 38% of the protocol patients were continuously free of disease for 24 months compared with only 17.4% of the historical control (P = 0.029). Since the historical control patients had a longer follow-up, and thus had expressed their true relapse potential, this difference could well have diminished in time, especially since the follow-up time on some treated patients was only 10 months. A comparison of patients receiving BCG with those not receiving BCG revealed no difference. The same applied to patients beginning with 50 mg/kg methotrexate compared with those beginning with 100 mg/kg.

Significant differences were observed between the treated group and the historical control. Four patients considered for inclusion in the treated group were excluded because chest tomography demonstrated pulmonary metastases while the chest roentgenogram was normal. These patients would have been amputated before lung tomography was routinely used and would therefore have been included in the historical control. If these four current protocol patients, excluded because of positive tomograms, are included in the analysis with a hypothetic recurrence at three months, then the difference between the treated group and the historical control patients loses statistical significance (P = 0.11).

Another difference between the two groups was in histologic grading. The adjuvant treated group had a slightly lower incidence of Grade III and IV lesions (33 of 39 as compared with the historical control [22 of 23]). When comparison was made only with patients with Grade III and IV lesions, then the difference in disease-free survival again lost statistical significance (P = 0.11).

Rosenberg concludes his paper with the statement that "in the absence of a prospective randomized study, it is impossible to be certain that the use of high-dose methotrexate has improved the disease-free interval in current protocol patients."

While a randomized comparison between doxorubicin and HDMTX has not been reported, the CALGB (Cortes et al, 1981) has performed a randomized study comparing adjuvant doxorubicin alone with doxorubicin plus HDMTX. Ninety-nine patients were randomized between December 1975 and April 1980. Doxorubicin alone was 30 $mg/m^2/d \times 3$, repeated every four weeks for a total of six courses. The combination included two cycles of doxorubicin as above followed by two courses of HDMTX 200 mg/kg three weeks apart. The two cycles were then alternated until six cycles of each of the components had been administered. Patients were excluded from this study who had parosteal osteosarcoma, chondrosarcoma, fibrosarcoma, radiation-induced osteosarcoma, soft-tissue sarcoma, and osteosarcoma after Paget's disease.

Fifty-five patients were randomized to receive doxorubicin alone and 45 to receive doxorubicin plus HDMTX. The number considered evaluable were 51 and 38 respectively. The results have been published with a mean follow-up of 34 months. At 36 months following primary surgery, the life table analysis estimates the percent to be 52% for the doxorubicin alone and 40% for the two-drug combination (P = 0.02). The advantage for doxorubicin is seen only in men and in those over 16 years of age.

When overall survival is examined, the life table estimates at 36 months indicate that 82% of patients in the doxorubicin group are survivors in contrast to 57% in the doxorubicin plus HDMTX group. In the single agent group, 18 patients failed with pulmonary metastases and 11 underwent thoracotomy and tumor resection. In the two-drug cohort, 18 also failed with pulmonary metastases but only six underwent therapeutic surgery.

This study indicates that the addition of HDMTX to doxorubicin gave an inferior result to doxorubicin alone. The disease-free survival of 40-52% is within the range of most of the series listed in Table I, and again is only meaningful if the historical controls are valid which is unlikely.

The Mayo Clinic (Edmonson et al, 1980) has reported the only randomized study comparing surgery alone with surgery plus adjuvant chemotherapy. A total of 37 patients were entered into the study. The estimated two-year survival for the whole group was 74% with an estimated 52% remaining disease-free at two years. All patients, except one, had an amputation to remove the primary tumor with 17 receiving no further therapy and 20 adjuvant drug treatment. The adjuvant therapy was vincristine, 2 mg/m^2 followed at 30 minutes by a six-hour infusion of HDMTX escalated from 3 to 7.5 g/m^2 in successive courses. To date, no differences exist between treated and control groups with the data being similar to the historical control of surgery alone between 1976 and 1978.

The best results of adjuvant chemotherapy have been reported by Rosen and his associates (Rosen and Nirenberg, 1982 at the Memorial Sloan-Kettering Cancer Center MSKCC) in New York. They have reported three sequential studies labelled T-4, T-7 and T-10. The first protocol (T-4) used HDMTX at a dose of 200 mg/kg, doxorubicin at a dose of 90 mg/m^2 and cyclophosphamide at a dose of 1200 mg/m^2. A total of 54 patients was treated and the overall continuous disease-free survival is 50% which is in the range of most of the results shown in Table I. In this study, 58% of the patients underwent a limb salvage procedure and the rest had amputations. Therefore, the local control approach (and one would presume the selection of patients), differed significantly from the earlier reported experiences with local control alone published by Marcove and used as the reference historical control.

The second protocol at MSKCC (T-7) incorporated the concept that HDMTX had to be given at a dose of 12 g/m^2 to achieve optimal results. In addition, the combination of bleomycin, cyclophosphamide and dactinomycin (BCD) replaced the single agent cyclophosphamide. In the T-7 protocol they have reported 82% continuous disease-free survival for all patients with a mean follow-up of 52 months. This study also included a significant number of patients receiving limb salvage with presurgical chemotherapy. Presurgical chemotherapy has allowed an evaluation of drug responsiveness in the primary tumor at the time of surgery. In their experience, all patients having a histologically-confirmed complete or near complete (>90%) response of the primary tumor following preoperative drug treatment became continuous disease-free survivors. Whether the superior results in the T-7 protocol, as compared with the T-4 experience, was due to the higher HDMTX dose, the addition of the BCD regimen, different local control, different case selection, or other factors is impossible to determine. It is unfortunate that MSKCC did not randomly compare T-7 with T-4 to clarify the situation.

In 1978, the T-10 protocol was initiated at MSKCC. This study called for the identification of poor responders to presurgical chemotherapy with HDMTX and the rest of the T-7 regimen, and the addition of cisplatin to their adjuvant chemotherapy protocol following surgery after HDMTX was discontinued. At last report, with 79 patients entered and a mean follow-up of 24 months, the continuous disease-free survival was over 90%.

It is difficult to interpret the MSKCC data which are dramatically superior to the results reported from all other major centers. Their approach is highly individualized, mixing rigid adherence to a very high HDMTX approach, preoperative chemotherapy and limb salvage, and evaluation of drug responsiveness of the primary tumor. It is impossible to relate it to any other center's approach except in the crudest sense of comparing literature reports. This does not allow any comparison of possible selection biases and exclusions. It is possible that the MSKCC has found the near optimal ways to treat this disease but the acceptance of this will require confirmation in other centers, ideally within a prospectively randomized comparison with more standard procedures.

Lange and Levine (1982) have asked why the MSKCC procedure's results were so far superior to those in other centers. They found it difficult to accept that increasing the dose of HDMTX to >8 g/m^2 in young children accounts for the difference. They discuss an unpublished study of the children's cancer study group that "has failed to demonstrate any advantage of high-dose methotrexate over moderate-dose methotrexate (1-Ertel, personal communication)". They put forth the alternate explanation of the selection of a particularly good risk group of patients for the T-7 and T-10 protocols.

Surgical Resection of Pulmonary Metastases

Metastatic disease is usually a manifestation of widespread dissemination of disease. Occasionally, metastatic disease is clinically limited to isolated pulmonary lesions. These types of lesions are amenable to surgical resection and osteosarcoma is one disease where this is being done with increasing frequency. Surgical resection of pulmonary metastases can be performed with either curative or palliative intent. It can be performed either alone or combined with a chemotherapy regimen.

The biologic assumption behind surgical resection of pulmonary metastasis alone for cure is tenuous. It presumes that the resected lesions are the only manifestations of disease existing within the host and that no foci of micrometastatic disease exists in other locations. This is biologically improbable and cannot be supported by either experimental or clinical evidence. Within a combined chemotherapeutic setting, the curative concept is identical to that presumed in traditional surgical adjuvant chemotherapy: surgery removes the clinically evident disease and chemotherapy eradicates the micrometastatic residual disease. This is conceptually feasible if the proper combination of residual tumor burden and cell kill effectiveness of chemotherapy is achieved. It is not unreasonable to assume that the micrometastatic burden existing at the time of clinical pulmonary metastatic disease is at least as great as existed at the time that the primary lesion was ablated. It can also be assumed that the metastatic foci have had a longer time to express the heterogeneity which results from the relative genetic instability of metastatic tumor cells. Included in this heterogeneity is the mutation to intrinsic resistance to chemotherapy which is the conceptual focus of the Goldie-Coldman Hypothesis now in vogue in circles advocating cancer chemotherapy. In addition, if these patients have been given adjuvant chemotherapy, there may be a further lack of chemosensitivity due to that exposure. All of this makes a relatively poor case for the surgical resection of pulmonary metastasis followed by chemotherapy as a curative approach, particularly in patients relapsing on or after primary adjuvant chemotherapy.

If palliation is the goal of pulmonary resection then the cost-benefit ratio must be carefully considered. The benefits must be viewed in terms of disease-free survival and overall survival prolongation. The disease-free survival is measured from the time of surgical resection. The zero points for the overall survival can be either from time of initial diagnosis or from time of diagnosis of metastatic disease. The costs must include:

- surgical morbidity of the thoracotomy and resection;
- surgical complications;
- surgical mortality;
- effects of chemotherapy.

There is great tendency in the reporting of these types of trials to play down the surgical morbidity of the therapy.

The two-year survival after resection of pulmonary metastases has been reported to range from 30-61% in five series for major cancer centers (Table III) with four of the five showing results of 50% or better.

Surgical treatment of pulmonary metastases has made overall survival a difficult end-point for adjuvant studies compared with earlier historical controls. The importance of secondary therapy making overall survival more difficult to interpret can be seen in the NCI Osteogenic Sarcoma Study (Rosenberg et al, 1979). In the original adjuvant chemotherapy

Table III. Resection of pulmonary metastases for osteogenic sarcomas

Group	No. of patients	% 2-year survival
Memorial Sloan-Kettering Cancer Center (Rosen et al, 1978)	31	50
Mayo Clinic (Telander et al, 1978)	28	55
NCI (Rosenberg et al, 1979)	18	61
Roswell Park Memorial Institute (Huang et al, 1978)	13	30
Stanford (Giritsky et al, 1978)	12	53

study with and without immunotherapy, relapse occurred in 22 of 39 patients. These patients had early aggressive resection of all metastatic disease whenever possible. Any patient developing new pulmonary lesions thought to represent metastatic foci underwent early thoracotomy except when lesions were too numerous to resect. Twenty of the 22 recurrences had pulmonary disease as their initial sign of recurrence. In 18 of 20 patients, this was seen on plain chest roentgenogram while in two it was diagnosed after routine tomograms. In 12 of 20 patients, a single unilateral lesion was first observed clinically with nine being confirmed as single at surgery. In 18 historical control patients with single pulmonary recurrence, only three had single unilateral lesions. Eighteen of 20 patients, relapsing after adjuvant treatment, underwent thoracotomy compared with only six of 18 in the historical control group. Eleven of the 18 protocol patients had resection of all known pulmonary metastatic disease. After resection, nine of 11 patients received additional chemotherapy. Four received doxorubicin plus cyclophosphamide and five a combination of vincristine, high-dose methotrexate, cyclophosphamide, doxorubicin, dacarbazine, melphalan and cisplatin.

Overall relapse had occurred in four of the 11 resected patients. These patients have had additional thoracotomies and are still disease free. The other seven are also still disease free beyond 12 months. With the combined treatment strategy of adjuvant chemotherapy followed by resection of pulmonary metastases where possible, 28 of 39 patients (71.8%) were disease free with a mean follow-up of 27 months. This represents a significant improvement in survival compared with that of the historical control with a P-value of <0.001.

What cannot be determined is how to explain the secondary salvage potential in the protocol group compared with the historical control. One potential explanation is that the adjuvant chemotherapy decreased the frequency of pulmonary metastases. A second explanation is that the screening for relapse with chest roentgenograms every three weeks was responsible for picking up recurrence at an earlier time. Rosenberg et al believe that the latter view is "a more likely explanation".

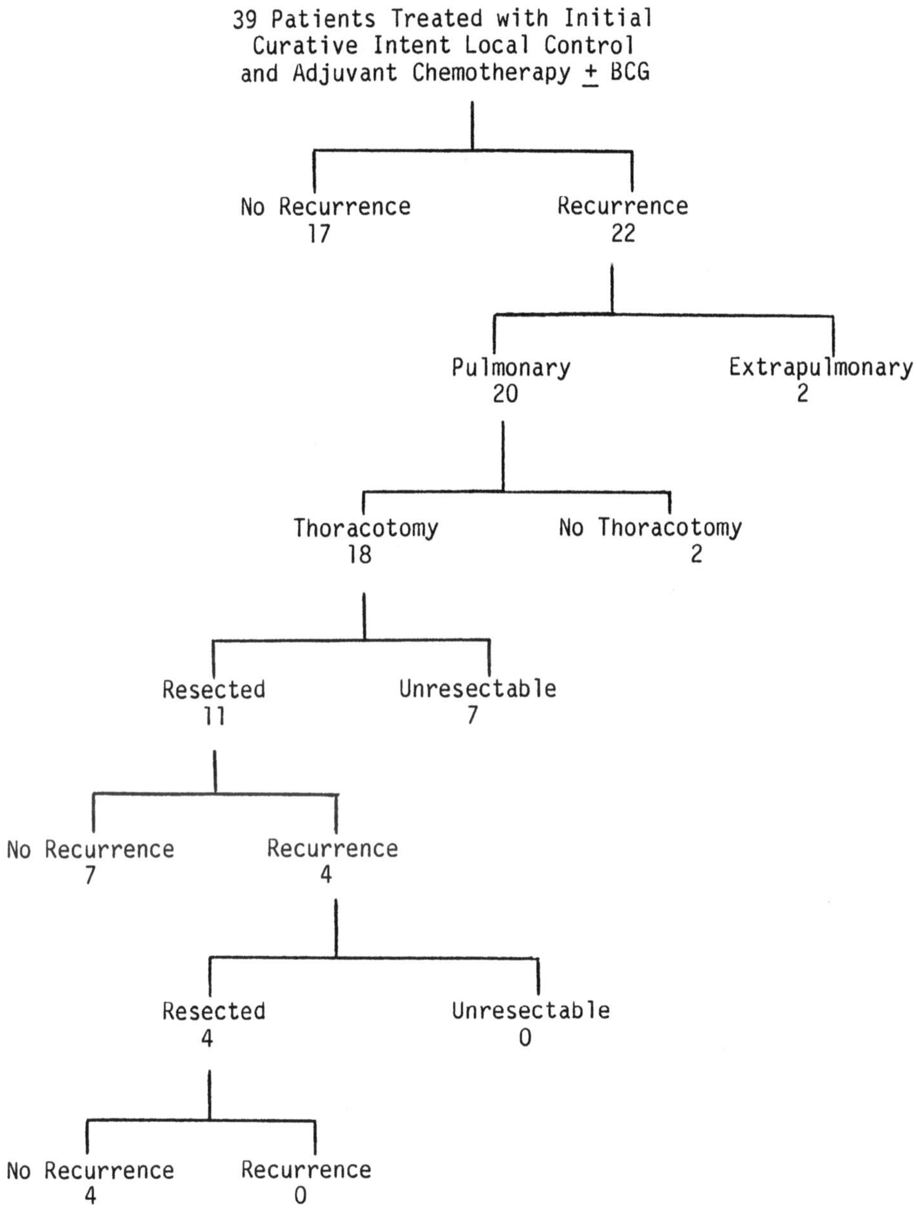

Fig. 1. Osteogenic sarcoma course in 39 patients treated in a protocol study at the NCI (Rosenberg et al, 1982).

In the NCI experience all protocol patients received chest roentgenograms every three weeks and lung tomography every three months. When pulmonary nodules were found, the patients were operated and an exploratory thoracotomy with wedge resections of metastases performed. The course of the 39 protocol patients at the NCI with regard to resection of pulmonary metastases is shown in Figure 1.

The true value of the surgical resection of pulmonary metastases cannot be determined yet. The case rests on retrospective analyses of selected populations without any attempt to use a matched historical control. What will be required is a prospectively randomized study, restricted to patients with metastatic disease deemed appropriate for surgery. This selected group should be randomized to surgery plus chemotherapy versus chemotherapy alone. The study should demonstrate that the patients treated by surgery have a longer RFS and OS and that this superiority outweighs the morbidity, risks, and costs of a surgical procedure. The assurance by selected surgical and medical oncologists that this approach is beneficial, without a randomized study, is not enough to justify its widespread use.

Conclusion

The advantage of adjuvant chemotherapy for osteogenic sarcoma has not yet been proven. The clinical research history of this approach in the last decade is a massive indictment of the historical control methodology as a failure. Osteogenic sarcoma, despite its low incidence, is an extremely heterogeneous disease. Unfortunately, low incidence does not mean homogeneity. Osteogenic sarcoma is as heterogeneous as breast cancer or the malignant lymphomas. In these latter diseases, nobody would accept an adjuvant trial of 10 to 50 patients compared to an earlier historical control as meaningful. Of 16 trials listed in Table I, four (25%) involved less than 20 patients and nine involved less than 50 patients. Osteogenic sarcoma cannot be viewed as a special case. In future, major cancer centers should put aside their desires for individual institutional achievements and pool their resources with other centers so that meaningful randomized trials can be undertaken. Data, to date, indicate strongly that a local control-only arm should be part of such a study. A presumed poor prognosis is no excuse for the administration of toxic therapies which have not been proven within an appropriate clinical research setting.

References

1. Carter, S.K.: Adjuvant chemotherapy of osteogenic sarcoma. In, Principles of Cancer Treatment. Eds: Carter, S.K., Glatstein, E., Livingston, R.B. McGraw-Hill, New York, 1982.

2. Cortes, E.P., Holland, J.F., Wang, J.J., et al: Amputation and Adriamycin in primary osteogenic sarcoma. N. Engl. J. Med. 291, 998, 1974.

3. Cortes, E.P., Necheles, T.F., Holland, J.F., et al: Adjuvant chemotherapy for primary osteosarcoma: a cancer and leukemia Group B experience. In, Adjuvant Therapy of Cancer III. Eds: Salmon, S.E., and Jones, S.E., Grune and Stratton, New York, 1981.

4. Edmonson, J.H., Green, S.J., Ivins, J.C., et al: Methotrexate as adjuvant treatment for primary osteosarcoma. New Engl. J. Med. 303, 1980.

5. Friedman, M.A., and Carter, S.K.: The therapy of osteogenic sarcoma - current status and thoughts for the future. J. Surg. Oncol. 4, 482, 1972.

6. Gehan, E.A.: Comparative clinical trials with historical controls: a statistician's view. Biomedicine 28, 13, 1978.

7. Giritsky, A.S., Etcubanas, E., and Mark, J.B.D.: Pulmonary resection in children with metastatic osteogenic sarcoma. J. Thorac. Cardiovasc. Surg. 73, 354, 1978.

8. Huang, M.N., Takita, H., and Douglass, H.O.: Lung resection for metastatic osteogenic sarcoma. J. Surg. Oncol. 10, 179, 1978.

9. Jaffe, N., Frei, E., Traggis, D., et al: Adjuvant methotrexate and citrovorum factor treatment of osteogenic sarcoma. N. Engl. J. Med. 291, 994, 1974.

10. Jaffe, N., Frei, E., Watts, H., et al: High-dose methotrexate in osteogenic sarcoma: a 5-year experience. Cancer Treat. Rep. 62, 259, 1978.

11. Lange, B., and Levine, A.S.: Is it ethical not to conduct a prospectively controlled trial of adjuvant chemotherapy in osteosarcoma? Cancer Treat. Rep. 66, 1699, 1982.

12. Marcove, R.C.: En bloc resection of osteogenic sarcoma. Cancer Treat. Rep. 62:225, 1978.

13. Rosen, G., Huvos, A.G., Mosende, C. et al: Chemotherapy and thoracotomy for metastatic osteogenic sarcoma: a model for adjuvant chemotherapy and the rationale for the timing of thoracic surgery. Cancer 41, 841, 1978.

14. Rosen, G., and Nirenberg, A.: Chemotherapy for osteogenic sarcoma: an investigative method, not a recipe. Cancer Treat. Rep. 66, 1687, 1982.

15. Rosenberg, S.A., Chabner, B., Young, R., et al: The treatment of osteogenic sarcoma. I. Effect of adjuvant high-dose methotrexate following amputation. Cancer Treat. Rep. 63, 739, 1979.

16. Rosenberg, S.A., Flye, M.W., Conkle, D., et al: The treatment of osteogenic sarcoma. II. Aggressive resection of pulmonary metastases. Cancer Treat. Rep. 63, 753, 1979.

17. Rosenberg, S.A., Suit, H., Baker, L.H., and Rosen, G.: Sarcomas of the soft tissue and bone. In, Principles and Practice of Oncology. Eds: DeVita, V.T. Jr., Hellman, S., Rosenberg, S.A., J.B. Lippincott Co., Philadelphia, Toronto, 1982.

18. Sutow, W.W., Sullivan, M.P., Ferbach, D.J., et al: Adjuvant chemotherapy in primary treatment of osteogenic sarcoma. Cancer 36, 1598, 1975.

19. Taylor, W.F., Ivins, J.C., Dahlin, D.C., et al: Trends and variability in survival osteosarcoma. Mayo Clinic Proc. 53, 695, 1978.

20. Telander, R.L., Pairolero, P.C., Pritchard, D.J., et al: Resection of pulmonary metastatic osteogenic sarcoma in children. Surgery 84, 335, 1978.

Adjuvant Interferon Therapy in Primary Osteosarcoma

H. STRANDER, T. APARISI, L.-Å. BROSTRÖM, S. EINHORN, S. INGIMARSSON,
U. NILSONNE, and G. SÖDERBERG

Introduction

Human interferons (IFN) are antiviral proteins classified into three types: α, β, and γ (Isaacs et al, 1957; Lengyel, 1982; Stewart, 1979; Strander, 1982f). Work in bacteria, including genetic analysis and purification by monoclonal antibodies, has shown that a least 15 types of human IFNs exist. The biologic implication of this fact remains to be established.

Until 1979, IFN-α produced by human leukocytes could only be obtained in impure form. It was then purified to homogeneity and the cloning of several human IFN-α subclones was achieved later (Weissman, 1982). Most investigators of the clinical antitumor trials in the world have used semipurified IFN-α (Borden and Hawkins, 1980; Cantell and Strander, 1977; Gutterman, 1981; Merigan and Friedman, 1982; Priestman, 1981; Strander, 1977; 1981a; 1982c; 1982d; 1983g; Strander and Einhorn, 1982a; 1982b; WHO report, 1982).

Using the impure preparations available up to 1979, a trial with a human leukocyte IFN as adjuvant treatment was initiated in classic osteosarcoma (Adamson et al, 1979a; 1979b; Aparisi et al, 1981; Broström, 1979; 1980d; Ingimarsson et al, 1980c; 1981; Nilsonne and Strander, 1979; 1981; Strander et al, 1977; 1979; 1981b; 1982a; 1982b; 1982e; 1982g). This study has been ongoing for 11 years at the Karolinska Hospital, in Stockholm, Sweden. Interim results have been reported from time to time while the trial is being carried out. It is too early to draw any firm conclusions on the basis of current results, but they have been encouraging enough to continue the trial. An updated report is being presented in this communication.

Materials and Methods

IFN-treated group: 51 patients with osteosarcoma without metastases on admission have been treated at Radiumhemmet and at the Division of Orthopaedic Oncology at the Karolinska Hospital. All patients have had primary classic osteosarcomas. Patients with parosteal and periosteal osteosarcoma have been excluded. The series consists of 32 men and 19 women, with a mean age of 24 ± 12 years (±SD).

Current Concepts of Diagnosis and Treatment
of Bone and Soft Tissue Tumors.

Nonadjuvant-treated groups: 30 patients, treated between 1972 and 1976 in other Swedish institutions, have been included as nonadjuvant-treated controls. During this time, no adjuvant treatment was given at all in Sweden. This group consisted of 18 men and 12 women with a mean age of 16 ± 8 years.

Chemotherapy group: after 1976, 21 patients were treated in Sweden, outside the Karolinska Hospital, with chemotherapeutic agents. They received either high doses of doxorubicin* or high doses of methotrexate. This group consists of 13 men and 8 women with a mean age of 18 ± 5 years.

Historical control group: this group consists of 35 patients, 22 men and 13 women with a mean age of 17 ± 9 years, all attending the Karolinska Hospital.

During the time of the trial, the patients included in this series constitute all the patients in Sweden suffering from osteosarcoma without signs of metastases on admission.

Treatment delay

Duration of symptoms was 4 ± 5 months before presentation in the IFN-treated group. The corresponding values for the nonadjuvant-treated group, the chemotherapy group, and the historical control group were 3 ± 2; 3.5 ± 2.5; and 3.5 ± 2.5 years, respectively.

Size

The size of the patients' tumors on radiograms were: in the IFN-treated group, 10 cm ± 3.5; in the chemotherapy group, 10 cm ± 5, in the non-adjuvant-treated control group 8.5 cm ± 3; and in the historical control group, 13 cm ± 6.5.

Histopathologic findings

In the IFN-treated group, 61% of the tumors were osteoblastic. In the chemotherapy-treated, nonadjuvant-treated, and historical control groups, the corresponding values were 52%, 53% and 83%. The percentage of chondroblastic tumors were 18%, 24%, 17% and 12%, respectively.

Tumor grade

In the IFN-treated group, 51% grade IV tumors were observed. The corresponding values for the chemotherapy-treated, nonadjuvant-treated, and historical control groups were 33%, 57% and 80%, respectively. The corresponding percentages of grade III tumors were 37, 57, 37, and 17.

----

* Adriamycin

Tumor sites

The tumors were situated in a proximal site in the IFN-group in 57%, in the chemotherapy-treated group in 62%, in the nonadjuvant control group in 60%, and in the historical control group in 66%. In the four groups, tumors located in the region of the knee occurred in 75%, 67%, 77% and 71%, respectively.

Type of tumor surgery

Local resections were performed on 33% of the patients in the IFN group. Corresponding values for the control chemotherapy-treated group, the nonadjuvant-treated group and the historical control group were 0%, 7% and 0%, respectively.

Interferon

Over the years, two types of preparations have been used; one type was semipurified (P-IF), while the first employed was extremely crude (C-IF). The properties of these preparations have been described previously in detail (Cantell et al, 1981). IFN was administered intramuscularly. During the first month, after an open biopsy was performed in all patients, 3 x 10^6U of IFN-α were given intramuscularly daily. Thereafter, the preparation was administered three times a week for a further 17 months.

Most serious adverse effects during C-IF administration were fever (62%); local pain (38%); shivering (38%); scalp hair loss (28%); itching and erythema (17%) (Ingimarsson et al, 1979). Better toleration was observed with the routine use of P-IF (Ingimarsson, 1980). Fever, however, persisted for some time in most patients. The erythrocyte sedimentation rate decreased significantly and the hemoglobin concentration rose significantly after the cessation of C-IF therapy (Ingimarsson et al, 1980b).

Results

Infections

Patients receiving IFN had fewer acute viral infections than their family members (Ingimarsson et al, 1980; 1980c). Serologic identification tests were carried out in all groups of patients. It is interesting to note that subclinical and clinical infections were detected during ongoing therapy in the IFN-treated group but only in those patients in whom metastases had developed. These viral infections included: herpes simplex, herpes zoster, rubella and measles. They did not develop in patients on IFN treatment who had no metastases.

Immunologic findings

IFN is known to affect the immune system (Berthold et al, 1981; Bloom, 1980; Einhorn, 1980; Einhorn et al, 1978b; 1980a; 1981; 1983). IFN

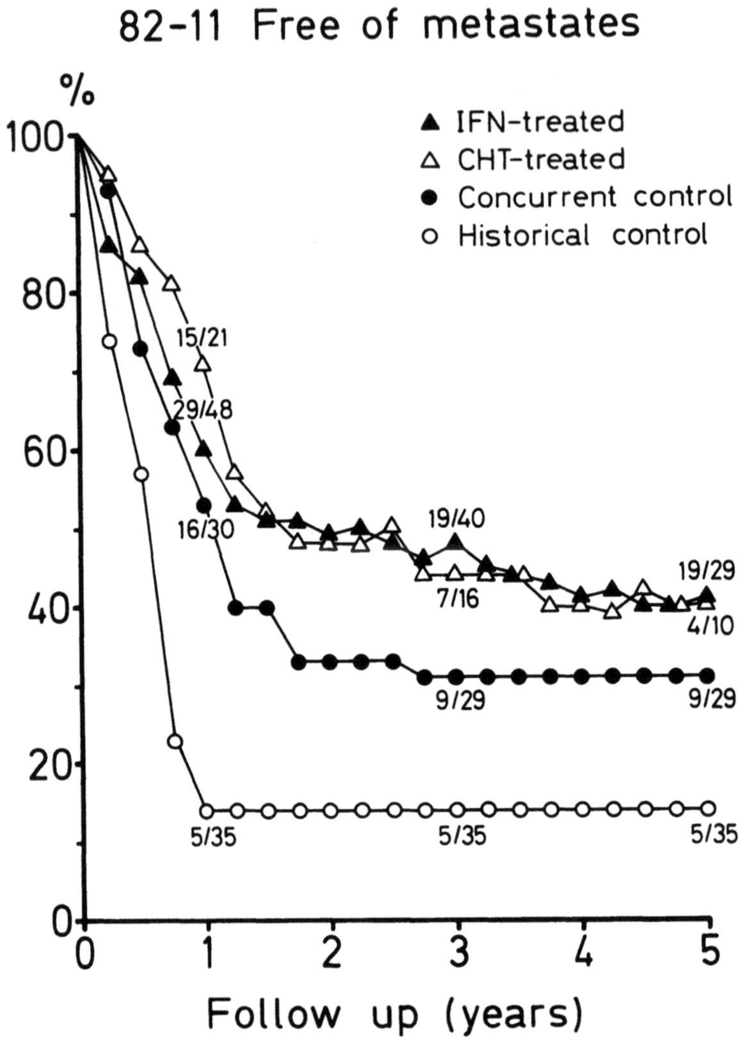

Fig. 1. Current metastasis-free rate in: IFN-treated patients (-▲-); chemotherapy-treated control group (-△-); nonadjuvant-treated control group (-●-); and historical treatment control group (-o-).

might, therefore, be assumed theoretically to act in patients with tumors by stimulating the action of lymphoid cells directed against tumors. IFN is able to inhibit some functions of lymphoid cells and stimulate other functions. In the patients receiving IFN-α, it can be shown that NK activity in peripheral blood is increased on continued therapy for a long time after initiating IFN therapy (Einhorn, 1980; Einhorn et al, 1980b; 1982a). Monocyte phagocytosis has been found to decrease during IFN therapy while we have been unable to find any change in the lymphocyte response to mitogens in IFN-treated patients (Einhorn et al, 1979; 1982b).

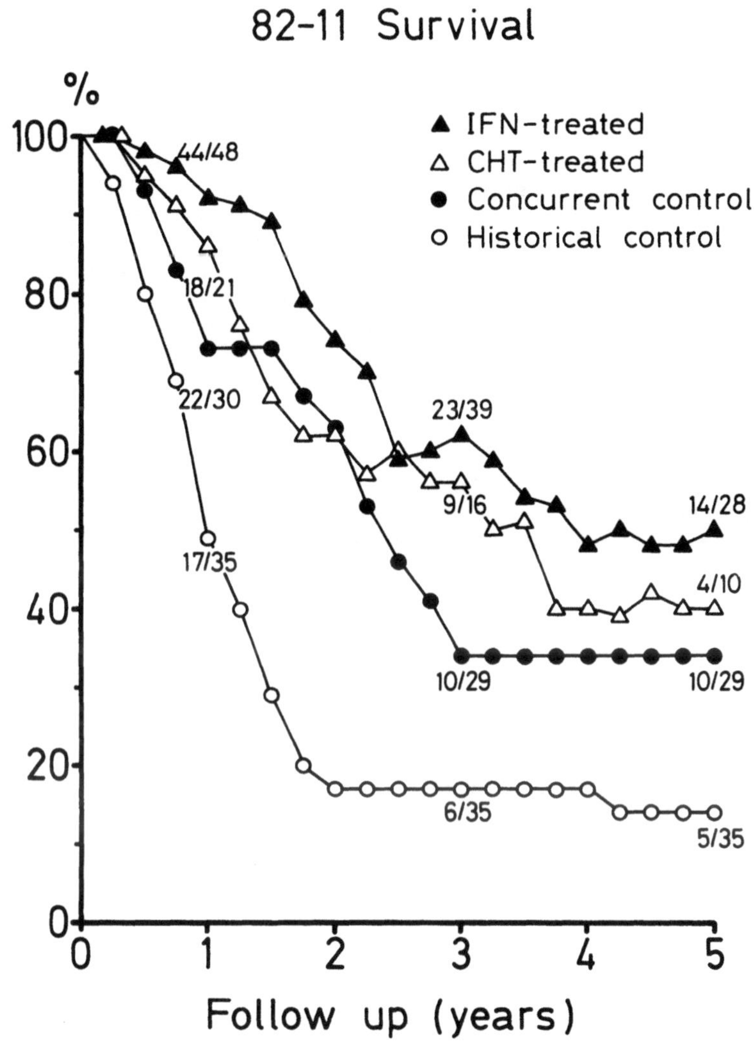

Fig. 2. Current survival rate in IFN-treated patients. (Symbols as in Fig. 1.)

Metastases and survival rate

After three years' follow-up, 46% of the patients are free of metastases in the IFN-treated group, 44% in the chemotherapy-treated group, 31% in the nonadjuvant-treated control group, and 14% in the historical control group. At five years, the percentage rates are 44%, 40%, 31% and 14%. Corresponding survival rates at three years are 61%, 56%, 34% and 17%. At five years, the survival rates are 45%, 40%, 34% and 14% respectively.

The results are encouraging but there is no significant difference between the IFN-treated group and the two control groups. The trial is continuing to determine whether or not the observed difference can be substantiated and confirmed (Fig. 1).

Treatment of metastases

In patients who developed metastases despite IFN treatment, we used high doses of methotrexate followed by high doses of doxorubicin according to recommended regimens. The results have not been encouraging. Over the last three years, we have tried to develop a new treatment for pulmonary metastases consisting of a combination therapy with daily IFN treatment, 3×10^6U intramuscularly, with irradiation, 20 Gy, to both lungs. It has been reported by a Japanese group that IFN might have an effect on pulmonary metastases in osteosarcoma (Ban et al, 1982). In the laboratory at the Radiumhemmet, Karolinska Hospital, we have found that the action of irradiation and IFN in tissue culture is either additive or potentiating. Two patients who have come to the hospital with pulmonary metastases with osteosarcoma have been treated with this type of therapy, so far successfully. The follow-up time is still too short to permit conclusions. It also has to be determined whether such a treatment will help patients who are included in the IFN trial and, hence, previously have received IFN therapy (Fig. 2).

Discussion

Treatment with IFN-α is currently performed on several tumor diseases, usually in phase 1 and 2 trials. Since 1972, we have been conducting a trial in patients with osteosarcoma using IFN as adjuvant treatment based on animal models showing that IFN therapy presumably has its greatest effect in experimental animals with small tumors. Hence, we decided to give IFN as adjuvant treatment for a trial period of 1-1/2 years.

It has been shown in experimental animals that IFN can be effective in osteosarcoma, and we have shown in tissue culture that IFN has an effect on typical human osteosarcoma cell lines (Einhorn and Strander, 1977; 1978a; Glasgow et al, 1978a; 1978b; 1979; Glasgow and Kern, 1981; Strander and Einhorn, 1977). Whether the results of our ongoing trial will be statistically significant is difficult to predict at this time. We are also continuing to analyze various prognostic factors in clinical situations to be able in future to assess which patients should be given IFN in this particular disease (Broström, 1979; Broström et al, 1980c; 1982).

At the present time it is impossible to predict in which situations IFN will have an antitumor effect and to explain its mode of action. It might act via the immune system, due to its antiviral action, or due to direct effects on the growth of tumor cells (Brouty-Boyé, 1980; Gresser, 1980; Strander, 1980). It might also act via a system which we have not yet considered at all. It will be interesting to observe the experimental systems where IFN is used. We will continue our research in patients with osteosarcoma on IFN therapy in the hope of enabling us to determine eventually which patients would benefit most from such therapy.

Summary

This report concerns a clinical trial with human leukocyte interferon (IFN) as adjuvant treatment for patients with osteosarcoma. The trial has been ongoing for 11 years at Karolinska Hospital, Stockholm, Sweden, in 32 men and 19 women with an average age of 24 ± 12 years (±SD).

The three concurrent control groups consist of one group of patients without adjuvant therapy, one group on chemotherapeutic agents, and one group on historical treatment. None of the patients had signs of metastases on admission to hospital.

Up to the present time, the difference between IFN-treated patients and the concurrent control groups is not statistically significant. The historical control group, however, has so far fared worse than the other groups.

The trial in patients with osteosarcoma is being continued to determine the mode of antitumor action of IFN and to establish predictive factors which of the patients will benefit from IFN treatment.

Three pathways for the mode of antitumor action of IFN are currently under consideration; the immune system, a specific antiviral action, or a direct effect on tumor cell growth.

Acknowledgements

The study was supported by grants from the Swedish Cancer Society and the Cancer Society of Stockholm.

References

1. Adamson, U., Aparisi, T., Broström, L.-Å., Cantell, K., Einhorn, S., Hall, K., Ingimarsson, S., Nilsonne, U., Strander, H., and Söderberg, G.: Interferon treatment of human osteosarcoma. Pontificiae Acad. Scient. Scripta Varia 43, 383, 1979a.

2. Adamson, U., Aparisi, T., Broström, L.-Å., Cantell, K., Einhorn, S., Hall, K., Ingimarsson, S., Nilsonne, U., Strander, H., and Söderberg, G.: Adjuvant interferon treatment of human osteosarcoma. In, Recent Results in Cancer Research. Eds.: G. Bonnadonna, G. Mathé and S.E. Salmon. Springer-Verlag, Berlin/Heidelberg, 68, 40, 1979b.

3. Aparisi, T., Strander, H., Broström, L.-Å., and Nilsonne, U.: Experience avec l'interferon dans le traitement de l'osteosarcome. Proc. of the Journées Annuelle de Chirurgie Orthopedique. Ed.: J. Duparc, 1981. (In press)

4. Ban, S., Yamagiwa, T., Yanagawa, T., Sakakida, K., Ito, H., Murakami, K., Matsuo, A., Imanishi, J., and Kishida, T.: Effect of human leukocyte interferon on metastatic lung tumor of osteosarcoma and primary lung cancer. In, The Clinical Potential of Interferon. Eds.: R. Kono and J. Vilček. University of Tokyo Press, pp 223, 1982.

5. Berthold, W., Grazia-Masucci, M.-G., Klein, E., and Strander, H.: Interferon treatment elevates the cytotoxic potential of lymphocytes and thereby increases the frequency of functional killer cells. Human Lymphocyte Differentiation 1, 1, 1981.

6. Bloom, B.R.: Interferons and the immune system. Nature 285, 593, 1980.

7. Borden, E.C., and Hawkins, J.J.: Interferons for human neoplastic and viral diseases. Comprehensive Therapy, Harvard MA, 6, 6, 1980.

8. Broström, L.-Å.: On the natural history of osteosarcoma. Aspects on diagnosis, prognosis and endocrinology. Acta Orthop. Scand., Suppl. No. 183, 1979.

9. Broström, L.-Å., Aparisi, T., Ingimarsson, S., Lagergren, C., Nilsonne, U., Strander, H., and Söderberg, G.: Can historical controls be used in current clinical trials in osteosarcoma? Metastases and survival in a historical and a concurrent group. Int. J. Rad. Oncol. Biol. Phys. 6, 1717, 1980a.

10. Broström, L.-Å., Aparisi, T., Ingimarsson, S., Lagergren, C., Nilsonne, U., Strander, H., and Söderberg, G.: Can historical controls be used in current clinical trials in osteosarcoma? Analysis of prognostic factors. Int. J. Rad. Oncol. Biol. Phys. 6, 1711, 1980b.

11. Broström, L.-Å., Ingimarsson, S., Strander, H. and Eklund, G.: Correlation between prognostic factors and blood variables in osteosarcoma. Acta Med. Scand. 207, 429, 1980c.

12. Broström, L.-Å., Ingimarsson, S., Strander, H., and Söderberg, G.: Adjuvant interferon treatment and late development of cerebral metastases in a patient with osteosarcoma. Acta Orthop. Scand. 51, 589, 1980d.

13. Broström, L.-Å., Strander, H., and Nilsonne, U.: Survival in osteosarcoma in relation to tumour size and location. Clin. Orthop. 167, 250, 1982.

14. Brouty-Boyé, D.: Inhibitory effects of interferons on cell multiplication. Lymphokine Reports 1, 99, 1980.

15. Cantell, K., Hirvonen, S., Kauppinen, H.L. and Myllylä, G.: Production of interferon in human leukocytes from normal donors with the use of Sendai virus. Methods in Enzymology (New York) 78, 29, 1981.

16. Cantell, K., and Strander, H.: Human leukocyte interferon for clinical use. In, Blood Leukocytes. Functions and Use in Therapy. Eds.: C.F. Högman, K. Lindahl-Kiessling, and H. Wigzell. Almqvist and Wichzell, pp 73, 1977.

17. Einhorn, S.: Influence of interferon on human lymphocytes. *In vitro* and *in vivo* studies on spontaneous cytotoxicity and cell surface receptors. Thesis, Sundt Offset, Stockholm, 1980.

18. Einhorn, S. and Strander, H.: Is interferon tissue specific? Effect of human leukocyte and fibroblast interferons on the growth of lymphoblastoid and osteosarcoma cell lines. J. Gen. Virol. 35, 573, 1977.

19. Einhorn, S. and Strander, H.: Interferon therapy for neoplastic diseases in man. In, Production of Human Interferon and Investigations of Its Clinical Use. A Symposium Workshop, May 19-20, 1977. Human Interferon. Eds.: Warren R. Stinebring and Paul J. Chapple. Plenum Publ. Corp. 159, 1978a.

20. Einhorn, S., Blomgren, H., and Strander, H.: Interferon and spontaneous cytotoxicity in man. I. Enhancement of the spontaneous cytotoxicity of peripheral lymphocytes by human leukocyte interferon. Int. J. Cancer 22, 405, 1978b.

21. Einhorn, S., Blomgren, H., Cantell, K., and Strander, H.: Effect of prolonged *in vivo* administration of leukocyte interferon on the mitogen responsiveness of human lymphocytes. Acta Med. Scand. 206, 345, 1979.

22. Einhorn, S., Blomgren, H. and Strander, H.: Effect of an intramuscular injection of human leukocyte interferon on blood leukocyte counts and proportions of lymphocytes forming E, EA and EAC rosettes. Int. Arch. Allergy Appl. Immun. 63, 139, 1980a.

23. Einhorn, S., Blomgren, H., and Strander, H.: Interferon and spontaneous cytotoxicity in man. V. Enhancement of spontaneous cytotoxicity in patients receiving human leukocyte interferon. Int. J. Cancer 26, 419, 1980b.

24. Einhorn, S., Blomgren, H. and Strander, H.: Influence of interferon on the immune system. Proc. XIth Cong. Europ. Allergolog. Clin. Immunology, Vienna, October 6-10, 1980. In, Clinical Immunology and Allergology. Eds.: C. Steffan and H. Ludwig, Elsevier/North Holland Biomedical Press. 1981.

25. Einhorn, S., Ahre, A., Blomgren, H., Johansson, B., Mellstedt, H., and Strander, H.: Enhanced NK activity in patients treated by interferon-α. Relation to clinical response. In, Natural Cell-Mediated Immunity Against Tumours. In, NK Cells and Other Natural Effector Cells. Ed.: R.B. Herberman, Academic Press, 1259, 1982a.

26. Einhorn, S., Blomgren, H., Jarstrand, C., Strander, H., and Wasserman, J.: Influence of interferon-α therapy on different functions of the human immune system. Proc. 2nd International TNO Meeting on the Biology of the Interferon System, Rotterdam, April 18-22, 1983. (In press)

27. Einhorn, S., Blomgren, H., and Strander, H.: *In vitro* and *in vivo* effects of interferon on the mitogen responsiveness of human lymphocytes. Clin. Exp. Immunol. 1982b. (In press).

28. Glasgow, L.A., Crane, J.L., Schelupner, C.J., Kern, E.R., Youngner, J.S., and Feingold, D.S.: Activation of reticuloendothelial system macrophages and enhancement of host resistance to a transplantable osteogenic sarcoma in mice by an extract of *Brucella abortus*. Cancer Treat. Rep. 62, 1931, 1978a.

29. Glasgow, L., Crane, J., and Kern, E.: Antitumor activity of interferon against murine osteogenic sarcoma cells *in vitro*. J. Natl. Cancer Inst. 60, 659, 1978b.

30. Glasgow, L., Crane, J., Schleupner, C.H., Kern, E., Youngner, J., and Feingold, D.: Enhancement of resistance to murine osteogenic sarcoma *in vivo* by an extract of *Brucella abortus* (Bru-Pel): Association with activation of reticuloendothelial system macrophages. Infect. Immun. 23, 19, 1979.

31. Glasgow, L. and Kern, E.: Effect of interferon administration on pulmonary osteogenic sarcomas in an experimental murine model. J. Natl. Cancer Inst. 67, 207, 1981.

32. Gresser, I.: Usefulness of the results of studies on the antitumor effects of interferon in animals to interferon therapy of patients. Recent Results of Cancer Research 75, 226, 1980.

33. Gutterman, J.U.: Clinical investigation of the interferons in human cancer. Bulletin du Cancer 33, 271, 1981.

34. Ingimarsson, S.: Studies on human leukocyte interferon in clinical use. Thesis, Sundt Offset, Stockholm, 1980.

35. Ingimarsson, S., Cantell, K., and Strander, H.: Side effects of long-term treatment with human leukocyte interferon. J. Infect. Dis. 140, 560, 1979.

36. Ingimarsson, S., Cantell, K., and Strander, H.: Acute infections in patients receiving interferon. In, Interferon - Properties and Clinical Uses. Eds: A. Khan, N.O. Hill, and G.L. Dorn. Wadley Institutes, 633, 1980a.

37. Ingimarsson, S., Bergström, K., Broström, L.-Å., Cantell, K., and Strander, H.: Effect of long-term treatment with human leukocyte interferon on various laboratory parameters. Acta Med. Scand. 208, 155, 1980b.

38. Ingimarsson, S., Cantell, K., Carlström, G., and Strander, H.: Virus infections and recurrence of osteosarcoma in patients receiving human leukocyte interferon. Int. J. Cancer 26, 395, 1980c.

39. Ingimarsson, S., Cantell, K., Carlström, G., Dalton, B., Paucker, K. and Strander, H.: Immune reactions and long-term therapy with human leukocyte interferon. Acta Med. Scand. 209, 17, 1981.

40. Isaacs, A., and Lindenmann, J.: Virus interference. I. The interferons. Proc. Royal Soc. Lond.B 147, 258, 1957.

41. Lengyel, P.: Biochemistry of interferons and their actions. Ann. Rev. Bioch. 51, 251, 1982.

42. Merigan, T.C. and Friedman, R.M.: Interferons. New York & London, Academic Press, 1982.

43. Nilsonne, U., and Strander, H.: Clinical experience of interferon treatment in osteosarcoma. Proc. 12th Int. Immunobiological Symp. Yugoslav; Academy of Sciences and Arts Symp. on Interferon, Zagreb. 137, 1979.

44. Nilsonne, U. and Strander, H.: Traitement de l'osteosarcome par l'interferon et la chirurgie différenciée. Revue de Chirurgie Orthopedique 67, 193, 1981.

45. Priestman, T.J.: Interferons and cancer: The end of the beginning or the beginning of the end. Clin. Oncol. 7, 271, 1981.

46. Stewart, II, W.E.: The Interferon System. Springer-Verlag , Wien & New York, 1979.

47. Strander, H.: Interferons: Anti-neoplastic drugs? Blut 35, 277, 1977.

48. Strander, H.: Mechanisms of cell growth inhibitory and antitumor action; genetics of the interferon system. Introductory remarks. Ann. New York Acd. Sci. 350, 168, 1980.

49. Strander, H.: Interferon (IFN) therapy of tumor disease. Proc. Int. Workshop on Regulation and Modulation of Immune Activities. Hasliberg-Reuti, Feb. 24-27, 1981. In, Mikroökologie und Therapie, 11, 135, 1981a.

50. Strander, H.: Clinical trials conducted with interferon on tumour patients in Europe. TNO and Erasmus University Meeting on the Biology of the Interferon System, Rotterdam, Apr. 21-24, 1981. In, The Biology of the Interferon System. Eds.: E. De Maeyer, G. Galasso, and H. Schellekens. Elsevier/North-Holland Biomedical Press. 1981b.

51. Strander, H., Einhorn, S., Ingimarsson, S., Aparisi, T., Broström, L.-Å., Nilsonne, U., and Söderberg, G.: Treatment of human osteosarcoma with HU LEU IFN alpha. Presented at The Annual International Congress of Interferon Research, Miami, Nov. 1-4, 1982a.

52. Strander, H., and Einhorn, S.: Interferon in cancer - faith, hope and reality. Am. J. Clin. Oncol. 5, 297, 1982b.

53. Strander, H.: Interferon therapy for tumour diseases in man - prospects for the near future. Proc. 3rd Deutsche Stiftung für Krebsforschung, Bonn, March 13-16, 1981. In, Interferon - Properties, Mode of Action, Production, Clinical Application. Eds.: K. Munk and H. Kirchner, S. Karger, Heidelberg. 227, 1982c.

54. Strander, H.: Interferon therapy of malignant disease. In, The Clinical Potential of Interferons. Eds.: R. Kono and J. Vilček. University of Tokyo Press. 179, 1982d.

55. Strander, H.: Antitumor effects of interferon and its possible use as an antineoplastic agent. In, Texas Reports on Biology and Medicine - The Interferon System: A Review to 1982. 41 (in press) 1982e.

56. Strander, H.: The Role of Interferon in Disease. Cambridge University Press (in press) 1982f.

57. Strander, H.: Adjuvant treatment of osteosarcoma - some personal views. International Conference on Advances in the Adjuvant Therapy of Cancer. London, June 17-18. 1982g.

58. Strander, H. and Einhorn, S.: Effect of human leukocyte interferon on the growth of human osteogenic sarcoma cells in tissue culture. Int. J. Cancer 19, 468, 1977.

59. Strander, H., and Einhorn, S.: Interferon alpha trials in human tumor diseases. In, Chemistry and Biology of Interferons: Relationship to Therapeutics. UCLA Symposia on Molecular and Cellular Biology. XXV. Eds.: Thomas Merigan, Robert Friedman and C. Fox. Academic Press, New York, 399, 1982a.

60. Strander, H. and Einhorn, S.: Interferon therapy in neoplastic diseases. Phil. Trans. R. Soc. London. B, 299, 113, 1982b.

61. Strander, H., Cantell, K., Ingimarsson, S., Jakobsson, P.A., Nilsonne, U., and Söderberg, G.: Interferon treatment of osteogenic sarcoma - a clinical trial. Conf. on Modulation of Host Immune Resistance in the Prevention or Treatment of Induced Neoplasias. December 9-11, 1974. Fogarty Int. Ctr. Proc. US Govt. Printing Office, Washington, DC 28, 377, 1977.

62. Strander, H., Adamson, U., Aparisi, T., Broström, L.-Å., Cantell, K., Einhorn, S., Hall, K., Ingimarsson, S., Nilsome, U., and Söderberg, G.: Adjuvant interferon treatment of human osteosarcoma. In, Recent Results in Cancer Research. Eds.: G. Bonnadonna, G. Mathé and S.E. Salmon. Springer-Verlag, Berlin-Heidelberg 68, 39, 1979.

63. Strander, H., Aparisi, T., Blomgren, H., Broström, L.-Å., Cantell, K., Einhorn, S., Ingimarsson, S., Nilsome, U., and Söderberg, G.: Adjuvant interferon treatment of human osteosarcoma. EORTC meeting, Paris 1980. In, Recent Results in Cancer Research, Springer-Verlag, Berlin-Heidelberg, 80, 103, 1982b.

64. Weissmann, C., Nagata, S., Boll, W., Fountoulakis, M., Fujisawa, A., Fujisawa, J.-I., Haynes, J., Henco, K., Mantei, N., Ragg, H., Schein, C., Schmid, J., Shaw, G., Streuli, M., Taira, H., Todokoro, K., and Weidle, U.: Structure and expression of human IFN-α genes. Phil. Trans. Royal Soc., London, B, 299, 7, 1982.

65. WHO Scientific Group Report: Interferon therapy. Technical Report Series 676, WHO, Geneva, 1982.

Chemotherapy by Infusion for Malignant Bone Tumors

F. R. EILBER, J. MIRRA, J. ECKARDT, and E. CAULKINS

Introduction

One of the important factors for determining the destruction of human cancer cells by chemotherapeutic agents is the proliferative state of the tumor cell population at the time of drug exposure. Two classes of chemotherapeutic agents are available: one is nonspecific or cell-cycle specific; the second is phase specific (Valeriote and Von Pulten, 1975). Phase-specific agents kill cells in one phase of the cell cycle and have little effect on cells in other phases, whereas cell cycle specific agents kill cells throughout the cell cycle and are less cytotoxic to cells in the nonproliferative state. These differences in cell cycle kinetics are responsible for the differences in tumor response to various anticancer agents. For example, little or no response would be noted with a cell cycle specific drug in a tumor with a large fraction of cells in the resting or G_0 phase (low grade or mitotically inactive tumors). If the drug were administered frequently, however, or if the exposure were prolonged, then more normal hemopoietic stem cells would be destroyed, causing increasing toxicity and intolerable side effects to nonmalignant cells that would be mitotically active.

Several investigators have used prolonged low-dose infusional chemotherapy to obviate some of these problems. With this method, all the different phases of the cell cycle can be exposed to a lower dose of the specific anticancer agent over a prolonged period. Theoretically, more cells are then killed as they move from a G_0 resting stage to the next phase of the cell cycle. Longer exposure may also permit a higher concentration of the drug within the tumor cells and this, too, appears to affect the cytotoxicity. For example, 5-FU and doxorubicin* may be more effective at higher concentrations in the liver, as may cisplatin in bone tumors. Several other drugs have been investigated which apparently have no additional cell-kill benefit with higher local drug concentrations.

Several methods for increasing local drug concentrations have been investigated. Stehlin, McBride and Krementz reported a marked increase in clinical responses from L-PAM, a relatively inactive agent against malig-

----

* Adriamycin

Current Concepts of Diagnosis and Treatment
of Bone and Soft Tissue Tumors.

nant melanoma provided the drug was given in higher concentrations by the isolated limb perfusion technique (Stehlin et al, 1975). Similar responses have been reported for perfusions or infusions, including those from infusions in the head and neck with 5-FU and methotrexate for sinus cancer, the intra-arterial administration of chemotherapeutic agents, such as 5-FU for metastases to the liver or for primary hepatoma. Most recently, Benjamin, Jaffe and Wallace reported on intra-arterial cisplatin for primary malignant bone tumors (Didolkar et al, 1978; Jaffe et al, 1980; Kraybill et al, 1977; Mavligit et al, 1981). In each of these reports systemic administration of these same drugs failed to induce a significant response, whereas a constant infusion enhanced the antitumor effect.

Doxorubicin. This antitumor antibiotic has a broad-spectrum antineoplastic activity in both leukemia-lymphemia and several solid tumors, including breast cancer and bone and soft tissue sarcomas. Doxorubicin's known side effects include alopecia, cardiomyopathy, mucositis, gastrointestinal problems (such as nausea, vomiting and diarrhea), and, finally, bone marrow toxicity leading to leukopenia and thrombocytopenia. Several investigators have reported a marked local toxicity if doxorubicin infiltrates the soft tissues during intravenous administration (Rudolph et al, 1976). Several studies have shown marked tissue necrosis, and in fact, ulceration can occur if high concentrations of the drug are allowed to extravasate from a vein. This local toxicity is not related to any cell cycle-specific kinetics because the normal tissue being destroyed is not mitotically active.

In 1974, we began a pilot study using doxorubicin by the intra-arterial route (Haskell et al, 1974). We based our study on the premise of potential advantages to delivering high concentrations of a drug known for its toxicity to tumor and normal tissues. Furthermore, systemic toxicity could be reduced if the same dose were delivered by a prolonged, three-day, constant infusion. Doxorubicin has a known affinity for nuclear DNA during its first pass; therefore, more drug would be concentrated in the tumor and less would escape into the systemic circulation. Slow administration would also obviate some of the cardiac toxicity associated with doxorubicin (Bristow et al, 1978). Benjamin and others suggest that at least a portion of the cardiac toxicity is caused by the bolus effect of high concentrations of doxorubicin given by vein when the heart is the first target organ (Benjamin et al, 1981; Lokich et al, 1983).

Finally, the intra-arterial route was chosen because doxorubicin is known to be a radiation sensitizer. Several investigators have observed a radiation portal flare in patients who were given systemic doxorubicin during or immediately after radiation therapy (Greco et al, 1976).

Our purpose was to deliver radiation immediately after the doxorubicin infusion to take advantage of these sensitizing effects with lower doses of radiation.

To date, a total of 153 patients with malignant soft tissue sarcomas and 104 patients with malignant bone tumors have been treated with a combination of intra-arterial doxorubicin followed immediately by radiation therapy. The radiation dose varied from 3500 rad given at 350 rad fractions times ten to 1750 rad, again at 350 rad dose fractions, from one to three weeks. The primary tumor was surgically excised following the preoperative chemotherapy and radiation therapy.

This communication will limit its discussion to the results achieved for patients with primary malignant tumors of bone.

Materials and Methods

Indwelling arterial catheters were placed by the radiologists after percutaneous puncture of a major vessel by the Seldinger technique. For the upper extremity, the catheter was placed in the axillary artery, and for the lower extremity it was placed in the common femoral artery for distal lesions or in the iliac artery for proximal lesions of the femur. The catheter was placed in the common iliac artery for lesions involving the ilium and sacrum. Patients were then transferred back to their rooms with an indwelling catheter. Catheter position was monitored with daily injections of fluorescein dye and a Wood's lamp. If the skin distribution of the dye varied from the primary tumor site, the catheters were repositioned until the proper skin distribution was obtained. Early in our experience, we tried to position the catheters based on the tumor's arterial blood supply at the time of angiography, but we found that this placement did not necessarily correlate with drug-flow patterns in the vascular system.

Doxorubicin was then infused at a dosage of 30 mg in 500 mL of saline per extremity for the lower extremity and 20 mg per extremity for the upper extremity delivered continuously by an Ivac infusion pump over a 24-hour period for each of three consecutive days. The daily fluorescein dye injections were used to confirm infusion of the proper skin area for evidence of erythema. If, during the infusion, areas of hot spots or skin erythema were noted, the catheters were repositioned to change their distribution pattern. No heparin was added to the system.

Early in our experience, the catheters were placed more distally in the superficial femoral artery or the popliteal artery for tibial lesions or lesions of the distal femur and into the brachial artery for forearm lesions. However, severe complications such as erythema of the skin and fibrosis of the subcutaneous tissue were noted. In retrospect, one of the obvious difficulties was placement of the catheter in a vessel with a diameter nearly equal to that of the catheter. Thus, the treatment acted more like a perfusion than an infusion, and at that dose level severe toxicity was encountered. Subsequently, all catheters were placed in a high-flow vessel. Initially, the catheter was placed on the same side as the lesion -- when the lesion was in the right distal femur, the catheter was placed in the right femoral artery. In several patients the catheter came out of the arterial puncture because too little length had been inserted. Subsequently, all catheters were placed in the opposite groin and fed across the aortic bifurcation down to the opposite side. Thus, more of the catheter could be inserted in the patient and the problems of catheter migration were avoided. In the first two patients, heparin was instilled into the line to prevent clotting; however, it became apparent that heparin precipitated doxorubicin. Subsequently, no heparin was used in the system.

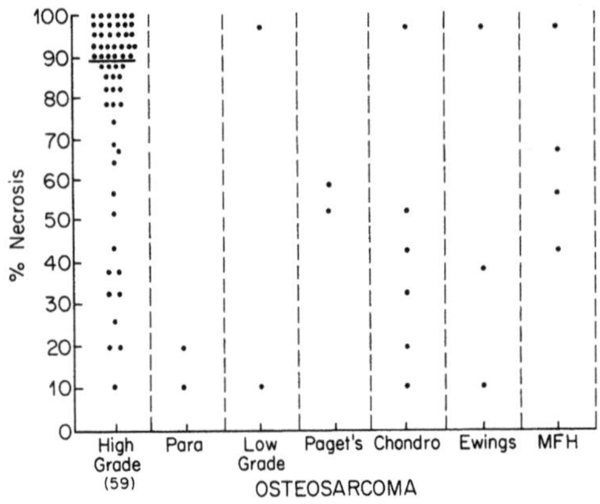

Fig. 1. Preoperative doxorubicin and radiation therapy.

At the completion of the 72-hour infusion, the catheter was removed, manual pressure was placed on the arterial puncture for 10 minutes and then pressure was applied with a sandbag for six hours, after which time the patient was allowed to ambulate. Radiation therapy usually started the next day but no later than five days after removal of the catheter. Radiation therapy was given by external beam with a cobalt-60 generator and was delivered daily by AP-PA ports at a dosage of 350 rad fractions times ten for 51 patients and 350 rad fractions times five for 40 patients. The entire involved bone and soft tissue component of the tumor was treated. In all patients, a strip of skin or a compartment was spared either medially or laterally opposite the tumor. All radiation was performed on an outpatient basis.

Surgical excision of the involved bone took place from three to six weeks after completion of the preoperative therapy. The delay was usually occasioned by waiting for bony replacement materials such as cadaver allografts or the specific manufacture of a metallic endoprosthesis.

The excised tumor specimen was examined by a surgical pathologist, Dr. Mirra. At least 15 sections were taken from the surgical margin and the central portion of each tumor. An estimation of tumor cell necrosis was based on the criterion for lack of nuclei. All 15 slides of each specimen were examined by at least 10 high-power fields per sample. Estimates of tumor cell necrosis were then given as a percentage of the total tumor mass evaluated.

Results

Ninety-one patients with malignant bone tumors were treated with this preoperative regimen from 1975-1983. Histologic tumor types included 73 high-grade osteosarcomas, two parosteal osteosarcomas, two low-grade osteosarcomas, and one osteosarcoma arising in Paget's disease. Additional histologic types included six chondrosarcomas, three Ewing's sarcomas, and four malignant fibrous histiocytomas. The tumor cell necrosis in the high-grade osteosarcomas ranged from 5 to 100%, with a mean of 90%. Of the parosteals studied, the range was from 2 to 30%, with a mean of 20%. The necrosis rate for chondrosarcomas was from 10 to 95%, with an average of 35%, and with malignant fibrous histiocytoma and Ewing's sarcoma mean necrosis was 80% and 95%, respectively (Fig. 1).

The different doses of radiation therapy -- 3500 rad versus 1750 -- were compared in patients with osteosarcoma. Forty-four patients with high-grade osteosarcoma were treated with 3500 rad , and the mean tumor necrosis in this group was 90%. Of the 29 patients treated thus far with 1750 rad of radiation after treatment with intra-arterial doxorubicin, the mean necrosis rate appears to be the same: 90%. If the total number of patients studied in each group is considered, at the present time we can find no dramatic differences in tumor cell necrosis with the smaller radiation dose.

Complications

Several complications were noted with this preoperative therapy regimen. In one patient, arterial thrombosis occurred which had to be resolved by amputation. In this case, the catheter had been placed in the distal superficial femoral artery for a distal lesion of the femur. At the time of surgery, the superficial femoral artery had thrombosed, and despite attempts of performing a thrombectomy and arterial bypass, after excising the distal femur, the viability of the extremity could not be maintained. Since that time there has been no incidence of arterial thrombosis. The other complication was wound slough in eight patients. All of these complications occurred early in our trial, when wounds were closed under tension. Since that time, the incisions have not been closed under tension and the muscle flaps have been mobilized. No further wound necrosis has been encountered.

Skin and subcutaneous fibrosis occurred in approximately 20% of the patients early in the series who received cadaver allografts. These complications were prevented with the metallic endoprostheses and earlier mobilization of the patients.

Discussion

This report summarizes our experience with the intra-arterially infused doxorubicin and radiation therapy in a series of 91 patients with malignant tumors of bone. The overall disease-free survival is 65% for the 91 patients and 50% for the 73 with high-grade osteosarcomas at a mean of 30 years' follow-up. Due to three incidents of local tumor recurrence within the treatment field, the local control rate in the patients undergoing limb salvage was 96.8%. The purpose of this study was to determine whether it was possible to achieve in situ tumor cell destruction to limit the surgical resections and permit limb salvage for a high percentage of patients with malignant bone tumors. This study was based on the premise that the most highly malignant bone tumors have a significant extramedullary component and that surgical attempts at limb salvage must of necessity come extremely close to the popliteal vessels, posterior tibial and perineal nerves. Therefore, we reasoned that if these structures were to be saved, it would be potentially beneficial to destroy the tumor at its closest surgical margin.

In situ tumor cell destruction was achieved in the majority of the patients. There was a range in tumor cell necrosis for each histologic type and within each histologic type. High-grade lesions such as osteosarcoma, Ewing's sarcoma, and malignant fibrous histiocytomas had a mean tumor cell necrosis of 90%. Although these high rates do not necessarily equate to biologic tumor cell death, the evidence of pathologic necrosis is truly impressive. It is unlikely that the necrosis is due to radiation therapy alone, because the dose of radiation is much less than is normally given (6000 r) to achieve much less tumor cell necrosis. It is also unlikely that this result can be credited to doxorubicin alone because earlier attempts at infusions of doxorubicin produced an average of 30% tumor cell necrosis. It therefore appears that there is synergism between doxorubicin and radiation. The biologic equivalent to the tumor cell necrosis for the patient, of course, is whether or not local tumor recurrence is controlled. The fact that only three patients in this entire series had local tumor recurrence appears to be further justification for the presurgical therapy.

Rosen et al, at Sloan-Kettering, are also using presurgical therapy for patients with malignant skeletal tumors (primarily osteosarcoma)(Rosen et al, 1982). Patients received preoperative adjuvant chemotherapy consisting of high-dose methotrexate and bleomycin/cyclophosphamide/actinomycin D before local surgical resection. With this T-7, or the first portion of the T-10, protocol, Rosen has found that approximately 50% of the tumor specimens have greater than 90% necrosis (Grade III-IV); however, at least 50% of the specimens show no significant necrosis. The patient's adjuvant chemotherapy is altered based on this finding. The overall limb salvage rate for lesions of the distal femur at the Memorial Sloan-Kettering Hospital is approximately 30%, whereas the overall limb salvage rate at University of California, Los Angeles (UCLA) is 80%.

Other investigators perform the en bloc excision without preoperative therapy, and in a small series their local recurrence rate is approximately the same as that with presurgical therapy.

The concepts of presurgical therapy and en bloc excision are appealing. This regimen permits an interpretation of its antitumor effects on tumor cells in situ. If little tumor cell necrosis were apparent, either the combination chemotherapy or the irradiation dosage could be altered to achieve the goal of in situ tumor cell destruction. Furthermore, Rosen's concept could be used for presurgical chemotherapy only, and if no tumor cell destruction were seen, then the adjuvant chemotherapy could be altered. In case of significant destruction, however, then presurgical chemotherapy would be maintained because if the primary tumor could be killed by this therapy occult metastases would similarly be killed.

Summary

A consecutive series of 91 patients with highly malignant bone tumors has been treated with intra-arterial doxorubicin and radiation therapy (3500 rad to 1750 rad). Surgical excision of involved bone was then undertaken, and pathologic evaluation of the excised specimen was done. In high-grade lesions, such as osteosarcoma, malignant fibrous histiocytoma, and Ewing's sarcoma, a mean of 90% tumor cell destruction was achieved. In lower grade lesions, such as parosteal sarcoma, low-grade osteosarcoma, and chondrosarcoma, the average tumor cell necrosis was only 20%. Local recurrence was noted in only three patients in this consecutive limb salvage series. It, therefore, appears that presurgical therapy with intraarterial doxorubicin and radiation therapy is a highly effective means to obtain in situ tumor cell destruction. The local recurrence rate of 3% is comparable to series of patients treated by primary amputation. These results suggest that presurgical therapy should be continued and expanded in patients with malignant bone tumors (Pratt, 1982).

Supported by Grants CA29605 and CA12582, awarded by the National Cancer Institute, DHHS.

References

1. Benjamin, R., Hegha, H., Mackay, B., Ewer, M., Wallace, S., Valdivieso, M., Rasmussen, S., Blumenschein, G., and Freireich, E.: Reduction of adriamycin cardiac toxicity using a prolonged continuous intravenous infusion. Proc. AACR 22, 179, 1981.

2. Bristow, M.R., Billingham, M.E., Menson, J., and Daniels, J.R.: Clinical spectrum of anthacyline antibiotic cardiotoxicity. Cancer Treat. Rep. 62, 873, 1978.

3. Didolkar, M., Kanter, P., Boffi, R., Swartz, H., Lopez, R., and Baez, N.: Comparison of regional versus systemic chemotherapy with adriamycin. Ann. Surg. 187, 332, 1978.

4. Greco, F., Brereton, H., Kent, H., Zimbler, H., Merril, J., and Johnson, R.: Adriamycin and enhanced radiation reaction in normal esophagus and skin. Ann. Int. Med. 85, 294, 1976.

5. Haskell, C.M., Silverstein, M., Rangel, D., Hunt, J., Sparks, F., and Morton, D.L.: Multimodality cancer therapy in man: A pilot study of adriamycin by arterial infusion. Cancer 33, 1485, 1974.

6. Jaffe, N., Chuang, V., Wallace, S., Ayala, A., Murray, J., Romsdahl, M., and Benjamin, R.: Osteosarcoma: Control of the primary tumor with intraarterial cisplatinum. Proc. AACR 21, 197, 1980.

7. Kraybill, W., Harrison, M., Sasaki, T. and Fletcher, W.: Regional intra-arterial infusion of adriamycin in the treatment of cancer. Surg. Gynecol. Obstet. 144, 335, 1977.

8. Lokich, J., Bothe, A., Zipoli, T., Green, R., Somneborn, H., Paul, S. and Philips, T.: Constant infusion schedule for adriamycin: A phase I-II clinical trial of a 30-day schedule by ambulatory pump delivery system. J. Clin. Oncol. 1, 24, 1983.

9. Mavligit, G., Benjamin, R., Pratt, Y., Jaffe, N., Chuang, V., Wallace, S., Murray, J., Ayala, A., Johnston, S., Hersh, E., and Calvo, D.: Intraarterial cis-platinum for patients with inoperable skeletal tumors. Cancer 48, 1, 1981.

10. Pratt, Y.: Intraarterial chemotherapy in cancer treatment -- selected abstracts. International Cancer Research Databank Program. National Cancer Institute, U.S. Department of Health and Health Services, September 1982.

11. Rosen, G., Caparros, B., Huvos, A., Kosloff, C., Nirenberg, A., Cacavio, A., Marcove, R., Lane, J., and Metha, B.: Preoperative chemotherapy for osteosarcoma: Selection of postoperative adjuvant chemotherapy based on the response of the primary tumor to chemotherapy. Cancer 49, 1221, 1982.

12. Rudolph, R., Stein, R., and Pattillo, R.: Skin ulcers due to adriamycin. Cancer 38, 1087, 1976.

13. Stehlin, J.S., De Ipolyi, P., Giovanella, B.C., Gutierrez, A., and Anderson, R.F.: Soft tissue sarcomas of the extremity: Multidisciplinary therapy employing hyperthermic perfusion. Am. J. Surg. 130, 64, 1975.

14. Valeriote, F., and Van Pulten, L.: Proliferation-dependent cytotoxicity of anticancer agents: A review. Cancer Res. 35, 2619, 1975.

Toxicity and Supportive Care Related to Chemotherapy

J. A. MAROUN

Chemotherapeutic agents for the treatment of cancer characteristically have narrow therapeutic margins leading to significant side effects. Their early recognition and appropriate management are major tasks in today's cancer treatment. Progress in supportive care has made an important contribution to cancer control over the last two decades. It is useful to discuss chemotherapy toxicity under four categories:

A) Immediate and local toxicities

B) Early toxicities

C) Specific organ toxicities

D) Long-term complications

A) Immediate and Local Toxicities

1. Hypersensitivity reactions

The occurrence of hypersensitivity reactions to many drugs is a widely recognized phenomenon, and a number of cytotoxic drugs have been reported to cause allergic reactions (Weiss and Bruno, 1981).

The polypeptide enzyme L-asparaginase has been frequently associated with allergic reactions. Although less frequently, other drugs have been associated with such reactions, eg, cisplatin, doxorubicin*, bleomycin, methotrexate, melphalan and VP-16.

2. Phlebitis

Phlebitis is frequently caused by cytotoxic drugs, but its occurrence is more prevalent with nitrogen mustard, anthracyclines, vinca alkaloids, and nitrosoureas.

A variety of methods has been used to prevent its occurrence but with limited success, such as adding heparin or steroids to the infusion.

----

* Adriamycin

Current Concepts of Diagnosis and Treatment
of Bone and Soft Tissue Tumors.

More effective methods consist of a higher dilution, slow rate of drug delivery, and the use of central venous lines. Once established, topical heat therapy and analgesics are indicated with avoidance of the involved vein for long enough to permit resolution.

3. Extravasation (Ignoffo and Friedman, 1980; Larson, 1982)

Despite the utmost care, extravasation accounts for approximately 2-5% of all adverse effects from antineoplastic drugs. Vesicant agents lead to the most severe reactions and include nitrogen mustard, doxorubicin, mitomycin-C, vinca alkaloids, and the nitrosoureas.

Prevention of such complication cannot be overstressed; however, once extravasation has occurred, appropriate treatment should be initiated immediately, consisting of local and surgical measures.

a) *Local measures:*

- Stop infusion.
- Aspiration of as much drug as possible.
- Administration of appropriate antidote to enhance drug inactivation, eg, sodium bicarbonate for doxorubicin.
- Administration of corticosteroids.
- Application of warm compresses.

b) *Surgical measures:*

- Local excision followed by plastic surgery should be considered early and can be most effective, especially in cases of doxorubicin extravasation.

4. Nausea and vomiting

These reactions are among the most common and most distressing of cytotoxic drugs. They are usually due to the central stimulation of the chemoreceptor trigger zone, which subsequently induces activation of the vomiting center in the brain stem. Some drugs (eg, cisplatin) have an additional emetic effect by stimulating the peripheral sites such as the pylorus. Treatment of nausea and vomiting can be quite difficult and a variety of agents can be used. The phenothiazines have been effective in several trials and the most useful are probably prochlorperazine and thiethylperazine (Moertel et al, 1963). The butyrophenones, haloperidol and droperidol, are also effective antiemetics (Plotkin et al, 1973), although with more powerful emetics, particularly cisplatin, the effectiveness of the above agents is limited. Large dose metoclopramide appears to be more effective in controlling vomiting due to cisplatin (Gralla et al, 1981).

Cannabis and its derivatives, including delta-9-tetrahydrocannabinol and nabilone, have in some controlled trials been superior to placebo and to prochlorperazine; however, they do not give complete protection against cisplatin and cause a high incidence of side effects including sedation, confusion, and unsteadiness, particularly in older patients (Heriman, 1979; Sallan et al, 1980). High doses of steroids have been demonstrated to reduce vomiting induced by highly emetic cytotoxics. Their mechanism of action is not understood, although they are believed to act through inhibition of prostaglandin release (Baker et al, 1980; Lee, 1981). In general, the control of vomiting due to cytotoxic drugs is unsatisfactory, particularly when highly emetic drugs are used. Combination of antiemetic drugs has received less attention and should be tried more widely. The addition of sedatives has also been reported to enhance the efficacy of other drugs (Maher, 1981).

B) Early Toxicities

1. Stomatitis

Stomatitis appears usually a few days to two weeks after treatment, leading to widespread ulcerations of the buccal mucosa as well as other parts of the gastrointestinal tract. This reaction can be quite severe, leading to malnutrition and local bacterial and fungal infections.

It is often associated with the use of cell cycle specific agents (eg, antimetabolites). Treatment is symptomatic with avoidance of irritant foods, topical viscous lidocaine, and appropriate treatment of infections. In general, stomatitis requires subsequent reduction of cytotoxic drug dosage.

2. Alopecia

Over 80% of the cells in the hair follicles are in an actively growing phase and are therefore susceptible to most anticancer agents. It is most severe with cyclophosphamide, doxorubicin and the nitrosoureas. Alopecia can start within two weeks of initiation of treatment, being maximal at one to two months. With prolonged chemotherapy, hair loss can affect the other parts of the body.

Several methods have been attempted to minimize hair loss due to cytotoxic agents. The use of scalp tourniquet or hypothermia have had a certain degree of success (Dean et al, 1979; Edelstyn et al, 1977). However, there is a risk of creating a tumor cell sanctuary in the scalp, particularly in widely disseminated diseases, such as leukemia, when these procedures should be avoided.

3. Bone marrow depression

From the oncologist's point of view, this is one of the most serious complications associated with cancer chemotherapy, in view of its significant morbidity and potential mortality. With the recent progress in supportive care, however, control of bone marrow depression has had a significant impact on cancer cures and overall survival.

The bone marrow contains the stem cell population for the various blood components; therefore, the kinetics of the various components will determine the clinical manifestation of toxicity. The rate of reduction for the various blood components depends on two factors:

i) *Their half life:* The rapidity of myelosuppression is inversely related to the half-life of the various elements, being slowest with red blood cells (half-life 120 days) and faster with platelets and granulocytes (half-life 7 days and 6 hours respectively).

ii) *The type of chemotherapeutic agent used:* Cell cycle and phase specific agents (eg, antimetabolites) lead to the most rapid myelosuppression and recovery, whereas cell cycle nonspecific agents (eg, nitrosoureas) are associated with the longest delay of myelosuppression (5-6 weeks).

Some agents do not produce clinically significant bone marrow depression and their addition to various combinations could be beneficial. These include: vincristine, bleomycin, L-asparaginase, hormones, including steroids, and cisplatin.

The most important elements of myelosuppression are neutropenia and thrombocytopenia which will be discussed briefly:

a) *Neutropenia:* The frequency of infection is inversely related to the absolute neutrophil count, increasing as the neutrophil counts fall below 1000/mm^3 (Bodey et al, 1966).

An increasing number of infections are caused by gram-negative bacilli, the most common organisms being *Escherichia coli, Pseudomonas aeruginosa,* and *Klebsiella* species. Compromised hosts are prone to gram-positive coccal infections as well as to nosocomial infection with rare pathogenic organisms such as *Serratia marcesceus, Clostridium septicum,* and *Bacillus cereus* (Bodey et al, 1978a).

In the management of these patients some guidelines should be emphasized:

- Classic signs and symptoms of infections may be absent due to the lack of an inflammatory response.

- Untreated infections may disseminate rapidly and terminate fatally.

- Not all antibiotics are equally effective in the neutropenic host.

In a neutropenic patient who develops fever, appropriate physical examinations, collection of culture specimen, and prompt therapy with combination of broad-spectrum antibiotics are indicated. Various antibiotics have been advocated for immediate therapy, preferably a penicillin effective against *Pseudomonas* species, such as carbenicillin, an aminoglycoside or a cephalosporin. When the results of cultures are known, appropriate antibiotics should be continued (Bodey et al, 1978b).

The use of prophylactic chemotherapeutic agents in afebrile neutropenic patients has been advocated, such as sulfamethoxazole and trimethoprim), and some encouraging results have been obtained in leukemic patients (Dekker et al, 1981; Weiser et al, 1981). However, the advantages of such an approach should be carefully weighed against the potential toxicity with routine use of those agents.

b) *Thrombocytopenia:* The risk of spontaneous hemorrhage, which could be serious or fatal, increases when the platelet count falls below $20{,}000/mm^3$. Prophylactic transfusions are then indicated. Platelet transfusions should also be considered in patients with moderate thrombocytopenia associated with bleeding.

C) Specific Organ Toxicities (Dorr and Fritz[a])

These reactions are usually cumulative and dose related, frequently due to the dose-limiting toxicities of the drug. A detailed discussion is beyond the scope of this review but a brief list of the various drugs is included:

Neurologic: vincristin, cisplatin, L-asparaginas, procarbazine, 5-fluorouracil, intrathecal drugs

Pulmonary: bleomycin, busulfan, nitrosourea (BCNU), cyclophosphamide, methotrexate

Genitourinary: cyclophosphamide, cisplatin, methotrexate

Hepatic: methotrexate, 6-mercaptopurine, L-asparaginase, mithramycin

Cardiac: anthracyclines

Dermatologic: busulfan, bleomycin, 5-fluorouracil (5-FU), methotrexate

D) Long-Term Complications (Dorr and Fritz[b])

With improved responses to treatment and prolonged survival, long-term complications are becoming more significant. They are subdivided into anatomic, genetic and oncogenic complications.

1. Anatomic complications

Anatomic complications are reported mainly in children, where growth retardation and osteoporosis are associated with prolonged use of steroids, radiotherapy and various other cytotoxic drugs.

2. Genetic complications

a) *Testicular suppression:* Azoospermia is often associated with cytotoxic drugs, particularly alkylating agents. Sterility following prolonged combination therapy has been reported in as many as 80% of patients (Sherins and DeVita, 1973). Potency is not affected. Men should consider sperm banking before initiation of cytotoxic chemotherapy.

b) *Ovarian suppression:* Amenorrhea is frequently associated with cytotoxic therapy but sterility is not as frequent as in men, accounting for about 30%. When it occurs, pregnancy is usually normal (Horning et al, 1981).

c) *Teratogenesis:* Cytotoxic drugs have a high potential for teratogenesis and mutagenesis. They should not be used during pregnancy, particularly during the first 60 days (Stutzman and Sokal, 1968).

3. Oncogenic complications

The onset of a second malignancy is the most serious late complication of cancer therapy and might prove to be the greatest risk of attempted cancer cure. The number of reported cases of second malignancies is increasing and consists mostly of acute nonlymphocytic leukemias.

The mechanisms underlying this phenomenon have not been fully elucidated, but based on experimental and clinical evidence three processes might be responsible:

a) *Direct oncogenesis of cytotoxic drugs,* well documented with alkylating agents (Karchmer et al, 1974; Reimen et al, 1977).

b) *Cooncogenesis of cancer drugs:* based on experimental evidence, some agents, such as methotrexate, can potentiate the effect of a carcinogen on the host (Harris, 1975).

c) *Induction of immunosupression* with an increased incidence of a second malignancy as demonstrated in patients treated for various tumors (Canellos et al, 1975; Penn, 1976), and following immunosuppressive therapy of allogeneic transplantation (Penn, 1975).

   Similar increase in incidence of malignancies has been demonstrated in primary immunodeficiency states (Penn, 1974).

It is important to emphasize that long-term complications represent a relatively small risk when weighed against the benefits derived from the treatment of advanced malignancies which carry a grave prognosis. Nevertheless, the long-term use of antineoplastic drugs, particularly in nonmalignant disorders or in patients with minimal residual disease, deserves careful reassessment with evaluation of the risk of carcinogenesis versus projected long-term survival.

References

1. Baker, J.J., Lockey, J.L., Price, N.A., Bowen, J., and Winokur, S.H.: Comparison of dexamethasone plus prochlorperazine to placebo plus prochlorperazine as antiemetics for cancer chemotherapy. Proc. Am. Soc. Clin. Oncol. 21, 339, 1980.

2. Bodey, G.P., et al: Quantitative relationships between circulating leukocytes and infection in patients with acute leukemia. Ann. Intern. Med. 64, 328, 1966.

3. Bodey, G.P., Rodriguez, V., Chang, H.Y., and Narboin, G.: Fever and infection in leukemic patients. A study of 494 consecutive patients. Cancer 41, 1610, 1978a.

4. Bodey, G.P.: Using drugs for the compromised patient: Management of bacterial infections. Drug Therapy 8 , 161, 1978b.

5. Canellos, G.P., DeVita, V.T., Arseneau, J.C., Whang, P.J., and Johnson, R.E.C.: Second malignancies complicating Hodgkin's disease in remission. Lancet i, 947, 1975.

6. Dean, J.C., Salmon, S.E. and Griffith, K.S.: Prevention of doxorubicin-induced hair loss with scalp hypothermia. N. Engl. J. Med. 301, 1427, 1979.

7. Dekker, A.W., Rozenberg-Arska, M., Sixma, J.J., and Verhoeff, J.: Prevention of infection by trimethoprim-sulfamethoxazole plus amphotericin B in patients with acute nonlymphocytic leukemia. Ann. Intern. Med. 95, 555, 1981.

8. Dorr, R.T., and Fritz, W.L. (a): Cancer Chemotherapy Handbook, Elsvier, N.Y., pp 116-135.

9. Dorr, R.T. and Fritz, W.L. (b): Cancer Chemotherapy Handbook, Elsvier, N.Y., pp 136-142.

10. Edelstyn, G.A., MacDonald, M., and MacRae, K.D.: Doxorubicin -induced hair loss and possible modification by scalp cooling. Lancet 2, 253, 1977.

11. Gralla, R.J., Itri, L.M., Pisko, S.E., et al: Antiemetic efficacy of high-dose metoclopramide: randomized trials with placebo and prochlorperazine in patients with chemotherapy-induced nausea and vomiting. N. Engl. J. Med. 305, 905, 1981.

12. Harris, C.C.: Immunosuppressive anticancer drugs in man: Their oncogenic potential. Radiology 114, 163, 1975.

13. Heriman, T.S., Einhorn, L.H., Jones, S.E., et al: Superiority of nabilone over prochlorperazine as an antiemetic in patients receiving cancer chemotherapy. N. Engl. J. Med. 300, 1295, 1979.

14. Horning, S.J., Hoppe, R.T., Kaplan, H.S. and Rosenberg, S.A.: Female reproductive potential after treatment for Hodgkin's disease. N. Engl. J. Med. 304, 1377, 1981.

15. Ignoffo, R.J., and Friedman, M.A.: Therapy of local toxicities caused by extravasation of cancer chemotherapeutic drugs. Cancer Treat. Rev. 7, 17, 1980.

16. Karchmer, R.K., Amare, M., Larsen, W.E., Mallouk, A.G., and Caldwell, G.G.: Alkylating agents as leukemogens in multiple myeloma. Cancer 33, 1103, 1974.

17. Larson, D.L.: Treatment of tissue extravasation by antitumor agents. Cancer 49, 1796, 1982.

18. Lee, B.J.: Methylprednisolone as an antiemetic. N. Engl. J. Med. 304, 486, 1981.

19. Maher, J.: Intravenous lorazepam to prevent nausea and vomiting associated with cancer therapy. Lancet i, 91, 1981.

20. Moertel, C.G., Reitemeier, R.J., and Gage, R.P.: A controlled clinical evaluation of antiemetic drugs. JAMA 186, 116, 1963.

21. Penn, I.: Occurrence of cancer in immune deficiencies. Cancer 34, 858, 1974

22. Penn, I.: The incidence of malignancies in transplant recipients. Transpl. Proc. 7, 323, 1975.

23. Penn, I.: Second malignant neoplasms associated with immunosuppressive medications. Cancer 37, 1024, 1976.

24. Plotkin, D.A., Plotkin, D., and Okum, R.: Haloperidol in the treatment of nausea and vomiting due to cytotoxic drug administration. Curr. Therap. Res. 15, 599, 1973.

25. Reimen, R.R., Hoover, R., Froumeni, J.F., Jr., and Young, R.C.: Acute leukemia after alkylating-agent therapy of ovarian cancer. N. Engl. J. Med. 297, 177, 1977.

26. Salan, S.E., Cronim, C., Zelen, M., and Zimberg, N.E.: Antiemetics in patients receiving chemotherapy for cancer: a randomized comparison of delta-9-tetrahydrocannabinol and prochlorperazine. N. Engl. J. Med. 302, 135, 1980.

27. Sherins, R.J., and DeVita, V.T., Jr: Effect of drug treatment for lymphoma on male reproductive capacity. Ann. Int. Med. 79, 216, 1973.

28. Stutzman, L., and Sokal, J.: Use of anticancer drugs during pregnancy. Clin. Obstet. Gynecol. 11, 416, 1968.

29. Weiser, B., Lange, M., Fialk, M.A., et al: Prophylactic trimethoprim-sulfamethoxazole during consolidation chemotherapy for acute leukemia: A controlled trial. Ann. Int. Med. 95, 436, 1981.

30. Weiss, R.B., and Bruno, S.: Hypersensitivity reactions to cancer chemotherapeutic agents. Ann. Int. Med. 94, 66, 1981.

Allograft Transplantation in the Management of Bone Tumors

H. J. MANKIN

Introduction

Bone allografts in the reconstruction of skeletal defects have a long and somewhat controversial history in orthopedic surgery. Until relatively recently, they were viewed with little enthusiasm, chiefly because of the unpredictability of the end result of the surgical procedure.

Clearly, the best type of graft which one can use in replacement for skeletal defects is the autograft which is known to be well accepted, serves as a scaffolding for bony ingrowth, stimulates healing and may, under appropriate circumstances, include some live cells (Boyd, 1943; Clark, 1959; Enneking et al, 1980; Enneking and Shirley, 1977; Johnson, 1978; Miller and Phalen, 1947; Wilson and Lance, 1975). Unfortunately, the supply and possible shapes of autografts are limited; moreover, the absence of an articulating surface often prevents reconstruction of massive defects resulting from tumor surgery. Xenograft (material obtained from another species) has been shown to be strongly antigenic in man and is now considered to be of no value. Allografts, on the other hand, have a number of advantages. They may be used for joint reconstruction without limit as to size, shape or quantity. There is no need to sacrifice normal structures and donor site morbidity is not encountered.

Background Data

Although sporadic reports of allograft transplants are recorded both in legends and early medical history, in modern times the first large series of allograft transplantations was that of Lexer, who in 1908 performed 23 whole and 11 hemijoint transplants about the knee (Lexer, 1908). He reevaluated the patients in 1923 and declared 50% of them to be successful (Lexer, 1925). A few case reports followed but it was not until the 1960's that several large series were published. The reason for the reawakening of interest in the procedure was the demonstration by Herdon and Chase and Curtiss et al that freezing the grafts reduced their immunogenicity (Curtiss et al, 1959; Herdon and Chase, 1954). Ottolenghi in Argentina (1972), and Parrish (1966) in the United States, reported on a small series of patients, but the larger series reported by Volkov (1970), and the subsequent report by Ottolenghi (1972) clearly demonstrated that this procedure was possible and that good results could be obtained in over 50% of patients. A long-term follow-up of patients by Parrish in 1973 showed that

Current Concepts of Diagnosis and Treatment
of Bone and Soft Tissue Tumors.

the grafts were subsequently at least partially incorporated and that joint function could be preserved for as long as 24 years after surgery.

To fully evaluate allograft transplantation as a possible method of treatment in orthopedics, it is essential to review current scientific knowledge regarding the procedure. It is obvious that much of the problem lies in the immune system and study of the immune response to allograft bone has clearly demonstrated that components of the graft are immunogenic and evoke a profound host response (Burwell, 1969; Langer et al, 1975; Muscolo et al, 1976). Although collagen is a weak antigen, the cells within the bone and the marrow provoke antibodies to the human lymphocyte antigen (HLA) system similar to those described for renal and skin grafts and both humoral and cellular antibodies can be detected by sophisticated immunologic techniques. The major factor which appears to ameliorate response to the foreign bone graft is freezing and thawing, presumably by not only killing the cells but by altering the cell membrane. Work in the 1950's by Herndon and Curtiss (Curtiss et al, 1959; Herndon and Chase, 1954) showed that freezing of the graft to -20°C markedly reduced the immune response. Studies by Friedländer et al (1976) demonstrated that a further reduction in an immune response could be achieved by freezing and drying. Evidence for this altered state is that fresh grafts provoke second set skin graft rejections while frozen or frozen and dried grafts do not. Despite the circulating antibodies to these grafts, however, even fresh material evokes only limited cytotoxic action, which led Langer and associates (1975) to postulate the presence of a blocking antibody which in some as yet unexplained manner decreases the likelihood of a rejection.

The cartilage of a composite bone-cartilage composite allograft behaves in a somewhat different manner. The cells and some of the proteins of the cartilage matrix are highly antigenic and several studies in animal models have demonstrated both humoral and cellular cytotoxic antibodies evoked by allogenic cartilage implantation (Elves, 1971; 1974; Langer and Gross, 1976). The matrix of articular cartilage is so dense, however, and the theoretic pore size is so small, that some of these antigens cannot leak out or antibodies gain ingress as long as the cartilage surface is intact. Therefore, cartilage is thought to be "immunologically privileged" but only as long as the cartilage surface is undamaged. If the cartilage becomes partially degraded or the surface injured, the pore size is markedly increased. It is, thus, theoretically possible that antigens will pass from the cartilage and evoke an intense response directed at the cells or the matrix components.

Over the years, numerous animal studies have been performed designed to assess the fate of experimentally implanted bone and cartilage allografts. The results to date have been inconclusive (Bonfiglio et al, 1955; Chalmers, 1959; Herndon and Chase, 1954; Hestop et al, 1960). The popular view held by many investigators is that the bone portion of the composite allograft behaves initially not unlike an autograft, evoking an acute local inflammatory reaction and an early repair response (Bonfiglio et al, 1955; Burchardt et al, 1975; Chalmers, 1959). Within a few days of transplant, however, when the host tissues begin to produce capillaries which invade the autograft as part of the process of creeping substitution (Burchardt et al, 1975), the allograft fails to evoke a similar response; in fact, the host bed shows modest to moderate inflammatory and fibroblast infiltration presumably in an attempt to wall off the graft. The vascular invasion believed to be necessary for incorporation and replacement of the graft with the host bone may under favorable circumstances appear quite late (months or even years after transplant). Then, over time it may carry out the creeping substitution with vascular cutting cones destroying the allograft and bringing in new host bone. Over several years (depending on the site, the bone, cortical thickness, and vascularity of the bed), the allograft is invaded by the host, bone is laid down, and a portion of the graft is converted to host bone without evidence of a major local inflammatory change (Chalmers, 1959; Hiki and Mankin, 1980; Mankin et al, 1976; Mankin and Schachar, 1977). This is the ideal response but under some circumstances, in clinical and experimental systems, a rapid dissolution of the allograft has been observed and in others the graft remains nonviable for a very long time. Either circumstance being considered less than optimal, rapid dissolution is of much greater consequence in the clinical situation.

As to the response of cartilage to joint transplantation, fresh articular cartilage implants remain viable with the cartilage surface intact for a number of months after implantation (Campbell, 1972; Herndon and Chase, 1954; 1959). A moderate degree of increased synovial reaction may be noted, even early in the procedure, and the synovial fluid shows an increased monocytic cellular response (Rodrigo and Lippiello, unpublished data; Yablon et al, 1977). Over time, a pannus is found to cover the cartilage surface and the matrix of the cartilage to be irregularly eroded and destroyed (Campbell, 1972). This process may take up to one year or longer. Thus, the ultimate result of even a well tolerated articular cartilage allograft may be a mildly to moderately osteoarthritic joint. This issue is difficult to fully assess, however, since the changes observed may in part be a function of the mechanical fit of the surface. It is likely that in surfaces where the graft is mechanically well matched to the host, the loss of the cartilage is much slower than in circumstances where the graft has a poor fit and the joint incongruous.

When our Orthopaedic Oncology Unit decided in 1971 to enter the field of allograft transplantation, we tried to define the optimal behavior of our allograft implants, principally based on our understanding of the above described immune mechanisms and the reported animal studies. As far as the bone is concerned, we wanted it to retain its tensile and compressive strength over a long period to provide structural stability to the skeleton. It seemed to us that, ideally, the allograft bone should unite to the host bone at the anastomotic site as rapidly as possible to diminish strain on the internal fixation device and free the patient from casts and braces. We hoped that the allograft bone would be replaced by host bone at a slow but steady rate, and that it would exert limited or no immunogenicity which might interfere with its eventual incorporation by the host. Since a second part of the system is the tendons and ligaments which serve as anchors for resected motors and/or joint stabilizers, we established similar criteria for these: they should be replaced by host collagen at an appropriate rate to maintain the stability of the joint and the function of important muscle groups, and exert limited or no immunogenic response. The viability of both the bone and tendinous structure would seem unimportant from a functional standpoint, and since evidence exists that frozen dead bone and ligament are less likely to exert an immunogenic response, we decided that these two segments should be treated by freezing.

The cartilage, however, is a considerably different system and it seemed best to attempt to use a preservative method which would maintain the cartilage in a functional state. Since chondrocyte viability is necessary for maintenance of the matrix of cartilage, we were faced with the problem of maintaining the chondrocyte in a viable state during the freezing and thawing process necessary to reduce the immunogenicity of the bone. If the surface remained intact, we would hope to maintain normal resiliency and joint gliding and at the same time avoid chemical or immunologic destruction of the cartilaginous tissue and the viable chondrocytes. It became apparent that if we were to be successful in developing an ideal system, we had to address the important issue of simultaneous cryoinjury to bone and cryopreservation of articular cartilage. The chondrocytes had to be maintained in a viable (although clearly hypometabolic) state during a process in which the adjacent bone is killed by freezing to -80°C and then thawing rapidly just before use.

Audrey Smith (1965) in England first described a system for preservation of isolated chondrocytes using exposure to dimethylsulfoxide (DMSO) before freezing. He noted a high percentage of viable cells after even prolonged exposure to low temperatures. Since then, investigators have proposed various protocols for the use of cryoprotectants for both isolated chondrocytes and the more difficult problem of the intact cartilage. Work in our laboratory has shown that either 8% DMSO or 10% glycerol will allow survival of a high percentage of chondrocytes in culture or in the intact organ (Schachar et al, 1977; Tomford and Mankin, unpublished data). The mechanism by which this occurs is presumably related to a decrease in the size of the ice crystals formed on the membrane and within the cell during the freezing and thawing procedure. A recent study by Tomford (unpublished) in our laboratory has suggested that the slower the thermal change between 0° and -4°C the more cells in the intact cartilage will remain viable and we have now adopted the technique of "slow freeze". Additional studies have demonstrated that the more rapidly the tissue is thawed the more cells will remain viable so that our optimal system has been defined as "slow freeze" and "rapid thaw" (Schachar et al, 1978).

Banking

Two major factors in an allograft program are the procurement of tissues and the maintenance of a bank. The more grafts available in the bank, the more likely we are to obtain an accurate fit for a patient who requires a resection for tumor. Because tumors must often be treated with some urgency, we must have sufficient stock in the bank to meet the demands with a minimal waiting period. Our banking program involves four separate and equally important steps: selection of donors; procurement of tissues; preservation and storage of the allograft segments; and accurate record keeping (Doppelt et al, 1981; Friedlaender and Mankin, 1979; 1981).

Donors are generally selected from individuals who are injured in automobile or motorcycle accidents and are usually shared with other transplant teams (kidney, liver, heart, eye and skin). According to our protocol, the donor should have been pronounced dead less than 12 hours previously and should have no historical or clinical evidence of tumor, infection, hepatitis or syphilis. We prefer individuals between the ages of 15 and

45 years so that we can exclude the possibilities of open epiphyseal plates (mechanically unsound) or occult neoplasm and osteopenia. To minimize the risk of diffuse infection, especially microabscesses in the marrow cavity, we reject potential donors who have been maintained on high-dose corticosteroids or a respirator for longer than one week. The next of kin are consulted at the time of death and a special permission form must be signed to proceed with the harvest.

The procurement or harvest procedure takes place in the operating room where the donor's body is subjected to a sterile preparation and draping as with any surgical procedure. Standard approaches are used to gain access to the bones and joints of the upper and lower extremities, and the bony segments are isolated and stripped of all muscle and periosteum. Tendons and ligaments, however, are carefully preserved with their length exceeding 2 cm so that they can later serve as reattachment sites for muscle and joint stabilizers. We incubate the cartilage in the cryoprotectant DMSO (8% in normal saline) for at least two hours to maintain chondrocyte viability during the freezing process. Since infection is one of our main problems, anaerobic and aerobic cultures are obtained from all possible sources of contamination (including heart blood, urine, pleural fluid, DMSO and the surface and medullary cavity of each of the bones). The grafts are then placed in polyethylene gas sterilized bags, wrapped in two sterile towels and an outside wrap labeled with the name, date and site. Following removal of the donor parts, we reconstruct the cadaver with wooden dowels and plaster of Paris. Before freezing, the graft and a sizing marker are roentgenographed in two planes and then refrigerated at 4°C for 18 hours, slowly frozen, and then stored at -70° to -80°C until needed.

Sizing of the graft is performed by comparing the dimensions of the recipient's bone (as determined by a standard 6-foot roentgenogram with a sizing marker in place) with those of the various donor segments available in the bank. When the bank contains sufficient parts (approximately 300), most normal size individuals can be fitted with a donor segment of approximately the correct dimensions. Due to the number of parts in the bank, the process of matching has become cumbersome and we are currently doing this by a computer program. At the time of surgery, the tissue is thawed in warm lactated Ringer's solution to which antibiotics are added after additional cultures are obtained. The warm Ringer's solution (40-45°C) brings the graft and particularly the cartilage to room temperature as quickly as possible.

To maintain an orderly bank, it is essential that accurate records be kept (Doppelt et al, 1981; Friedlaender and Mankin, 1979; 1981). Each donor has a separate file sheet (now maintained by computer) on which all important information regarding the individual's health and death are recorded. We record the blood type, tissue type, the culture reports, and the results of serum tests for hepatitis antigens. For each of the parts obtained, data regarding size, presence or absence of appropriate ligaments, analysis of the roentgenogram and culture tests are recorded in a separate file.

Surgical Procedure

Since 1971, the Orthopaedic Oncology Unit at the Massachusetts General Hospital has performed almost 180 allograft transplantations, principally for malignant or aggressive bone tumors (Eriksson et al, 1980; Hiki and Mankin, 1980; Mankin and Schachar, 1977; Mankin et al, 1976; 1982; 1983). For the most part, the segments have been procured as described above, and have been implanted following a wide (or, occasionally, marginal, depending on the nature of the tumor) en bloc resection of the bone containing the tumor, adjacent muscles, tendons, ligaments and the skin of the prior biopsy tract, all in continuity. An allograft, closest in size to the host part, is selected from the bank. After rapid thawing of the allograft in warm Ringer's lactate, it is cut to proper length and inserted and fixed to the host bone, most frequently with plates and screws but occasionally with an intramedullary rod. The ligaments, capsule, and tendons are carefully approximated with nonabsorbable sutures; drains are inserted and the wound closed in standard fashion. In the immediate postoperative period, the wound is drained for two to three days to decrease excessive dead space. Intravenous antibiotics are administered for approximately one week and oral antibiotics are given for three to four months. The limb is initially placed in a cast in an appropriate position and maintained until the soft tissue has healed. Depending on the degree of soft tissue repair and the stability of the segment, immobilization lasts from six weeks to three months for lower extremity grafts and four weeks to two months for those in the upper extremity. After removal of the plaster cast, a cast-type brace is applied and the patient begins a rehabilitation program principally consisting of active and passive exercises. Unlimited weight bearing is delayed until the anastomosis site is healed and the joint is considered stable.

Clinical Series

As of February, 1983, 179 patients were included in the series. Sixty-nine are men and 110 women with an average age of 31.8 years and an age range of 11 to 69 years. Follow-up for this group of patients ranges from four to 140 months. The lesions for which the procedures were performed are shown in Table I. As can be noted, most are tumors, the highest percentages being low-grade, with giant cell tumors and chondrosarcomas heading the list. We have, however, treated a significant number of high-grade sarcomas such as central osteosarcomas, high-grade chondrosarcomas, and a malignant fibrous histiocytoma. In addition, we have treated a small number of patients with benign conditions where a large segment of bone had been lost or destroyed. This group includes patients with Gaucher's disease, pigmented villonodular synovitis, fibrous dysplasia and a series of post-traumatic skeletal defects or osteonecrosis about the knee.

The operations performed on the patients are divided into three groups: 133 individuals in whom we implanted an osteoarticular graft (bone and a segment of articular cartilage such as the distal femur proximal humerus or the proximal tibia); 35 patients in whom the graft was intercalary (a bony segment between two parts of the host bone, such as the shaft of the femur or shaft of the tibia); and 11 patients who received an allograft combined with a prosthesis. This last group includes either total hip

Table I. Allograft transplantation (179) Diagnosis in 143 patients

Tumors	
Giant cell tumor	64
Chondrosarcoma	32
Parosteal osteosarcoma	13
Central osteosarcoma	11
Adamantinoma	7
Fibrosarcoma & MFH	7
Metastatic carcinoma	5
Recurrent chondroblastoma	4
Aggressive osteoblastoma	4
Desmoplastic fibroma	3
Lymphoma	2
Ewing's sarcoma	2
Plasmocytoma	1
Recurrent chondromyxoid fibroma	1
Destructive ABC	1
Hemangioma	1
	158
Nontumorous conditions	
Gaucher's disease	2
PVNS	2
Fibrous dysplasia	2
Failed THR	2
Paget's disease	1
Post-traumatic skeletal defects and osteonecrosis	12
	21

Table II. Allograft transplantation -- sites for 133 osteoarticular grafts

Site	Number
Distal femur	49 (23 hemi)
Proximal tibia	28 (16 hemi)
Proximal humerus	14 (1 hemi)
Distal radius	12
Hemipelvis	10 (4 with prox. femur)
Proximal femur	10
Distal humerus	4 (1 with prox. ulna 1 hemi)
Proximal ulna	3
Distal tibia	1

Table III. Allograft transplantation -- sites for 35 intercalary grafts

Site	Number
Tibial shaft	11
Femoral shaft	9
Humeral shaft	7
Proximal tibia	3
Distal radius	1
Ulnar shaft	1
Hemipelvis	1
Distal femur - proximal tibia (arthrodesis)	2

replacements or Austin Moore endoprostheses substituted for the femoral heads in patients with pelvic or proximal femoral grafts.

An analysis of the sites of the osteoarticular grafts is shown in Table II. It is evident that most are located in the distal femur, proximal humerus, distal radius, proximal tibia and hemipelvis. Of the 77 allografts of the distal femur and proximal tibia, 39 included only one or the other condyle and were therefore designated as hemigrafts. The sites for the intercalary grafts are shown in Table III. As can be noted, most of these are in the tibial and femoral shafts.

Results

All patients were carefully followed throughout the years since their procedures were performed and none has been lost to follow-up. Serial white counts, sedimentation rates, enzymes, and immunoelectrophoretic studies have all been normal. Patients who showed abnormalities either had an infection or hepatitis. Serial bone scans became normal four to eight months postoperatively and the patients' activities increased with time. A sustained high titre of anti-HLA antibody was noted in all but three patients out of the first 83. The quantity of antibody present, however, did not seem to correlate with the end results. Bone biopsies in 10 patients showed new cortical bone formation and creeping substitution after eight months. Cartilage biopsies showed varied results, some showing viable but a distinctly abnormal cartilage and others showing acellular tissue. The nature of these abnormalities has not yet been defined either in our human studies or in our cat model.

To assess our end results, we have developed a grading system based on function in which the patients were scored as *excellent, good, fair,* or *failure*. Results were rated *excellent* in those patients who had no evident disease, had normal function of the grafted part, and were able to return to normal

Table IV. Allograft transplantation. Current status of 77 patients followed two or more years.

Allograft	No.	Results: Excellent	Good	Fair	Failure Res. or amp.
Osteoarticular	75	29	23	11	6
Intercalary	22	17	-	3	2
Allograft and prosthesis	8	5	1	-	2
Totals*	105	51 (49%)	24 (23%)	14 (13%)	16 (15%)
		72% (Excellent + Good)			
* If the tumor failures (4 deaths and 3 recurrences) are deleted, the adjusted values are:					
Totals	97	51 (52%)	24 (25%)	13 (13%)	10 (10%)
		77% (Excellent + Good)			

activities with minimal limitations. A result was considered to be *good* in an individual who had no evident disease or pain but had reduced function of the grafted part. They did not need a brace or support and were able to return to most daily activities. A *fair* result was adjudged for an individual who had no evident recurrence of the tumor but a significant functional deficit which required a brace or support. A *failure* was adjudged for any patient who required resection of the graft for recurrence, infection, or fracture or who subsequently required an amputation or died of pulmonary metastases.

On February 20, 1983, 105 patients had been followed for over two years and were thus available for review. Of this group, 75 patients had osteoarticular grafts, 22 intercalary, and eight allograft plus prosthesis. As can be noted by examining Table IV, for the 75 osteoarticular grafts 52 were either good or excellent, 11 were fair and the remaining 12 either required a resection or amputation or had died. Of the 22 intercalary grafts, 17 were excellent, three were fair, and two were failures. Of eight allografts and protheses, six were good or excellent and two required resection. The total values show that in 49% of the patients we had obtained excellent results and 23 % had good results for a total of 72% in the acceptable range. In 13% the results were fair, and 15% were considered to have failed.

If, however, we consider the tumor failures (consisting of five deaths and three recurrences) as resulting from the tumor treatment rather than the allograft procedure and thus delete them from the values, the picture changes somewhat. Of 97 grafts without tumor recurrence or metastasis, 77% of the results were good or excellent, 13% fair and 10% were failures.

Table V. Allograft transplantation. Early complications in 147 patients followed for at least one year

Complication	Number
Deaths	0
Cardiac arrest	1
Wound hemorrhage	4
Arterial laceration	3
Nerve palsy (permanent)	2
Nerve palsy (transient)	3
Skin slough	9
Hepatitis	4

Most of the allograft failures resulted from complications. The early complications in 77 patients followed for at least two years are shown in Table V. No deaths occurred attributable to the surgical procedure. One patient had a cardiac arrest but was resuscitated. We had four massive wound hemorrhages in the series, one of which required a reoperation to tie the profunda femoris. Three arterial lacerations occurred which required repair. Three nerve palsies were transient and two partial nerve injuries resulting from pelvic surgery were permanent. Nine skin sloughs were encountered which caused major problems since many of these patients developed subsequent infections. Hepatitis occurred in four patients and although this is not excessive for a series such as this, it is cause for concern since it is possible that the virus was transmitted with the graft.

Of much more significance is the late complications in this group of patients (Table VI). Five patients died as a result of their tumor and three patients had recurrent local tumors, amounting to an overall tumor failure for the series of less than 10% which probably reflects that we were frequently dealing with low-grade tumors and that the wide or marginal surgery was adequate in the management of these patients. Deep infection occurred in 13 of the patients for a total of 12.4%. This is clearly an unacceptable rate and is far in excess of our infection rate for other surgery performed in our institution. Presumably, the high infection rate is in part due to the extent of the surgery but may be an expression of the immune reaction. Allograft fractures occurred in 15 patients and were treated surgically with successful salvage in over 75%. Non- or delayed union occurred in 10 patients and was easily treated by autograft implantation at the anastomosis site. Joint instability occurred in only five patients and has not caused great difficulty. One patient suffered a perforated colon on a sharp segment of the allograft. She is also included as having a deep infection and it should be pointed out that a number of the patients had more than one complication so that one cannot add up the figures shown in Tables V and VI to arrive at a total complication rate for the series.

Table VI. Allograft transplantation. Late complications in 105 patients followed for at least two years*

Complication	Number
Death as a result of tumor	6 (5.7%)
Recurrent local tumor	3 (2.9%)
Deep infection	13 (12.4%)
Allograft fracture	15 (14.3%)
Non- or delayed union	10 (9.5%)
Joint instability	5 (4.8%)
Perforated colon	1 (1.0%)

* Since a number of the patients had two or more complications, the figures noted in Tables V or VI cannot be added to arrive at the total number of patients in the series who developed a complication.

Table VII. Allograft transplantation. End results in patients who developed infections, fractures or nonunions*

	Infection (13)	Fracture (10)	Nonunion (8)	Uncomplicated cases (56)	Total series (87)
Excellent or good	1 (7.7%)	6 (60%)	7 (87.5%)	55 (98.2%)	69 (79.3%)
Fair	4 (30.7%)	3 (30%)	1 (12.5%)	1 (1.8%)	9 (10.3%)
Failure	8 (61.5%)	1 (10%)	0	0	9 (10.3%)

* Excluding patients in whom procedure failed as a result of recurrence or metastases

It is interesting to assess the effect of the complications on the outcome of the procedure. Table VII is a composite study of the first 87 patients classified according to excellent or good, fair or failure status and the results compared with their complications. Thirteen patients of this group had an infection and as can be noted, only one was in the excellent or good group, four in the fair group and eight were failure. On the other hand, of 10 patients with fractures, six (60%) were salvaged enough to eventually

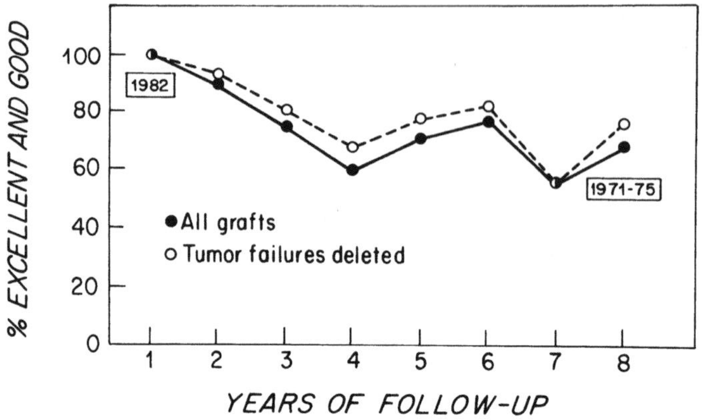

Fig. 1. Diagram showing the relationship between year of performance of allografts and percentage which remain in good or excellent results. Data for the entire series are shown with solid lines. When the tumor failures are deleted, the numbers remain at approximately 75%.

be classified as excellent or good. Nonunions occurred in eight patients and did not materially influence the end results. The 56 patients who were fortunate not to have complications had 98.2% good or excellent end results.

The most important test of this allograft system is how the patients fare over time. If a slow deterioration occurs with advancing years, the procedure has less merit than if the grafts, once having passed the hurdles of infection or allograft fracture, remain stable. An analysis of this particular issue is shown in Fig. 1 which graphically displays the results by year of treatment. As can be noted, the major fall in the good or excellent result category occurs in the first two years. Following this crucial period during which all infections and most fractures occur, the patient goes on to a steady existence with the graft. If the tumor failures are deleted (dotted line), the approximate number of such good and excellent results hovers around 75% and remains at that level throughout the duration of the study. What will happen to the patients after the 10-year mark, as more and more reach that point, is unknown but one hopes that they will continue to do well.

Discussion

In final consideration, it is clear to us that the system remains experimental, but the early results suggest that continued sustained research activities are indicated. One of the major areas of further exploration should be the problems of revascularization of bone. Despite the extent of our experience, we are not sure whether slow or rapid revascularization is more favorable. It is clear that under the current system it is unpredictable. We must continue to study the immune mechanism and its effect on this issue and try to arrive at a system that will provide us with an predictably optimal revascularization. A second major issue is improved survival of cartilage and joint structure. If we are to have a working joint, the cartilage must survive and we must develop a simple system for maintaining a high degree of cartilage viability during the freezing process. We must remain concerned with certain technical problems of the surgery which obviously must include control of infection, mechanical fixation of the grafts, and repair of ligaments and tendons. We are continuing to improve the quality of the surgery in each of these areas and hope that over time an optimal method will evolve.

A major project currently under consideration is correlation of the long-term follow-up data with some assessment of humeral evidence of the immune response. It is hoped that obtaining such data will provide us with a means of selecting the best graft from the bank for each patient. It is the only way, at least theoretically, to eliminate the failures which we now consider to be a result of the immune reaction.

Donor procurement and storage of allograft parts remain a major problem. We must educate physicians and the public as to the need for this important program. Finally, we must join with our colleagues in other areas of transplantation in the continued study of immunobiology in the hope of developing a series of safe systems to alleviate the immune response and to provide us with a more universally applicable system for patients with orthopedic problems.

Acknowledgement

Supported in part by grant #AM21896 from the National Institutes of Health.

References

1. Bonfiglio, M., Jeter, W.S. and Smith, C.L.: The immune concept: its relation to bone transplantation. Ann. N.Y. Acad. Sci. 59, 417, 1955.

2. Boyd, H.B.: The treatment of difficult and unusual nonunions with special references to the bridging of defects. J. Bone Joint Surg. 24, 535, 1943.

3. Burchardt, H., Busbee, G.A. and Enneking, W.F.: Repair of experimental autologous grafts of cortical bone. J. Bone Joint Surg. 57, 814, 1975.

4. Burwell, R.G.: The fate of bone grafts. In, Recent Advances in Orthopaedics. Apley, G.S. (Ed.) Churchill, London, 1969.

5. Campbell, C.J.: Homotransplantation of a half or whole joint. Clin. Orthop. 87, 146, 1972.

6. Chalmers, J.: Transplantation immunity in bone homografting. J. Bone Joint Surg. 41(B), 160, 1959.

7. Clark, K.: A case of replacement of the upper end of the humerus by a fibular graft reviewed after 29 years. J. Bone Joint Surg. 41(B), 365, 1959.

8. Curtiss, P.H., Jr., Powell, A.E. and Herndon, C.H.: Immunological factors in homogeneous-bone transplantation. III. The inability of homogenous rabbit bone to induce circulating antibodies in rabbits. J. Bone Joint Surg. 41(A),1482, 1959.

9. Doppelt, S.H., Tomford, W.W., Lucas, A. and Mankin, H.J.: Operational and financial aspects of a hospital bone bank. J. Bone Joint Surg. 63(A), 1472, 1981.

10. Elves, M.W.: Immunological studies of osteoarticular allografts. Proc. Royal Soc. Med. 64, 644, 1971.

11. Elves, M.W.: A study of transplantation antigens on chondrocytes from articular cartilage. J. Bone Joint Surg. 56(B), 178, 1974.

12. Enneking, W.F., Eady, J.L. and Burchardt, H.: Autogenous cortical bone grafts in the reconstruction of segmental skeletal defects. J. Bone Joint Surg. 62(A), 1039, 1980.

13. Enneking, W.F. and Shirley, P.D.: Resection-arthrodesis for malignant and potentially malignant lesions about the knee using an intermedullary rod and local bone grafts. J. Bone Joint Surg. 59(A), 223, 1977.

14. Eriksson, A.I., Schiller, A.L. and Mankin, H.J.: The management of chondrosarcoma of bone. Clin. Orthop. 153, 44, 1980.

15. Friedlaender, G. and Mankin, H.J.: Guidelines for the banking of musculoskeletal tissues. Newsletter, Am. Assoc. Tissue Banks 3, 2, 1979.

16. Friedlaender, G. and Mankin, H.J.: Bone banking: Current methods and suggested guidelines. AAOS Instructional Course Lecture Series 30, 36, 1981.

17. Friedlaender, G.E., Strong, D.M. and Sell, K.W.: Studies on the antigenicity of bone. I. Freeze dried and deep frozen bone allografts in rabbits. J. Bone Joint Surg. 58(A), 854, 1976.

18. Herndon, C.H. and Chase, S.W.: The fate of massive autogenous and homogenous bone grafts including articular surfaces. Surg. Gynec. Obstet. 98, 273, 1954.

19. Herndon, C.H. and Chase, S.W.: Experimental studies in the transplantation of whole joints. J. Bone Joint Surg. 34(A), 1482, 1959.

20. Hestop, B.F., Zeiss, I.M. and Nisbet, N.W.: Studies on transference of bone. I. A comparison of autologous and homologous bone implants with references to osteocyte survival, osteogenesis and host reaction. Brit. J. Exper. Path. 41, 269, 1960.

21. Hiki, Y. and Mankin, H.J.: Radical resection and allograft replacement in the treatment of bone tumors. J. Jpn. Orthop. Assn. 54, 475, 1980.

22. Johnson, J.R.: Reconstruction of the pelvic ring following tumor resection. J. Bone Joint Surg. 60(A), 747, 1978

23. Langer, F., Czitrom, A., Pritzker, K.P. et al: The immunogenicity of fresh and frozen allogeneic bone. J. Bone Joint Surg. 57(A), 216, 1975.

24. Langer, F. and Gross, A.E.: Immunogenicity of allograft articular cartilage. J. Bone Joint Surg. 56(A), 297, 1976.

25. Lexer, E.: Die Verwendung Der Freien Knochenplastic Nebst Versuchen Uber Gelenkversteifung and Gelenk-transplantation. Arth. Klin. Chir. 86, 939, 1908.

26. Lexer, E.: Joint transplantation and arthroplasty. Surg. Gynec. Obstet. 40, 782, 1925.

27. Mankin, H.J., Doppelt, S.H., Sullivan, T.R. and Tomford, W.W.: Osteo-articular and intercalary allograft transplantation in the management of malignant tumors of bone. Cancer 50, 613, 1982.

28. Mankin, H.J., Doppelt, S. and Tomford, W.: Clinical experience with allograft implantation: the first ten years. Clin. Orthop. 174, 69, 1983.

29. Mankin, H.J., Fogelson, F.S., Thrasher, A.Z. and Jaffer, F.: Massive resection and allograft replacement in the treatment of malignant bone tumors. N. Engl. J. Med. 294, 1247, 1976.

30. Mankin, H.J. and Schachar, N.S.: The role of allograft transplantation in the treatment of malignant bone tumors. Annual meeting (Boston), Musculoskeletal Council of the American Association of Tissue Banks. Rockville, Maryland, pp 95-107, 1977.

31. Miller, R.C. and Phalen, G.S.: The repair of defects of the radius with fibular bone grafts. J. Bone Joint Surg. 29, 629, 1947.

32. Muscolo, D.L., Kawai, S. and Ray, R.D.: Cellular and humeral immune response analysis of bone allografted rats. J. Bone Joint Surg. 58(A), 826, 1976.

33. Ottolenghi, C.E.: Massive osteo and osteo-articular bone grafts: technique and results of 62 cases. Clin. Orthop. 87, 156, 1972.

34. Parrish, F.F.: Treatment of bone tumors by total excision and replacement with massive autologous and homologous grafts. J. Bone Joint Surg. 48(A), 968, 1966.

35. Parrish, F.F.: Allograft replacement of part of the end of a long bone following excision of a tumor: Report of twenty-one cases. J. Bone Joint Surg. 55(A), 1, 1973.

36. Rodrigo, J.J. and Lippiello, L.: Unpublished data.

37. Schachar, N.S., Henry, W., Wadsworth, P.L., Castranova, F.P., Jr. and Mankin, H.J.: A feline model for the study of frozen osteoarticular allografts. I. Quantitative assessment of cartilage viability and bone healing. Trans. Orthop. Res. Soc. 3, 130, 1978.

38. Schachar, N.S., Tomford, W.W. and Mankin, H.J.: The effect of cryopreservative agents on the viability of frozen human articular cartilage. Annual Meeting (Boston), Musculoskeletal Council for the American Association of Tissue Banks. Rockville, Maryland, pp. 108-109, 1977.

39. Smith, A.U.: Survival of frozen chondrocytes isolated from cartilage of adult mammals. Nature 205, 782, 1965.

40. Tomford, W.W. and Mankin, H.J.: Unpublished data.

41. Volkov, M.: Allotransplantation of joints. J. Bone Joint Surg. 52(B), 49, 1970.

42. Wilson, P.D., Jr. and Lance, E.M.: Surgical reconstruction of the skeleton following segmental resection for bone tumors. J. Bone Joint Surg. 47(A), 1629, 1975.

43. Yablon, I.G., Brandt, K.D. and DeLellis, R.A.: The antigenic determinants of articular cartilage - their role in homograft rejection. Trans. Orthop. Res. Soc. 2, 90, 1977.

Reconstruction of Skeletal Defects Following En-Bloc Excision of Bone Tumors

A. GROSS, N. MCKEE, I. FARINE, A. CZITROM, and F. LANGER

Introduction

A comprehensive bone and cartilage transplant program was established at the Mount Sinai Hospital, University of Toronto, in 1971. Previous research carried out at the Mount Sinai Institute and at the Royal National Orthopaedic Hospital in London (under the supervision of Prof. Geoffrey Burwell) had established the following facts concerning the immunogenicity of bone and cartilage.

Chondrocytes are immunogenic but when surrounded by a healthy matrix are protected from circulating antibodies (Langer and Gross, 1974). Bone is immunogenic when fresh, but freezing markedly reduces this property (Langer et al, 1975). Yet, fresh bone allografts do heal, thus suggesting that the immune response is not harmful. Osteochondral allografts, therefore, can be performed as fresh grafts with live cartilage which should survive. This has been the basis for the small fragment grafts that are performed for old plateau fractures, traumatic osteonecrosis and osteoarthritis in the young patient (Friedlander and Mankin). When dealing with reconstruction after en bloc excision of bone tumors, the clinical situation is more urgent and waiting for a suitable donor may be impractical. Deep frozen (-70°C) banked allograft tissue is therefore used for the more urgent tumor cases, but if waiting for a fresh donor is deemed safe, as in some recurrent giant cell tumors, then fresh tissue is used. If the allograft is to consist of bone only, for example a segmental graft, banked tissue is used. Since 1978, we have also employed the microvascularized fibular autograft, either alone or combined with an allograft depending on the size of the bone being reconstructed. The program, therefore, has the capacity to provide allografts (fresh or frozen) and microvascularized fibular autografts (used alone or with allografts) (Gross et al, in press).

The transplant program is basically divided into small fragment grafts (done for old plateau fractures, traumatic osteonecrosis and osteoarthritis), and large fragment grafts performed following en bloc excision of bone tumors. Only the large fragment grafts will be discussed in this paper. As of January, 1983, 45 reconstructions have been performed on 42 patients.

Current Concepts of Diagnosis and Treatment
of Bone and Soft Tissue Tumors.

Logistics of a Transplant Program

The transplant team consists of two orthopedic surgeons, one being responsible for the en bloc excision and the other for the reconstruction; a pathologist; a microvascular surgeon; and an immunologist who takes part in the clinical aspects and maintains an ongoing research program. Donors are obtained by the Metropolitan Toronto Organ Retrieval Program. Tissue typing and blood grouping is carried out but only as a retrospective study. Harvesting is done under sterile operating room conditions after the kidneys have been taken and life support has been terminated. Fresh grafts are stored at 4°C in normosol with added antibiotics (cefazolin sodium and gentamycin). Fresh grafts are carried out within 24 hours of harvesting in a clean-air operating room. Frozen grafts are stored at -70°C. No cryopreservation is used. The stored grafts are irradiated for sterility.

Technical Aspects

Dynamic compression plate (DCP) fixation is used and the plates left in as they contribute to the strength of these grafts. When the graft involves a joint, reconstruction of ligaments is carried out as meticulously as possible. The recipient's ligaments are left long and stapled to the graft. The donor ligaments are also left long in case the recipient's ligaments have been sacrified as part of the surgery. If the recipient's ligaments are available, the donor ligaments are plicated over them or simply used for anchorage. For knee reconstruction, the collateral ligaments, the posterior capsule, and anterior cruciate ligament are repaired.

Large fragment grafts are augmented with autologous iliac crest bone chips. If the allograft has a large diaphyseal component, a vascularized fibular autograft is used to reinforce it and bone chips are also placed in the groove between them. The vascularized fibular autografts when done around the knee are dissected with a vascular pedicle and transposed to the site not requiring a reanastomosis. When done at a distance from the knee, microanastomosis is, of course, carried out.

Cefazolin is given one hour before surgery and continued for three to five days when cephalexin is started for another week. Articulating grafts are immobilized for six weeks. Long leg ischial bearing braces are used for 12 to 18 months in all lower limb grafts.

Classification

Since May 1975, 45 reconstructions have been carried out on 42 patients. In 26 patients with joint involvement, reconstruction was carried out to maintain joint function. We call this procedure *articulating grafts*. In ten other patients in whom joints were involved, the joint was *fused* by means of the graft because of the extent of soft tissue excision. Nine grafts did not involve joints and are referred to as *segmental* (Table I).

Table I. Classification of the surgical procedures for large skeletal defects following en bloc excision of bone tumors.

Surgical reconstruction	Graft type
Articulating	Allograft
	Allograft & autograft
	Allograft & implant
Fused	Allograft & autograft
	Autograft
Segmental	Allograft
	Autograft
	Allograft and autograft

Different types of grafts were used depending on the requirements of the reconstructions, ie, allografts alone, autografts alone, combinations of allografts and autografts, and combinations of allografts and implants (Table II).

Table II. Classification of 45 surgical procedures according to the type of reconstruction and the type of graft.

Surgical reconstruction	Graft type				Total
	Allograft	Autograft	Allograft & autograft	Allografts & implants	
Articulating	22	--	2	2	26
Fused	2	4	4	--	10
Segmental	5	2	2	--	9
Total	29	6	8	2	45

Clinical Material

Articulating grafts

Twenty-six articulating grafts have been carried out. Twenty-two were allografts. Two allografts were reinforced by vascularized fibular autografts, and two were a combination of allograft and implant (Fig. 1, 2).

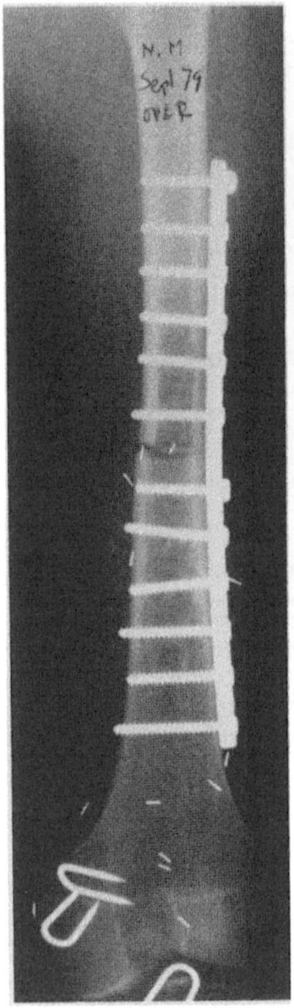

Fig. 1a

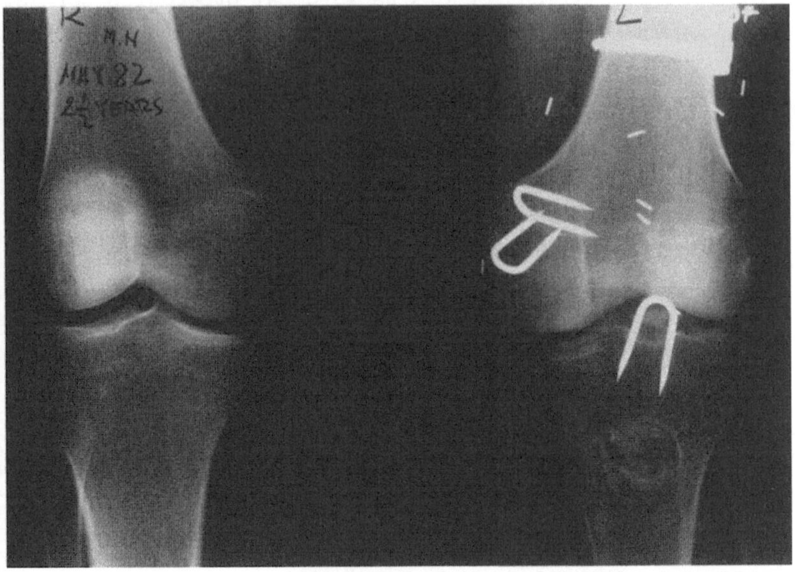

Fig. 1b

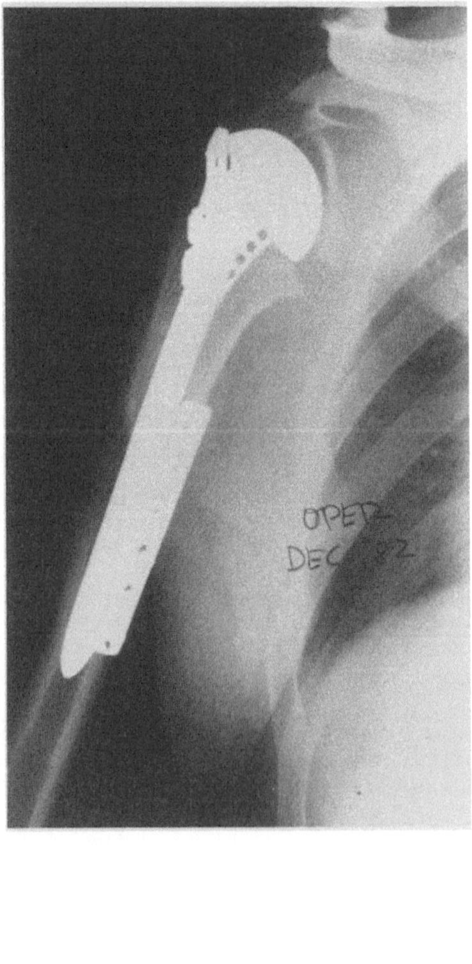

Fig. 2

Fig. 3

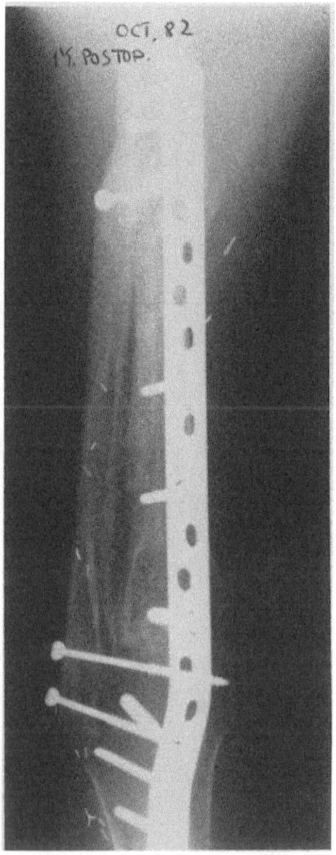

Fig. 4

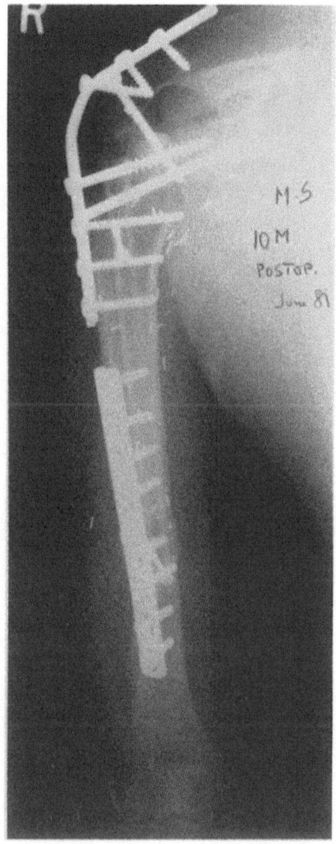

Fig. 3. Knee fusion with allograft and vascularized, fibular autograft following en bloc excision of distal femur for osteosarcoma in a 27-year old man. This roentgenogram is one year postoperative.

Fig. 4. Fusion of the shoulder with allograft and vascularized fibular autograft following en bloc excision of upper humerus for chondrosarcoma in a 31-year old man. The allograft allows for plate fixation while the autograft achieves the arthrodesis. Roentgenogram reveals solid fusion at ten months.

Fig. 1. a) Fresh articulating allograft following resection of distal femur for recurrent giant cell tumor in a 21-year old woman, immediately postoperatively. b) Two and a half years postoperative, the joint space is well preserved.

Fig. 2. Articulating reconstruction of a shoulder with allograft and implant following en bloc excision of upper humerus for an angiosarcoma in a 25-year old man, immediately postoperatively.

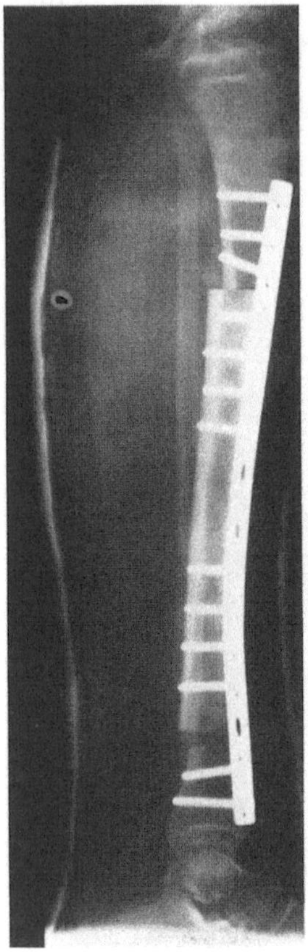

Fig. 5a

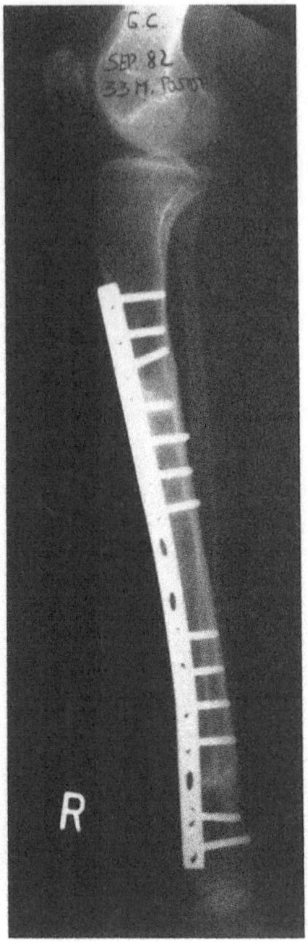

Fig. 5b

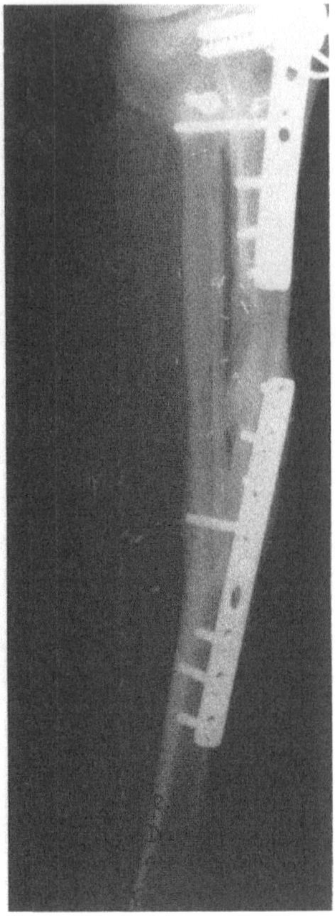

Fig. 6

Fig. 5. a) Segmental allograft following resection of the tibia for adamantinoma in an 11-year old child. This roentgenogram was taken immediately postoperatively. b) Thirty-three months postoperative -- solid union and remodelling of the graft.

Fig. 6. Segmental reconstruction of the tibia with an allograft and vascularized fibular autograft in a 17-year old boy with osteosarcoma. The allograft provides early strength and a means of plate fixation. The vascularized autograft provides a living graft to supplement the allograft which would remodel much slower in an adult than in a child as in the case in Fig. 5, where a fibular autograft was not necessary. This roentgenogram was taken two years postoperatively.

Graft fusions

Ten fusions have been carried out. Seven were primary fusions because of extensive lesions involving the soft tissues. One of the fusions was performed with a vascularized fibular autograft alone (wrist). The other six fusions were performed with allografts combined with vascularized fibular autografts, four of them as a one-stage procedure and two as two-stage procedures (Fig. 3, 4).

Three fusions were performed as secondary salvage procedures for infected allografts. These allografts had not been sterilized by irradiation. Vascularized fibular autografts were performed as delayed procedures following removal of infected allografts in these three cases.

Segmental grafts

Nine segmental grafts have been performed with five being allografts alone, two vascularized fibular autografts alone, and two a combination of allograft and vascularized fibular autograft (Fig. 5, 6).

Results

As of January 1983, 45 reconstructions have been carried out on 42 patients. In three patients, infected allografts were removed with delayed reconstruction by vascularized fibulae, thus accounting for the three extra reconstructions. The follow-up as of January, 1983 is an average of 34-1/2 months with a mean of 30 months and a range from three to 93 months.

Table III. Distribution of pathologic findings and surgical reconstruction.

Pathologic finding	Surgical reconstruction			Total
	Articulating	Fusion	Segmental	
Giant cell tumor	15	1	-	16
Chondromyxoid fibroma	1	-	-	1
Desmoplastic fibroma	-	-	1	1
Desmoid	-	-	1	1
Adamantinoma	1	-	2	3
Osteosarcoma	4	5	2	11
Parosteal sarcoma	1	-	-	1
Choldrosarcoma	2	1	1	4
Fibrosarcoma	1	-	1	2
Angiosarcoma	2	-	-	2
Total reconstructions	27	7	8	42

42 patients; 45 operations (3 revisions for infection)

The patients' ages ranged from 15 to 62 years, including 14 men and 29 women. Reconstructions were carried out around the knee in 25 patients, in the shoulder in seven patients, ankle in two patients, hip in one patient and wrist in one patient. Nine reconstructions were segmental -- one in the radius, one in the ulna, one in the femur, and six in the tibia. The usual distribution of pathologic findings was observed (Table III). The most common tumors were 16 giant cell tumors, and 11 osteosarcomas.

Complications

Recurrence. Three local recurrences were encountered. One patient had a segmental allograft for an osteosarcoma of the middle third of the femur, with a local recurrence at one year. An amputation was carried out but the patient died three years later of metastatic disease. Another patient had a local recurrence 18 months after an articulating proximal tibial allograft was performed for a fibrosarcoma. Amputation was performed and the patient is disease free three years later. The third patient had a segmental vascularized fibular autograft following en bloc excision of a desmoid tumor of the ulna. A local recurrence at one year was dealt with by graft removal and further excision of soft tissue. This patient is desease free two years after this procedure.

Metastatic disease. Two patients with osteosarcomas have died of metastatic disease, one of whom had an amputation for a local recurrence. Another patient with an osteosarcoma had metastatic disease and is still alive.

Amputations. Four amputations had to be carried out. Two were carried out for local recurrence, one due to an osteosarcoma and one due to a fibrosarcoma as already mentioned above. One amputation was carried out for recurrent stress fractures of an articulating proximal tibial allograft which was done for a giant cell tumor. The fourth amputation was carried out because the diagnosis was changed to a more aggressive tumor.

Stress fractures. Stress fractures occurred in three patients. One fracture occurred in a vascularized fibula with solid healing at three months. One proximal humeral articulating allograft which was supplemented by a vascularized fibular autograft also healed. The third patient with a stress fracture was seen at a centre in another city. The surgeons elected to treat her by amputation.

Infection. Three graft infections occurred. All three grafts were contaminated before transplantation. Two of these grafts came from the same donor and were done as fresh grafts. Unfortunately, we were unaware that the donor had an infected tracheostomy and eventually the allografts grew the same pathogens. Both patients were treated by graft excision and delayed reconstruction with vascularized fibular autografts and knee arthrodesis. The third infection occurred in a knee arthrodesis performed with an allograft and a vascularized fibular autograft. This was managed by removal of the allograft and delayed reconstruction using a nonvascularized fibula, thus leaving the patient with a knee arthrodesis using one vascularized and one nonvascularized fibula.

Degenerative arthritis. Severe degenerative changes occurred in two allografts, both in distal tibias and transplanted fresh. Both **had** undergone ankle fusion. One of the grafts was considered a poor size match at the time of surgery.

Joint instability. All the articulating grafts of the knee have some ligament laxity, nine of 18 being graded as severe. These patients do not participate in vigorous physical activities and, therefore, do not require bracing after 24 months. All patients were ambulatory without aids or bracing.

Joint stiffness. Only one patient with an articulating graft did not achieve a functional range of motion.

Nonunion and delayed union. We encountered no nonunions. One patient with a distal femoral articulating allograft fractured her plate at three months during partial weight bearing and had to be replated and bone grafted and went on to union. In this case the plate had been too small.

In this group of 42 patients, no significant complications were observed in 32. Thirty-three patients still have their original reconstruction in place; 37 patients still have their limb; 39 patients are disease free; 39 patients have had no local recurrence; and 40 patients are alive.

Discusssion

A comprehensive transplant program has been described. Three types of reconstructions are performed: articulating, fusion, and segmental. These reconstructions are carried out using allografts, autografts, allografts combined with autografts and, on occasion, allografts combined with implants.

The surgical treatment of the tumor is decided after routine tumor workup. When a limb salvage procedure is felt to be possible, then the type of reconstruction is decided according to the site (segmental or involving a joint) and the extent of soft tissue excision deemed necessary. If a significant amount of soft tissue is part of the en bloc excision, a fusion is carried out; otherwise, an articulating graft is used. When the tumor is slow growing and involves a joint, we wait for a donor so that a fresh articulating graft can be carried out. When the tumor is more aggressive, a deep frozen graft is performed. On two occasions, allografts have been combined with implants, ie, total hip prosthesis combined with allograft reconstruction of the pelvis.

The best indication for the fresh articulating graft is recurrent giant cell tumor. Fifteen of our 26 articulating reconstructions have been for this diagnosis with no recurrences to date (Table III). In our opinion, graft success is more dependent on the fate of the bone than the fate of the cartilage. If the bone fragments or fractures, instability, joint incongruity and malalignment result. The donor cartilage, if fresh, may survive, and we have evidence of live hyaline cartilage in two small fragment grafts at eight years. We have used fresh tissue in 18 articulating allografts. We feel that the cartilage has the potential to survive and, if possible, fresh tissue should be used as long as the wait for a donor does not compromise the treatment of the tumor. The long-term survival of the cartilage is part of an ongoing study to be published later. If the cartilage does not survive, it may be replaced by fibrocartilage which we have also observed in a few patients.

The small fragment grafts are ideally replaced by uniform creeping substitution before bone failure occurs. The large grafts are surrounded on their periphery by host bone which may grow into the graft but cannot possibly replace it in the patient's lifetime. In eight patients, we have fortified the allograft with a vascularized fibular autograft. All grafts are fortified with autograft iliac crest bone.

We believe that the vascularized fibular autograft plays a significant role in tumor surgery. We have used it alone as the only graft following en bloc excision of forearm bones on three occasions, combined with allografts on eight occasions, and to fuse joints after removal of infected allografts on two occasions. The vascularized graft survives in previously irradiated tissue and could be irradiated if necessary. This graft is also more resistant to infection; therefore, it was used in the three infected allografts.

The vascularized fibular autograft is combined with the allograft to provide immediate strength. Additional internal fixation can be achieved with a plate. The fibula will provide strength as it hypertrophies. It is interesting that in three stress fractures of allografts, the two that healed were supplemented by vascularized fibular autografts. We have, therefore, used the vascularized fibula on 13 occasions and feel strongly that it should be available as part of the biologic alternative in tumor surgery. The vascularized fibular autograft is, therefore, combined with an allograft under two conditions. The first is an arthrodesis of a large joint where the allograft provides a means of plate fixation and early strength while the fibular autograft along with iliac bone chips achieves the arthrodesis. The second is when a long segmental defect is to be reconstructed in an adult. The two grafts are combined with the allograft providing a means of fixation and early strength while the vascularized fibular autograft hypertrophies. In a child, a segmental allograft does not have to be supplemented by the vascularized fibular autograft. A rough guide for the choice of graft material is outlined in Table IV.

When deciding whether to do an articulating graft, the joint involved must also be considered. For example, the shoulder articulating grafts do well because stability does not present a problem and it is a relatively unloaded joint. The knee articulating grafts also do well despite some laxity which may or may not cause instability. If the patient is going to place heavy physical demands on the reconstruction, then a knee fusion rather than an articulating graft should be done. Our two ankle articulating reconstructions have developed severe degenerative changes and had

Table IV. Recommendations for choice of graft material

Graft	Indication
Allograft alone (with iliac bone autograft)	Articulating, fresh or frozen
	Articulating, in combination with implant (ie, shoulder)
	Segmental – in children
	Segmental – short segments in adults
Vascularized fibular autograft alone	Segmental in small long bones (ie, forearm)
	Fusion, in small joints (ie, wrist)
Allograft in combination with vascularized fibular autograft	Articulating, large - to supplement the diaphyseal component
	Fusion, large joints (ie, knee)
	Segmental, large in adults

to be fused. This is a relatively small load-bearing joint that does not tolerate any incongruity.

Conclusions

This is a small series of carefully selected patients for limb salvage surgery. Our follow-up is not long, but the results are gratifying to date. Of 42 patients, 40 are still alive, 39 have had no local recurrence, 39 are still disease free, 37 have their limbs, 33 have their original reconstruction, and 32 have had no significant complications. We have used a team approach, one team being responsible for the en bloc excision and the other for the reconstruction. This permits ample planning for both, and also enables both teams to remain alert and to work in their own areas of expertise. Our program started on a small scale 11 years ago, providing fresh small fragment grafts in a few selected patients a year. This program has expanded and now we perform approximately ten small fragment grafts and ten large fragment grafts per year.

In addition, we have an extensive research arm and also a bone bank that provides deep frozen bone to other surgeons in our region. We feel, however, that the need exists for such a program to be undertaken regionally.

References

1. Friedlaender and Mankin (eds): Surgical techniques and clinical experience with articular allografts at the knee. In, Osteochondral Allografts, Biology, Banking and Clinical Applications. Little, Brown & Co., Boston, Mass.

2. Gross, A.E., McKee, N.H., Pritzker, K.P.H., and Langer, F.: Reconstruction of skeletal deficits at the knee: A comprehensive osteochondral transplant programme. Clin. Orthoped. (in press).

3. Langer, F. and Gross, A.E.: Immunogenicity of allograft articular cartilage. J. Bone Joint Surg. 56(A), 297, 1974.

4. Langer, F., Czitrom, A., Pritzker, K. and Gross, A.E.: The immunogenicity of fresh and frozen allogeneic bone. J. Bone Joint Surg. 57(B), 216, 1975.

The Role of Resection and Reconstruction in the Treatment of Bone Tumors

M. CAMPANACCI, F. GHERLINZONI, and P. PICCI

Introduction

Ablative surgery has historically been the preeminent treatment of the majority of the musculoskeletal sarcomas. However, in recent years, early diagnosis, better surgical techniques, chemotherapy, and radiotherapy have improved the prognosis of these tumors and emphasized the practicality of conservative and reconstructive **surgical techniques.**

To standardize surgical procedures, the Musculoskeletal Tumor Society has proposed a surgical staging system (Enneking, 1980) that incorporates the most significant prognostic factors and delineates progressive stages of the disease. All the findings present specific implications for surgical management and provide guidelines to the use of adjunctive therapies.

The surgical staging system classifies the musculoskeletal sarcomas into three stages: grade of biologic aggressiveness; anatomic site; and presence of metastasis. The three stages are:

I - low-grade tumors

II - high-grade tumors

III - presence of metastasis.

Each stage may be subdivided into:

A - intracompartmental lesion

B - extracompartmental lesion

The compartment is an anatomic region with clearly delineated natural boundaries to arrest tumor invasion: in bone are cortical bone and articular cartilage; in joints are articular cartilage and joint capsule. Table I lists the classification of the intracompartmental and extracompartmental regions.

A lesion is extracompartmental (B) if it arises primarily in an extracompartmental anatomic region or if it secondarily extends into it from an original intracompartmental location. Furthermore, a poor surgical procedure such as incomplete removal of the tumor or a wrong biopsy with

Current Concepts of Diagnosis and Treatment
of Bone and Soft Tissue Tumors.

Table I. Surgical stages

Stage	Grade	Site	Metastases
IA	Low	Intracompartmental	None
IB	Low	Extracompartmental	None
IIA	High	Intracompartmental	None
IIB	High	Extracompartmental	None
III	Low or high	Intra- or extra-compartmental	Present

Table II. Surgical sites*

Intracompartmental	Extracompartmental
Intraosseous	Soft tissue extension
Intra-articular	Soft tissue extension
Superficial to deep fascia	Deep fascial extension
Paraosseous	Intraosseous or extrafascial
Intrafascial compartments	Extrafascial planes or spaces
ray of hand or foot	mid- and hindfoot
posterior calf	popliteal space
anterolateral leg	groin-femoral triangle
anterior thigh	intrapelvic
medial thigh	midhand
posterior thigh	antecubital fossae
buttocks	axilla
volar forearm	periclavicular
dorsal forearm	paraspinal
anterior arm	head and neck
posterior arm	
periscapular	

* Enneking, 1980

dissemination of the tumor may convert a lesion from intra- to extracompartmental. The surgical stages and their subdivisions are listed in Table II.

From the standpoint of surgical planning, the second step is to clearly define the surgical margins. The four types of surgical margins, again as defined by the Musculoskeletal Oncological Society, are the following (Table III):

Table III. Surgical margins*

Type	Plane of dissection	Results
Intralesional	Piecemeal debulking or curettage	Leaves macroscopic disease
Marginal	Shell out en bloc through pseudocapsule or reactive zone	May leave either satellite or skip lesions
Wide	Intracompartmental en bloc with cuff of normal tissue	May leave skip lesions
Radical	Extracompartmental en bloc with the entire compartment	No residual tumor or lesion

* Plane of dissection used to achieve a particular margin is shown as well as the result of that margin in terms of residual lesion remaining in the wound (Enneking, 1980).

1. *Intralesional*. The surgical dissection passes within the lesion and leaves macroscopic foci of tumor. An example is an incisional biopsy or a curettage for a benign lesion.

2. *Marginal*. The surgical dissection passes through the pseudocapsule or the reactive zone surrounding the tumor and leaves microscopic foci of disease. An example is the en bloc excision for benign tumors.

3. *Wide*. The surgical dissection comprises a cuff of normal tissue surrounding the tumor and passes into normal tissue. The wide resection may leave "skip" metastases. A wide margin is a correct surgical procedure for Stage I lesions.

4. *Radical*. The lesion is removed en bloc with the entire compartment. The surgical dissection comprises the entire bone or muscle involved and does not leave any residual tumor. A radical margin is definitive surgery for Stage II lesions.

Table IV lists the surgical procedures according to the margins and in Fig. 1 are shown the various types of amputations.

Table IV. Surgical procedures*

Margin	Local	Amputation
Intralesional	Curettage or debulking	Debulking amputation
Marginal	Marginal excision	Marginal amputation
Wide	Wide local excision	Wide through-bone amputation
Radical	Radical local **resection**	Radical disarticulation

* Classified as to type of margin achieved and whether obtained by a local or ablative procedure (Enneking, 1980).

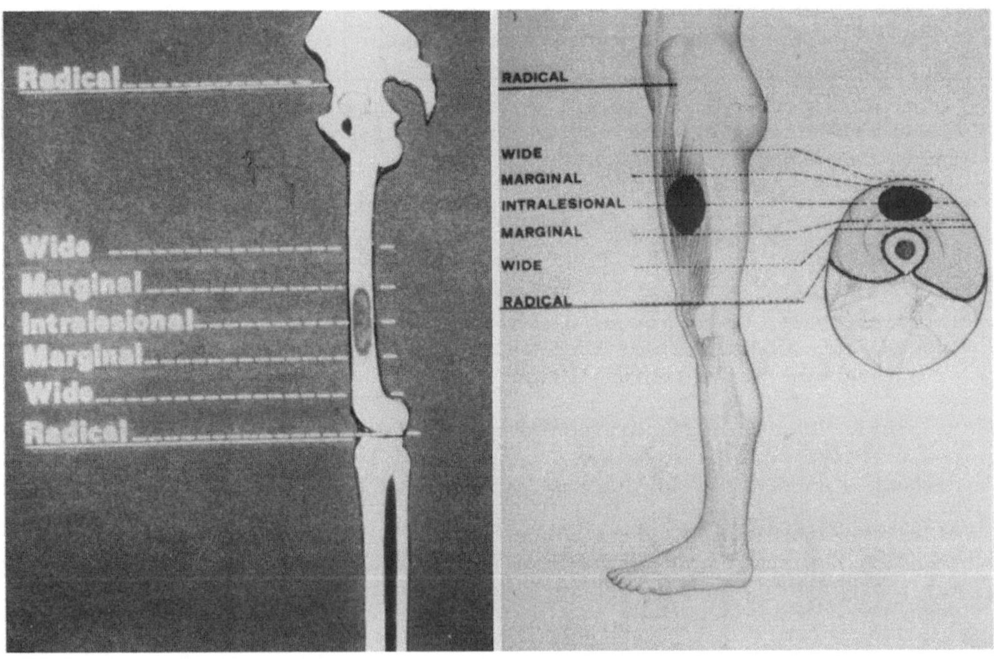

a b

Fig. 1. The surgical margins in accordance with the Surgical Staging System: a) for bone tumors; b) for soft tissue tumors.

Resection and Temporary Stabilization with Küntscher Rod (or Plate, or External Fixation) and Acrylic Cement

In high-grade tumors, when adjuvant chemotherapy is used, we do not implant bone grafts primarily. We rather stabilize the bone segments by means of

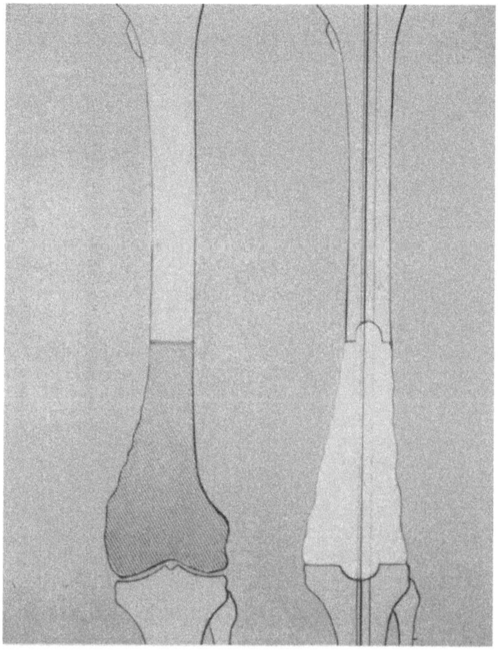

Fig. 2. Surgical resection of the distal femur and temporary stabilization with Küntscher rod and acrylic cement.

a Küntscher rod, or more rarely a plate, and a mass of acrylic cement moulded to replace the resected bone (Fig. 2). After the chemotherapy has been completed, provided the patient has no evidence of disease, the cement can be replaced by autogenous bone grafts.

With this procedure the patient can walk several days after the first operation without requiring plaster immobilization. Substitution of cement for the grafts may be delayed as long as the Küntscher rod does not break.

The reasons justifying these two-stage reconstructions after segmental resection for high-grade tumors are the following:

1. We save a lot of surgery and much time with plaster immobilization in patients with a high risk of local recurrence and metastases. Approximately 50% of them will die in the first five years.

2. We avoid surgical shock in these patients, and large blood loss that might decrease host defence against the metastatic spread.

3. Chemotherapy increases the risk of infection and adversely influences the incorporation of the grafts.

A second indication for temporary stabilization with cement and external fixation is in resections for benign or malignant tumors of bone where a previous biopsy has been done and a wound infection may or has already occurred.

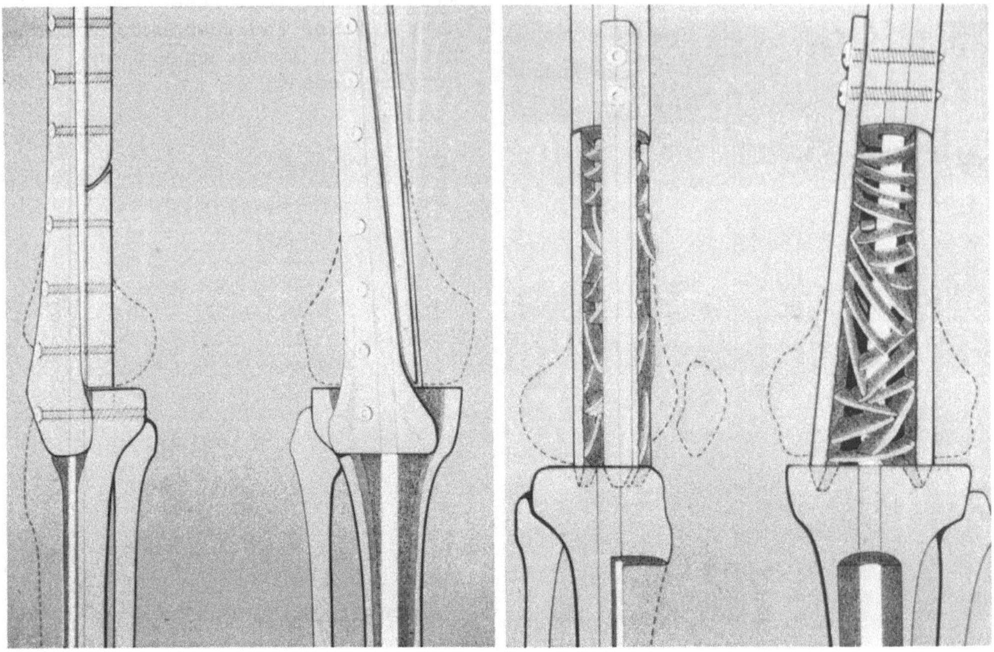

a) b)

Fig. 3. Surgical schema of an arthrodesis after resection of a tumor of the distal femur. a) The Putti-Juvara-Merle d'Aubigné technique. b) The Enneking modified technique.

Autograft Arthrodesis for Resection of the Distal Femur or Proximal Tibia

The most frequent indications for this operation are:

1. Giant cell tumor (radiographically active or aggressive, very expanded, with no healthy bone between the tumor and the articular cartilage, or recurrence).

2. Tumors with low-grade malignancy (central chondrosarcoma or parosteal osteosarcoma).

3. Tumors with high-grade malignancy (osteosarcoma and others). In this case, however, we apply the autografts only secondarily, using acrylic cement as a temporary stabilization (see previous section).

The technique of the arthrodesis is sketched in Fig. 3 and has been previously described (Campanacci and Costa, 1979). It is a modification of the Merle D'Aubigné technique (1958) which in turn derives from Putti (1928) and Juvara (1921). We immobilize our patients in a plaster cast for six to 12 months, and weight bearing is delayed one to two years after surgery.

Recently, we have modified this technique because it has, in our experience, some disadvantages:

- a high risk of infection;
- the hemicylindric resection of the femur and tibia damages the blood supply of the diaphyseal portions of these bones at the sites where they are needed as a bed for grafts and reduces the length of the untouched medullary canal where the rod can be tightly held.
- for all the above-cited reasons, the stability of the system can be weakened, causing nonunion or telescoping.

We actually prefer a technique similar to that described by Enneking (1977). With this technique most grafts are taken from the counterlateral leg (tibia or fibula), and from the iliac wing (lower risk of infection in the operated limb); and the diaphyseal segment residual to the resection, like the opposite epiphysis, are left undistrubed without damaging the blood vessels.

Enneking (1977) used a special, fluted rod with much greater strength for better stability of the system, but we have not yet applied it. Other techniques for the treatment of tumors of the skeleton of the knee are allograft and hinged prosthesis. We have no experience with the first and we apply the second only in elderly patients.

Patellar Autograft for Resection of One Femoral or One Tibial Condyle at the Knee

The indications for this surgical procedure are giant cell tumors involving only one-half of the metaepiphysis of the knee and where a curettage has only a slight chance of success. Occasionally, it might be used in a wide resection for a very small and circumscribed malignant tumor.

This procedure is considered indicated only when the resected sagittal plane crosses approximately the middle of the transversal diameter of the joint. For this reason, preoperative evaluation using computerized tomography is essential.

The articular surface of the patella was used as articular surface in all the resections of a tibial condyle; for the femoral reconstruction, we used more frequently, instead, the anterior nonarticular aspect of the patella as articular surface (Fig. 4).

A groin-to-ankle cast is applied for three weeks. It is important to start mobilization and rehabilitation at this time. In two patients, at six and ten weeks, manual mobilization under anesthesia was done with good results. Weight bearing is not allowed for at least one year.

Endoprostheses and Arthroprostheses for Resection of the Proximal Femur

Endoprostheses or arthroprostheses are almost universally used in the reconstruction after resection of tumors of the proximal femur. During

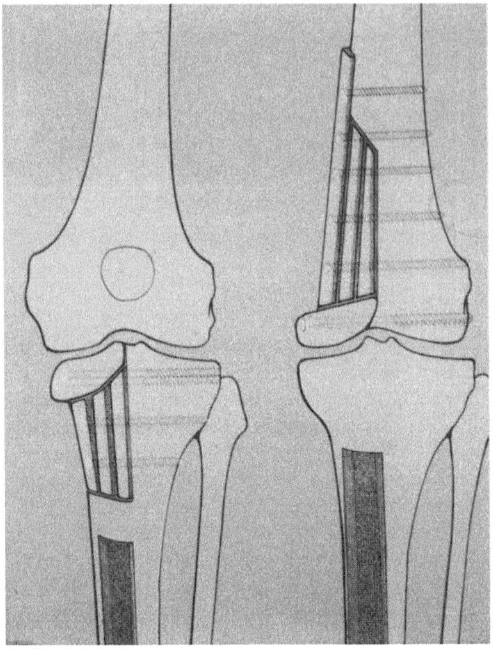

Fig. 4. Schema of surgical technique for resection of one femoral condyle and repair using the autogenous ipsilateral patella.

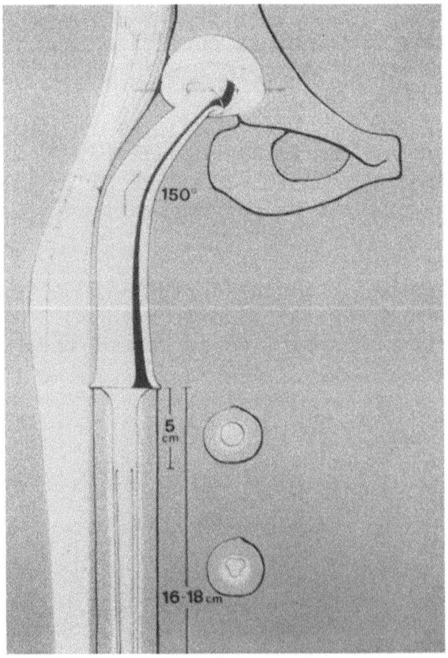

Fig. 5. Surgical schema of reconstruction with arthroprosthesis after resection of the proximal femur. The prosthesis has an angle of the neck of 30° and the acetabular component must be applied more horizontally than the normal prosthesis and in marked anteversion to avoid dislocation of the arthroprosthesis. The abductor muscles are sutured in tension to the deep aspect of the fascia lata. In the inferior part of the figure the sections of the stem of the prosthesis are shown.

the last ten years we have used different types of prostheses, all custom made and all with a medullary cemented stem. The angle of the neck was 45° in some and 30° in others. Sometimes it was used as an endoprosthesis with double cup, sometimes as an arthroprosthesis with a Charnley cemented cup.

In our experience, the major problem is dislocation of the prosthesis; therefore, to prevent the tendency of these prostheses to dislocate, we apply the acetabular component horizontally (10° - 20° obliquity in the transversal plane), in marked anteversion (30° - 40° in the frontal plane), and the femoral component 1 cm longer than the resected bone segment (Fig. 5).

After a four-week postoperative total immobilization, we change to a short cast to permit mobilization from knee to ankle for another two to four weeks.

We hope that, in future, the materials, the design and the anchorage of the prosthetic implants will be improved. We hope, furthermore, that a multicompomponent prosthesis will become available that can be assembled in the operating room according to individual requirements.

Endoprostheses for Resections of the Proximal Humerus

This technique represents a simple, practical and appropriate reconstruction system for many types of tumors, high or low-grade, primary or metastatic, of the proximal humerus.

In these wide intra-articular resections, the capsule is cut close to the glenoid, the muscles are sectioned at some distance from their humeral insertions. Not infrequently, the deltoid is largely or completely resected en bloc with the tumor, and it may be necessary to sacrifice the axillary nerve. The endoprosthesis acts as a spacer and the remaining muscles are sutured together but cannot be attached to the prosthesis. Thus, the active function of the shoulder is severely and permanently impaired and abduction is almost zero.

The prosthesis we use (Campanacci et al, 1982)(Fig. 6) has a multicomponent metallic body with a cemented intramedullary stem, and a polyethylene spheric head. There is a free rotatory connection between the body and the head which prevents dislocations, reduces the rotatory stresses at the interfaces of the stem, and allows for wide rotation of the arm.

This prosthesis should be improved in design and materials with the following principles to be retained:

- the rotatory hinge
- the medullary stem
- the multicomponent assemblage system.

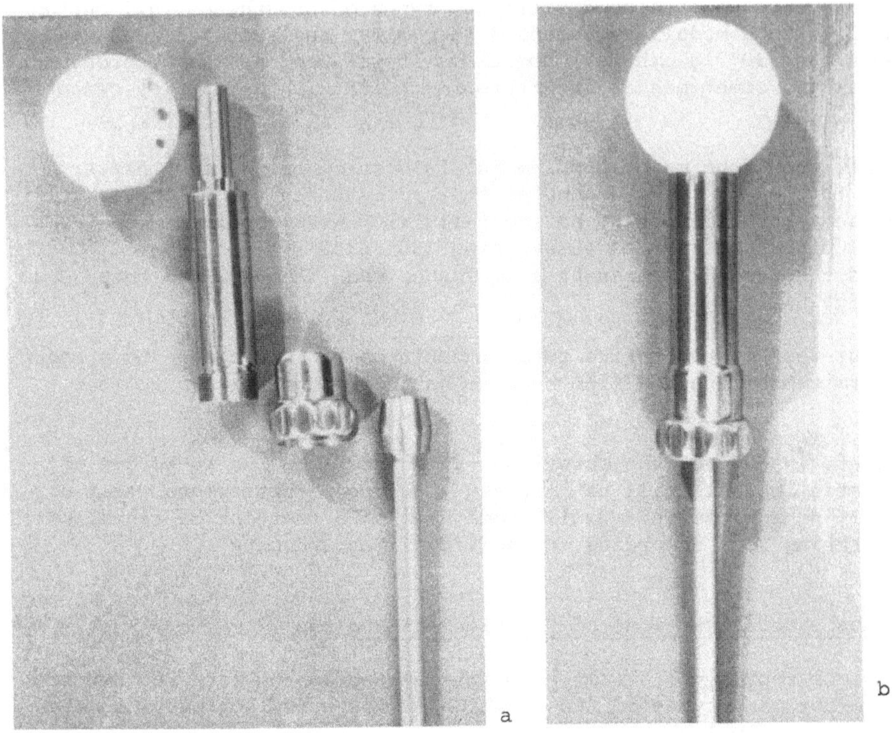

Fig. 6. a) The modular prosthesis for reconstruction after resection of the proximal humerus in use in our Institute: the head in polyethylene, the body, the connection, and the stem. b) The assembled prosthesis.

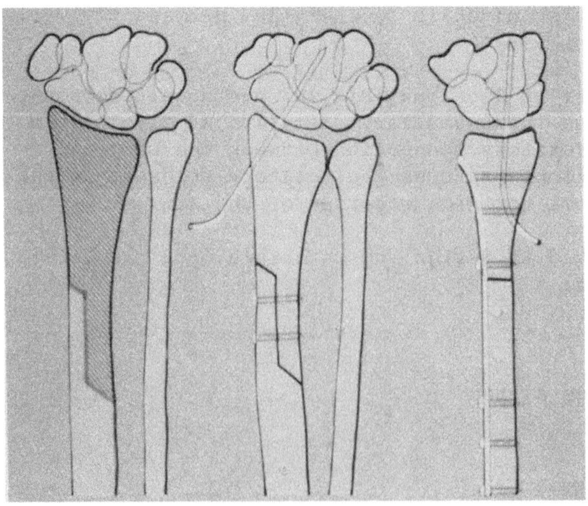

Fig. 7. Surgical schema of resection of the distal radius and reconstruction with fibular autograft.

Fibular Autograft for Distal Radius Resections

This procedure is almost exclusively indicated for giant cell tumors. We have experienced practically no complications and the results are satisfactory. The wrist is usually close to normal and no patients have complained of pain or tenderness.

We use the proximal ipsilateral fibula, subperiosteally excised with an intact epiphysis. The collateral ligament and biceps tendon are sutured to the peroneal muscles and/or to the tibia. The graft is fastened to the remaining radius by means of a step connection with two screws and a volar flat plate. The most difficult part of this simple procedure is to calculate exactly the length and the orientation of the graft. An attempt is made to obtain a symmetric articular contact between carpal bones and the facet of the fibula. A good cosmetic appearance of the wrist is thus obtained. Reduction is maintained by means of a Kirschner wire obliquely inserted through the graft and carpus (Fig. 7).

We apply a plaster which is removed five weeks later together with the Kirschner wire and gentle mobilization is started.

References

1. Campanacci, M., Cervellati, C., Gherlinzoni, R., Capanna, R.: Indoproteri d'omero: descrizione di un nuovo modello e sua applicazione. Giorn. It. Ortop. Traumat. 59, 1982.

2. Campanacci, M., and Costa, P.: Total resection of distal femur or proximal tibia for bone tumors. J. Bone Joint Surg. 61(B), 455, 1979.

3. Enneking, F.W. and Shirley, P.D.: Resection-arthrodesis for malignant and potentially malignant lesions about the knee using an intramedullary rod and local bone grafts. J. Bone Joint Surg. 59(A), 223, 1977.

4. Enneking, F.W., Spanier, S.S., and Goodman, M.A.: A system for the surgical staging of musculoskeletal sarcoma. Clin. Orthop. 153, 106, 1980.

5. Juvara, B.: Procedé de resection de la partie supérieure du tibia avec substitution, à la partie enlevée, d'une greffe prélevée sur le fémur. Presse Médicale 29, 241, 1921.

6. Merle D'Aubigné, R. and Dejouany, J.P.: Diaphyseal-epiphyseal resection for bone tumor of the knee. J. Bone Joint Surg. 40(B), 385, 1958.

7. Putti, V.: A study of two cases of tumor of the femur. In, The Robert Jones Birthday Volume. Oxford University Press, London, 1928.

Segmental Replacement of Long Bones Using Fiber Titanium Composites

F. H. SIM, E. Y. S. CHAO, and J. O. GALENTE

Introduction

Although the management of bone tumors continues to be one of the most challenging tasks in oncology, significant advances have been made. In recent years, interest has increased in en bloc resection (Wilson and Lance, 1965), not only for aggressive benign and low-grade tumors but also for limb-salvage procedures involving high-grade lesions (Jaffe et al, 1978; Marcove and Rosen, 1980). Recent improvements in clinical staging, particularly by the use of computed tomography, have enhanced our ability to assess the local extent of involvement of the tumor accurately and to decide which lesions are amenable to limb-saving resection.

In limb-salvage surgery, the first important goal is to achieve local control of the tumor by an adequate surgical procedure. This usually necessitates creating large osseous and soft tissue defects. The second goal is the reconstruction of the extremity to restore as much useful function as possible. The reconstruction of long bones after massive resection for tumors presents a complex surgical problem.

Tumors affecting the femur or tibia, particularly when the knee joint is involved, pose difficult reconstructive problems. Various techniques for reconstruction may be considered. Salvage of a mobile joint may be attempted using either an allograft or a custom-made prosthetic replacement. Allograft replacement (Mankin et al, 1976; Parrish, 1973) remains an experimental procedure with a risk of complications and limited availability. While prosthetic reconstruction has inherent advantages and the early results have been satisfactory, long-term fixation is a serious concern (Sim and Chao, 1979; Sim et al, 1979). Segmental reconstruction or knee arthrodesis can be performed using an allograft segment or a resection arthrodesis as popularized by Campanacci and Costa (1979) and Enneking and Shirley (1977) can be done. Either technique results in a complication rate varying from 5 to 20%. With the increased interest in limb-saving resection for primary bone tumors, efforts must continue to improve our techniques of oncologic reconstruction.

Reconstruction with Fiber Titanium Composites

In developing a prosthesis to substitute for a lost skeletal segment it is necessary to provide certain mechanical properties and compatibility. The prosthesis should simulate the lost segment and provide the mechanical pro-

Current Concepts of Diagnosis and Treatment
of Bone and Soft Tissue Tumors.

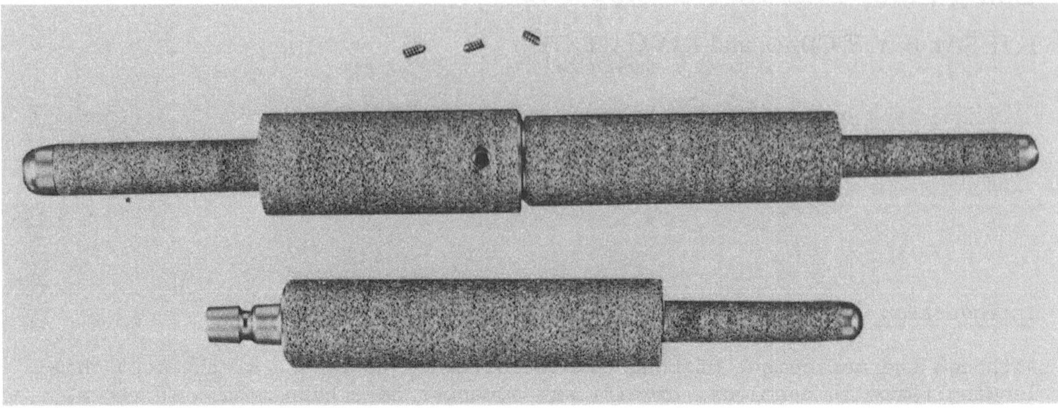

Fig. 1. Titanium fibermetal segmental bone prosthesis currently in use. Intermedullary stem fixation is used, and the two components are assembled through the self-locking conical coupling joint. Set screws are used to enhance the joint locking strength.

perties that are unique to the bony system. This is difficult to accomplish not only because of the large defects but also because of the need to provide adequate tensile and compressive strength to withstand the body's normal loading forces. Moreover, the prosthesis must be fixable both proximally and distally at the ends of the resected bone to obtain biologic incorporation. In addition to the desired mechanical properties, the prosthesis should have adequate surface pore size, and these pores must be interconnected to provide adequate bony ingrowth. Experimental work by Andersson et al (1978) has demonstrated bone ingrowth into a titanium alloy composite prosthesis. In dogs and rabbits, peripheral bone formation around the implant was noted by two weeks, with penetration by three weeks. By 17 days, the depth of ossification was 800 μ, and this was doubled by six weeks. In skeletal segmental replacement in baboons, massive consolidation of the grafts was noted at three months (Andersson et al, 1982). Skeletal reconstruction with a titanium fibermetal-coated prosthesis has been used at the Mayo Clinic and at Rush-Presbyterian Hospital since 1976 (Fig. 1). The implant consists of a solid core made of Ti-6Al-4V alloy with a porous layer of a titanium fiber composite on the surface.

In preparing the prosthesis, titanium wires are kinked and then molded to a density of 50% and sintered into sleeves that are subsequently sintered to the solid titanium alloy core. The diameter of the pores of the composite ranges from 250 to 400 μ. The initial designs provided for intermedullary stem projections at both ends, which were limited in length by the minimal amount of bone distraction at the time of implantation. However, a newer design has a coupling joint in its center portion with a self-locking conical fit, which allows for longer intermedullary stem projections and proper selection of prosthetic size and length to achieve better fixation. Additional set screws are used to enhance the locking strength at the joint. Ancillary plate fixation may be used to achieve initial fixation. The prosthesis has adequate tensile strength and is

Table I. Types of tumor

Type	Number of patients
Giant cell tumor	5
Osteosarcoma	4
Chondrosarcoma	4
Metastatic	2
Fibrosarcoma	1
Desmoplastic fibroma	1
Total	17

relatively inert because of its metallic composition. The large compliance of the porous layer, which compares favorably with that of trabecular bone, and the low modulus of elasticity of the solid core material permit uniform transmission of stresses to the bone and prevent stress-shielding bone resorption.

Clinical Series

Between 1976 and 1982, 18 patients underwent reconstruction of large defects in the femur or tibia using a fiber titanium composite prosthesis: nine patients at the Rush-Presbyterian Medical Center and nine at the Mayo Clinic. Seventeen of the 18 patients had a minimal follow-up of one year, and one patient died three months after prosthetic replacement. The ages of the 17 patients (6 men and 11 women) ranged from 20 to 79 years. Giant cell tumor, chondrosarcoma, and osteosarcoma predominated (Table I). Nine lesions were in the femur and eight were in the tibia, their sizes ranging from 1.5 to 16.5 cm. Eight patients had knee arthrodesis, four of whom had lesions in the proximal tibia and four in the distal femur. The nine other patients had diaphyseal segmental resections with preservation of knee function: four of the femur and five of the tibia. Partial weight bearing was allowed as early as four weeks, but full weight bearing was achieved in three to 11 months.

Illustrative cases

Case 1. A 31-year old woman experienced the gradual onset of pain in her left leg in February 1976. Roentgenograms revealed an extensive central cartilaginous lesion involving the midtibia (Fig. 2 A and B). Open biopsy revealed a grade 1 chondrosarcoma. In November 1976, 16.5 cm of the tibia were resected, and a temporary aluminum spacer was inserted. In December 1976, a custom-made fiber titanium prosthesis was inserted. Bone grafts from a single iliac donor site were placed. The patient became ambulatory while in a cast brace and short leg brace for six months. In December

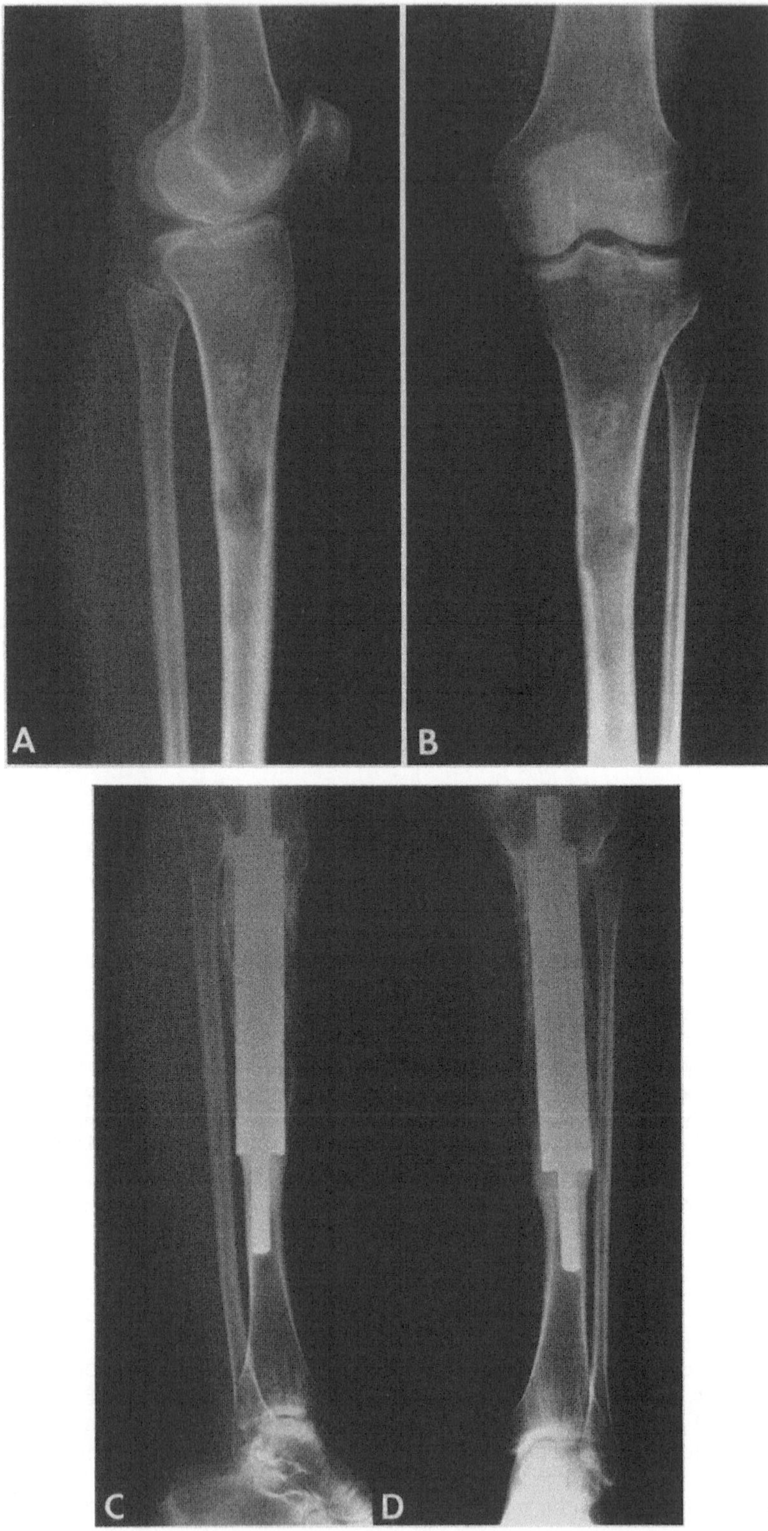

Fig. 2

Fig. 3. Case 2: A and B: anteroposterior and lateral views of tibia showing recurrent osteosarcoma. C and D: after segmental resection and reconstruction with a fibermetal prosthesis. Healing of the bone grafts is evident.

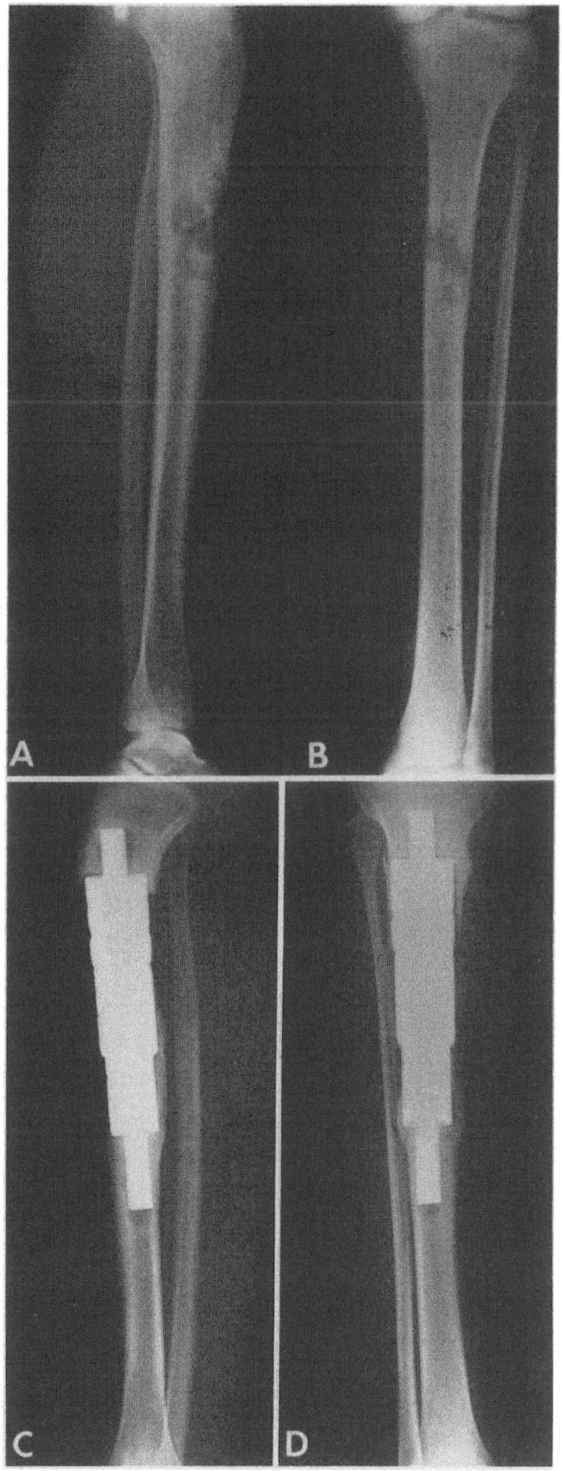

◄ Fig. 2. Case 1: A and B: anteroposterior and lateral views of tibia showing extensive grade 1 chondrosarcoma. C and D: after prosthesis implantation. The implant has solid fixation, with reconstitution of the cortex.

1977, bone grafting was repeated because of deficiency of the bone grafts in the midportion of the prosthesis. Exploration revealed that the implant was solidly fixed at both ends. The cortex was reconstituted, and the patient remains asymptomatic and unlimited in her activities (Fig. 2 C and D).

Case 2. A 21-year old woman was referred to the Mayo Clinic in September 1971 because of recurrent tumor in the left tibia (Fig. 3 A and B). She had undergone local excision of the lesion 10 months earlier. Histologic review revealed a grade 4 osteosarcoma. She underwent en bloc resection of 13 cm of the tibia. A titanium spacer was inserted, and on February 13, 1979 a delayed reconstruction with a fiber titanium composite prosthesis was inserted. Massive iliac bone grafts were applied. She remains fully active without limitations. Roentgenograms show healing of the bone grafts with consolidation of the cortex (Fig. 3 C and D).

Results

The results were rated as excellent, satisfactory, or failure. An excellent result was rated when full weight bearing and alignment of the limb with solid union and bony bridging was accomplished; 11 patients had excellent results. A satisfactory result was rated when full weight bearing, alignment, and a solid union was achieved but additional bone grafting was required to achieve healing. Of the five patients with satisfactory results, four required additional bone grafts and one required repositioning of an unsatisfactory knee arthrodesis. One failure occurred in a 66-year old woman who had persistent infection after knee arthrodesis for a proximal tibial giant cell tumor. There were two local recurrences, one was secondary to a giant cell tumor of the femur and the other was secondary to a fibrosarcoma of the tibia. Three patients had pulmonary metastasis: one from a benign metastasizing giant cell tumor, one from a fibrosarcoma, and one from an osteosarcoma. One patient with a chondrosarcoma of the femur subsequently required a quadricepsplasty 18 months after segmental replacement. One patient who had evidence of complete bone healing and function died of a myocardial infarction 11 months after surgery.

Discussion

The reconstructive problems in this series were difficult because of the large segment of bone involved. The mechanical requirements in replacing large defects are very demanding, and our experience indicates the necessity of careful attention to biomechanical principles. Immediate press-fit fixation and close apposition of the intermedullary stem with the cancellous bone are essential. Our experience also indicates that the procedure is probably best done in two stages. This provides the advantage of accurate

---

Fig. 4. A: diagram of modular design of proximal femoral replacement incorporating titanium fibermetal coating for bone ingrowth. B: diagram of modular design of custom-made proximal humeral replacement incorporating the fibermetal prosthesis. ►

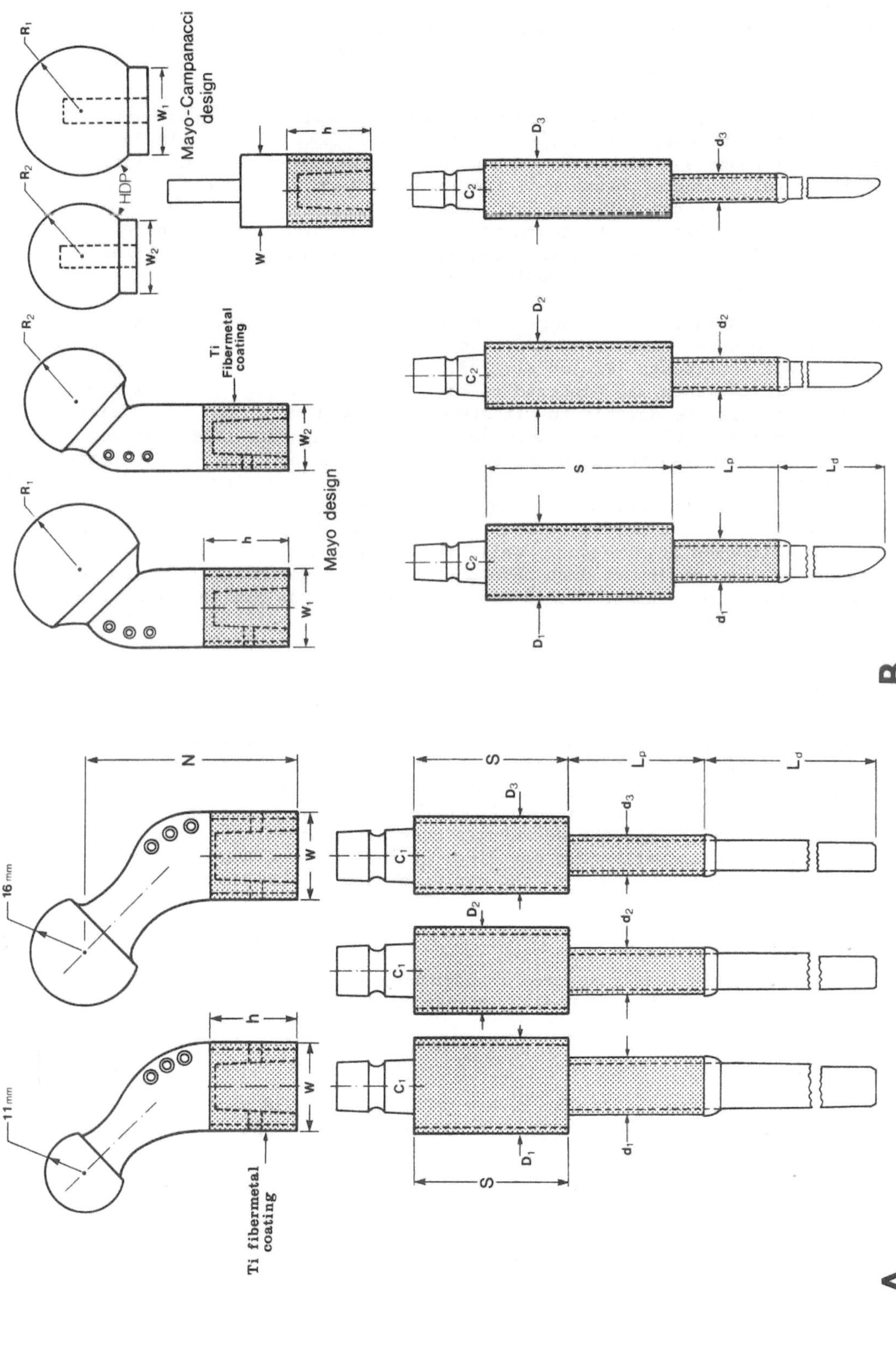
A
B
11 mm
16 mm
N
h
W
Ti fibermetal coating
S
C_1
D_1
D_2
D_3
d_1
d_2
d_3
L_p
L_d
Mayo design
Mayo-Campanacci design
R_1
R_2
W_1
W_2
HDP
Ti Fibermetal coating
C_2

custom-made fitting and allows for more massive bone grafts from both iliac crests to provide adequate bridging across the missing segment. In patients in whom bilateral posterior bone grafts were used, union and excellent results were obtained without any further surgery.

A potential problem of some concern is that the implant, when adequately incorporated in bone, may be very difficult to remove if it fails for mechanical reasons. Moreover, some concern exists over the compatibility of the metal products because of the large surface area exposed to the viable tissues of the patients. Fortunately, titanium is highly resistant to corrosion. However, the concentrations of titanium ions in the serum are currently being measured in these patients, both at the Mayo Clinic and at the Rush-Presbyterian Medical Center.

With increased interest in limb-saving resection in recent years, more patients are undergoing reconstruction of the hip, knee, and shoulder with custom-made prostheses (Sim et al, 1979). And with improved cancer treatments, more patients are expected to survive the disease. These implants must, therefore, be durable to withstand the long-term stresses. Fixation is a potential problem that determines long-term success. Clinical and experimental studies indicate the feasibility of using titanium fibermetal composites to provide mechanical fixation allowing bone ingrowth. Moreover, the fibermetal composite with its high compliance and low modulus of elasticity acts as a buffer between the bone and the prosthesis, with uniform distribution of stresses. While this system has been used only for lesions of long bones, it does have potential as a method of fixation for the components of a tumor prosthesis, and such designs are currently under investigation (Fig. 4).

Summary

Efforts continue to improve the techniques of oncologic reconstruction. Prostheses of fiber titanium composite have been successful in achieving full weight bearing status in 16 of the 17 patients studied. The principles of achieving adequate soft tissue coverage and adequate bone grafts are essential to achieve satisfactory reconstruction. The procedure is in an experimental stage, and the advantages over alternative methods of management, such as allografts, are not clear at this time because of the limited clinical experience. However, the early results are promising.

Acknowledgement

This investigation was supported in part by Research Grant CA-23751 from the National Institutes of Health, Public Health Service.

References

1. Andersson, G.B.J., Gaechter, A., Galante, J.O., and Rostoker, W.: Segmental replacement of long bones in baboons using a fiber titanium implant. J. Bone Joint Surg. (Am.) 60, 31, 1978.

2. Andersson, G.B.J., Lereim, P., Galante, J.O., and Rostoker, W.: Segmental replacement of the femur in baboons with fiber metal implants and autologous bone grafts of different particle size. Acta Orthop. Scand. 43, 349, 1982.

3. Campanacci, M. and Costa, P.: Total resection of distal femur or proximal tibia for bone tumours: Autogenous bone grafts and arthrodesis in twenty-six cases. J. Bone Joint Surg.(Br.) 61, 455, 1979.

4. Enneking, W.F. and Shirley, P.D.: Resection-arthrodesis for malignant and potentially malignant lesions about the knee using an intramedullary rod and local bone grafts. J. Bone Joint Surg.(Am.) 59, 223, 1977.

5. Jaffe, N., Watts, H., Fellows, K.E., and Vawter, G.: Local en bloc resection for limb preservation. Cancer Treat.Rep. 62, 217, 1978.

6. Mankin, H.J., Fogelson, F.S., Thrasher, A.Z. and Jaffer, F.: Massive resection and allograft transplantation in the treatment of malignant bone tumors. N. Engl. J. Med. 294, 1247, 1976.

7. Marcove, R.C. and Rosen, G.: En bloc resections for osteogenic sarcoma. Cancer 45, 3040, 1980.

8. Parrish, F.F.: Allograft replacement of all or part of the end of a long bone following excision of a tumor: Report of twenty-one cases. J. Bone Joint Surg.(Am.) 55, 1, 1973.

9. Sim, F.H. and Chao, E.Y.S.: Prosthetic replacement of the knee and a large segment of the femur or tibia. J. Bone Joint Surg.(Am.) 61, 88, 1979.

10. Sim, F.H., Pritchard, D.J., Ivins, J.C., and Shives, T.C.: Total joint arthroplasty: Applications in the management of bone tumors. Mayo Clin. Proc. 54, 583, 1979.

11. Wilson, P.D., and Lance, E.M.: Surgical reconstruction of the skeleton following segmental resection for bone tumors. J. Bone Joint Surg. (Am.) 47, 1629, 1965.

Principles of En-Bloc Excision of Limb Sarcomas

F. LANGER

Introduction

The classic surgical approach to sarcomas of the extremity has been amputation. The obvious goal of tumor surgery is local control, but in many cases this can be achieved by subamputative resection with preservation of the limb. Success depends on complete excision of the primary tumor which has been demonstrated by several clinical studies (Simon and Enneking, 1976). The basic theory guiding en bloc excision is that the gross anatomic setting of the sarcoma and not its histologic type should be the determining factor in the decision between amputation or en bloc excision (Bowden and Booher, 1958). The principles guiding this approach have been best enunciated in several presentations by Enneking (Enneking and Dunham, 1978; Enneking et al, 1981; Simon and Enneking, 1976).

1. Sarcomas do not have a true capsule, therefore "shelling-out" fails. At the growing edge of the tumor, the tumor cells compress the normal structures as they meet physical resistance to their growth. The cells of normal structure are also compressed and a reactive inflammatory zone usually exists peripheral to this zone of compression. Although this makes the tumor appear clearly circumscribed, shelling out of the tumor through this pseudocapsular plane inevitably leads to an incomplete excision leaving tumoral tissue behind. This occurs particularly in high-grade sarcomas because microextensions of tumor penetrate this zone and extend to varying distances.

2. Sarcomas tend to grow along planes of least resistance and remain confined to the structures within which they arise, particularly in the presence of a physical barrier to egression. This growth within anatomic compartments has been demonstrated by Enneking to be of singular importance in defining the location of the tumor and therefore the operative approach (Enneking et al, 1981), for these compartments can be delineated clinically and radiologically (especially by CAT scan) and then excised.

Examples of compartments in the lower extremity include muscles with fascial envelopes, eg, iliopsoas, gastrosoleus, gluteus maximus and rectus femoris, and functional muscle groups such as the quadriceps, adductors, lateral and medial hamstrings and the anterior tibial muscles. The bones and joints are also compartments for tumor growth. Within these muscle compartments, growth tends to be mainly longitudinal towards the origin and insertion of the respective muscles. Fascia is not transgressed until

Current Concepts of Diagnosis and Treatment
of Bone and Soft Tissue Tumors.

late. Primary bony sarcomas, especially high grade, on the other hand, often broach the bone cortex and form extraosseous soft tissue masses. This kind of tumor breakout usually occurs at sites of muscle attachment, blood vessel penetration, and capsular and ligament attachment.

Based on this concept of compartmentalization of tumors, Enneking developed a staging to facilitate the surgical management of extremity sarcomas (Enneking et al, 1980). It is based on the following three aspects:

i. *Grade*

From the surgeon's point of view, lesions are either low grade or high grade. Grade refers to the potential for metastases -- low being 10%. This potential is judged primarily by the histologic, clinical and radiologic findings.

ii. *Intra- or extracompartment*

Intracompartment means the tumor is confined to the anatomic compartment. A sarcoma may be extracompartmental in origin, eg, popliteal fossa, femoral triangle, or may extend extracompartmentally by natural growth or as a result of previous surgery. Especially in high-grade lesions, poorly executed surgery, eg, an excisional biopsy, can result in more radical definitive surgery by making an intracompartmental lesion extracompartmental. In addition, postoperative hematomas must be considered contaminated by tumor cells.

iii. *Presence or absence of regional and distal metastases*

In general, regional lymph node metastases are rare.

<u>Enneking Surgical Staging of Sarcomas</u>

Stage I A	Low grade	Intracompartmental
IB	Low grade	Extracompartmental
Stage IIA	High grade	Intracompartmental
IIB	High grade	Extracompartmental
Stage III	I or II but with metastases	

The possible surgical margins and procedures for either resection or amputation are:

i. *Intralesional* - Resection or amputation: operative margin is through the tumor.

ii. *Marginal* - Resection or amputation: the operative margin is through the **pseudocapsule**, ie, microscopic tumor is left behind. This is the usual result of an excisional biopsy.

iii. *Wide* - Resection or amputation: the tumor is removed with surrounding reactive tissue and a wide margin of normal tissue. Danger: especially in high-grade lesions, microscopic satellite deposits may be left behind. An equivalent amputation is through the bone, eg, AK amputation for a tumor of the vastus medialis, but in bone sarcoma the risk of skip lesions remains.

iv. *Radical* - Tumor is resected with the entire anatomic compartment; for example, with a bony lesion the entire bone is excised or a disarticulation is done. For a sarcoma of the tibialis anterior, the entire anterior tibial compartment is resected, or a disarticulation of the knee is performed.

Practical Considerations

According to Enneking, the biologic aggressiveness of the tumor determines the margin required, ie, type of excision required for local control, but the anatomic setting determines how that margin may be achieved (Enneking et al, 1981).

Most soft tissue sarcomas occur in the thigh and buttock. Stage I-A and II-A sarcomas in the anterior compartment (quadriceps), gluteus maximus, and ilipsoas can usually be resected. Sarcomas of the medial compartment (adductor muscles) usually are poorly contained. They tend to extend rather quickly towards the pelvis; they are hard to define. Unless the tumor is low grade or well delineated, an appropriate amputation (modified hindquarter) is the safer approach. The popliteal fossa and femoral triangles are anatomic spaces, but it is not feasible to resect these due to the femoral vessels in the former. Furthermore, the popliteal vessels and nerves in the latter cannot be excised without creating major functional problems. Therefore, in these areas we recommend amputation as the preferred management, at least in high-grade lesions. Basically, we feel that resection which includes the sciatic or posterior tibial nerve is contraindicated.

What is the role of chemotherapy and radiation in local control? In a low-grade lesion, radical resection should be curative. In a high-grade sarcoma, we usually follow a radical resection with a postoperative course of radiation and with systemic adjuvant chemotherapy. This is especially recommended when one margin only is wide or in cases where previous surgery had been carried out. In a low-grade sarcoma, wide resection is followed by radiation; and in circumstances where a border of the excision is marginal, eg, where the tumor abuts a major nerve or vessel, we always follow with radiation.

Radiation course, 5000 rad over 4 weeks.

Preoperative radiation is carried out where resection follows an inadequate excision, eg, excisional biopsy, and where the tumor is very large and proximal. In addition, when the tumor is threatening major vessels or nerves or in cases of extracompartmental extension, but where radical resection seems feasible, preoperative radiation is carried out combined with infusion of chemotherapeutic drugs to assure a successful en bloc excision.

Chemotherapy and irradiation will not compensate for insufficient surgery, but occasionally can aid in cleansing a marginal border of microscopic disease, at least in low-grade tumors (Suit et al, 1973). The best chance for cure, however, is still the total excision of the tumor, leaving no microscopic residual disease.

Table I. Chemotherapy and irradiation in local control after resection

Stage	Preoperative	Operation	Postoperative
I A	nil	wide or marginal	irradiation
	nil	radical*	nil
I B	occasional irradiation	wide or marginal	irradiation
	" "	radical*	irradiation
II A	nil	radical*	irradiation & adjuvant systemic chemotherapy
		wide	" " "
II B	irradiation & chemotherapy, systemic and/or infusion	wide or radical*	irradiation & adjuvant systemic chemotherapy

* Preferred surgical procedure

Radical resection of bone sarcomas can seldom be realized because possibilities of excision of entire long bones are rare. Almost all resections are wide, ie, through the bone at a distance of 5 cm at least from the edge of the primary tumor. Skip lesions are uncommon. Tumor breakout is managed by wide resection of the involved soft tissue en bloc with the osseous lesion. At the distal femur or proximal tibia this often involves removing most of the surrounding muscle, collateral ligaments, capsule, and the cruciate ligaments. Sarcomas will extend through the cruciates and also through the epiphyseal plate. Occasionally, erosion of the subchondral bone will result in a pathologic fracture and intra-articular spread. Similarly, sarcomas of the proximal fibula may extend through the proximal tibiofibular joint attachments to the tibia and vice versa.

Preoperatively, with high-grade bone sarcomas we usually administer at least one course of intra-arterial **cisplatin.** We do not use radiation (Chueng et al, 1981; Rosen et al, 1982).

Finally, a word about biopsy. Unless the lesion is small and obviously benign, biopsy should be incisional. The approach, including the skin incision, should be planned so that subsequent definitive surgery will not be compromised. Poorly executed biopsies are a major cause of problems in sarcoma surgery. They can cause wound healing difficulties and may necessitate a more extensive surgical procedure.

References

1. Bowden, L. and Booher, R.J.: The principles and techniques of resection of soft parts for sarcoma. Surgery 44, 963, 1958.

2. Chuang, V.P., Wallace, S., Benjamin, R.S., Jaffe, N., Ayala, A., Murray, J., Zornoza, J., Patt, Y., Mavligit, G., Charnsangavej, C., and Soo, C-S.: The therapy of osteosarcoma by intraarterial cis-platinum and limb preservation. Cardiovasc. Intervent. Radiol. 4, 229, 1981.

3. Enneking, W.F. and Dunham, W.K.: Resection and reconstruction for primary neoplasms involving the innominate bone. J. Bone Joint Surg. 60A, 731, 1978.

4. Enneking, W.F., Spanier, S.S. and Goodman, M.A.: The surgical staging of musculoskeletal sarcoma. J. Bone Joint Surg. 62A, 1027, 1980.

5. Enneking, W.F., Spanier, S.S. and Malawer, M.M.: The effect of the anatomic settling on the results of surgical procedures for soft parts sarcoma of the thigh. Cancer 47, 1005, 1981.

6. Rosen, G., Caparros, B., Huvos, A.G., et al: Preoperative chemotherapy for osteogenic sarcoma: Selection of postoperative adjuvant chemotherapy based on the response of the primary tumor to preoperative chemotherapy. Cancer 49, 1221, 1982.

7. Simon, M.A. and Enneking, W.F.: The management of soft tissue sarcomas of the extremities. J. Bone Joint Surg. 58A, 317, 1976.

8. Suit, H.D., Russell, W.O. and Martin, R.G.: Management of patients with sarcoma of soft tissue in an extremity. Cancer 31, 1247, 1973.

Treatment of Pathologic Fractures

J. M. LANE, R. R. McCORMACK, N. SUNDARESAN, B. HURSON, and P. BOLAND

The purpose of this article is to discuss the diagnosis and treatment of pathologic fractures secondary to metastatic disease. Fractures related to primary tumors are discussed in depth under their specific sections in other papers of this volume.

Metastatic disease is extremely common in the skeleton. Approximately 40% of all individuals at some time in their life will develop cancer. Depending on the etiology, approximately 50% of the tumors will not respond to primary care and lead to ultimate death. Diffuse metastasis in terminal cancer has been identified by Jaffe to occur in at least 70% of these patients who will have significant skeletal metastases (Jaffe, 1958).

The most common forms of metastatic disease leading to pathologic fractures are breast cancer, hypernephroma, lung cancer and multiple myeloma. These tumors account for 80% of the skeletal metastases leading to bone infraction (Galasko, 1974; Lane et al, 1980). Fracture incidence is not directly related to the frequency of metastases to the skeleton. This discrepancy results from the variable longevity of individuals with different forms of metastatic tumor. Lung cancer, which frequently metastasizes to bone, leads to a rapid death of the individual long before the bone can be significantly weakened and compromised. Breast cancer and hypernephroma, which can be extremely insidious, result in a higher incidence of pathologic fractures. In fact, breast cancer accounts for well over 80% of pathologic fractures of the hip in women at a time when lung cancer has achieved a higher mortality rate for women. Studies actually suggest that one out of 20 women with a newly developing breast cancer will have a pathologic fracture about the hip during the course of the disease process.

The distribution of skeletal metastases is in part related to the primary tumor and site (Galasko, 1972; 1974; Thomson, 1924). Generally, the metastases occur most frequently in the vertebral body, pelvic bone, skull, and the long bone proximal to the elbow and knee. All bones have been reported to be affected by metastases including such unusual locations as the patella and digits. Usually, lesions are more commonly found in areas rich in red marrow, probably related to the increased vasculature. Anatomic sites close to the drainage of specialized tumors have a higher incidence. Examples would be the pelvic rim in relationship to prostate cancer and the ribs in relationship to mammary carcinoma.

Current Concepts of Diagnosis and Treatment
of Bone and Soft Tissue Tumors.

The pathophysiology of skeletal metastases is indeed complex (Springfield, 1982). Initially, the primary tumor must give rise to separated metastatic tumors or microclumps. These are then carried either through the venous or the lymphatic drainage to the general circulation. To achieve distal seating, they must be capable of passing through the lung, either primarily or as a secondary focus from a lung metastasis. In the systemic circulation, they have to reach a distant capillary, pass through its wall, achieve a root in distant tissue and reestablish a nutrient source of blood supply to maintain their viability.

Skeletal metastases can present as either lytic (bone resorbing) or blastic (bone forming). Many tumors can present as a mixed blastic lytic pattern and this occurs especially in over 60% of metastatic mammary carcinoma (Galasko, 1972; Jaffe, 1958). Purely lytic metastases are usually associated in descending order of frequency with lung, kidney, breast, thyroid, gastrointestinal tract, and neuroblastoma. Blastic metastases occur most frequently with prostate, breast, bladder and stomach tumors. Pathologic fractures are more commonly associated with lytic lesions in contrast to blastic lesions. Prostate cancer, which is over 97% associated with blastic lesions, has a low incidence of pathologic fractures.

The diagnosis of bone metastases causing pathologic fracture is based on a careful history identifying usually preexisting cancer, radiographic and scan evidence of other metastases (bone, liver, lung, etc.), and tissue diagnosis. Most metastases will occur within two years of the primary tumor; however, periods over 10 years have been reported for breast and renal carcinoma. Metastatic disease would be suspected in case of multiple skeletal lesions in an individual over the age of 30. Isolated lesions in the younger age groups should be suspected of being a primary tumor. Although solitary lesions are statistically more frequently metastatic in the elderly, primary lesions can exist. The most sensitive diagnostic method for revealing metastases is the technitium diphosphonate bone scan (Galasko, 1972). Bone scans are in fact approximately three to five times more effective in identifying tumors than the skeletal survey. At least 30% of bone must be destroyed before routine roentgenographs can identify osseous abnormalities.

Fractures occurring in persons past age 30 without a clear history of trauma should be studied carefully. Lysis unaccounted for or periosteal elevation within five days of the fracture, recent history of cancer, abnormalities seen on bone scan and skeletal lesions could be suggestive of a pathologic fracture.

Vertebral fractures secondary to osteoporosis usually occur primarily at the apex of the kyphus and at the transition of the thoracolumbar spine (Saville, 1973). The pedicle is generally not involved. Consequently, vertebral fractures in the upper thoracic area (T-1 through T-4) and/or with pedicle involvement should be suspect of metastatic tumor.

Clinical symptoms associated with pathologic fractures frequently include pain before recognition of a slightly displaced fracture. Night pain is most commonly reported and is frequently unaffected by position. Two classes of fractures are identified in the area of metastatic disease:

impending and frank fractures (Beals, 1971; Harrington, 1981; Lane et al, 1980; Sherry et al, 1982). Impending fractures are lesions in which osseous destruction has occurred, compromising the integrity of the bone and approaching the point where conservative measures may not lead to successful healing. Frank pathologic fractures have passed that point and lead to obvious destruction of the osseous structure. Impending fractures often are difficult to determine and are in a clinical gray zone. Impending fractures often have the hallmark of an intramedullary defect, at least 50% of the cross-sectional diameter, or a cortical defect equal in length to the cross-sectional diameter, in which both sites caused continued pain in spite of radiotherapy and/or conservative protection (Lane et al, 1980). Borderline impending fractures may respond to radiotherapy for a transient period and then give rise to a frank fracture several months later. The treatment of impending fractures is complex and will be discussed below.

The goals of treating pathologic fractures include diagnosis of unrecognized metastatic tumor, stabilization of the whole bone and/or joint rather than treatment of a limited portion of a multiple involved bone, replacement of all defective bone, decreased hospitalization for the patient, pain relief, and restitution of significant function (Harrington et al, 1976; Lane et al, 1980; Sherry et al, 1982). The attainment of such goals depends on a realistic evaluation of the patient's potential for survival, degree of patient dependence, evaluation of extent of pain and determination of social support mechanism. Typically, patients with hip fractures have a mean projected survival between five to ten months depending on the type of tumor (Harrington et al, 1976; Lane et al, 1980). In metastatic breast cancer and hypernephroma, 20% of the patients are expected to live one year and 10% two years. Conversely, 10-20% of patients presenting with pathologic fractures may not survive two months. Lytic prostate cancer and lung cancer have extremely poor survival rates bordering on an average of three months in contrast to the more insidious nature of breast cancer and renal carcinoma.

Adjuvant support in the treatment of pathologic fractures centers on chemotherapy and radiotherapy (Schocker and Brady, 1982). Chemotherapeutic failures can lead to osseous destruction; in such cases, one should switch to a more effective mode of therapy. Similarly, radiation at the level of 2,000 rads may allow a sensitive tumor to heal in an impending fracture. It should be recognized that in a number of series radiation in the order of 2,000 rads has frequently led to a 20% local reemergence of the tumor, while radiation over 3,000 rads leads to less than 10% recurrence. Unfortunately, radiation is a serious inhibitor of osseous repair and is susceptible to delayed fatigue fractures of a borderline impending lesion. Early bone metastases are frequently treated with radiation in the order of 1,500-2,500 rads. Most of these lesions are controlled sufficiently during the course of the patient's life and never lead to pathologic fractures. Most orthopedists will be forced to treat pathologic fractures after radiation which can lead to destruction of normal tissue and increased fibrosis.

General principles of treating pathologic fractures should include a careful preoperative medical evaluation as to potential to survive surgery. Pulmonary and liver disease may have a critical effect on the

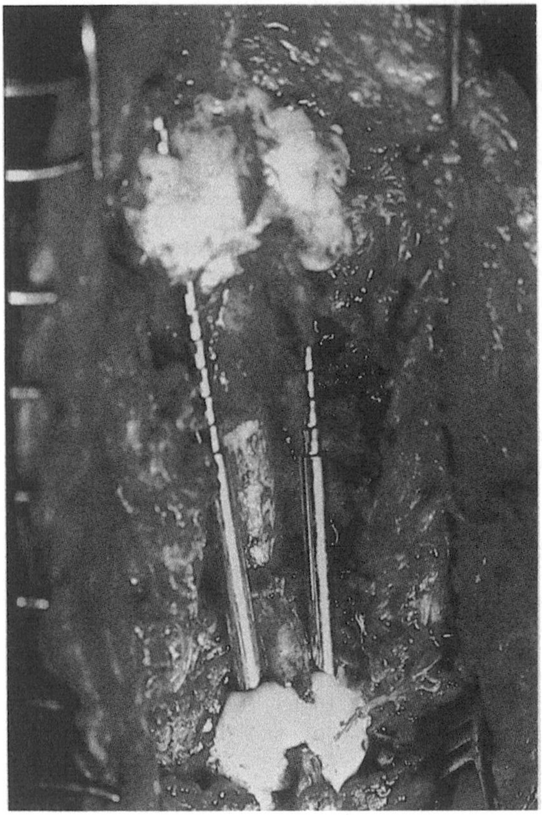

a

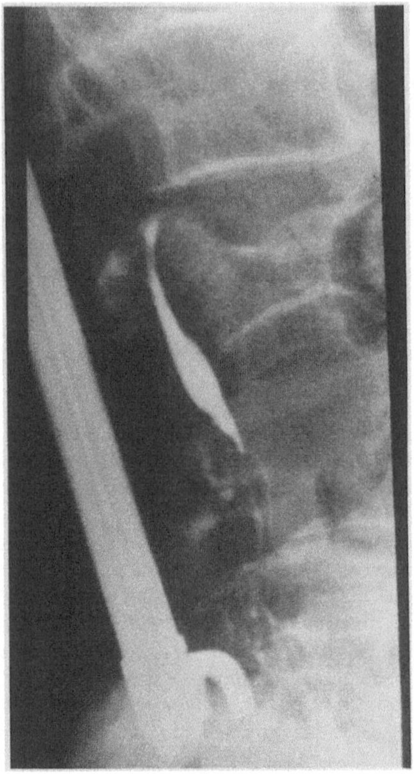

b

Fig. 1. Posterior decompression and Harrington rod stabilization of lumbar pathologic fracture. a) Photograph of operative field demonstrating thorough posterior decompression down to dura including resection. Spine has been stabilized by 2 Harrington rods inserted 2 levels above and below the laminectomy and secured with PMMA. b) Lateral radiograph demonstrating passage of myelogram dye and correction of lumbar gibbus with Harrington rods.

type of anesthesia. Significant blood transfusions should be prepared, particularly in case of hypernephroma (Lane et al, 1980). Large access lines should be established and a central venous catheter and/or arterial lines can be quite helpful in a debilitated patient. Foley catheters should permit gauging the effect of renal function, especially during a long procedure when copious blood may be lost. Arteriogram of the involved metastatic lesion before treatment and selective arterial thrombosis has been used experimentally and may be helpful such as in the treatment associated with hypernephromas.

Generally speaking, the principles of orthopedic management should include stabilization of whole bone, internal fixation and use of methylmethacrylate. Fractures will usually not heal and fixation must outlive the patient.

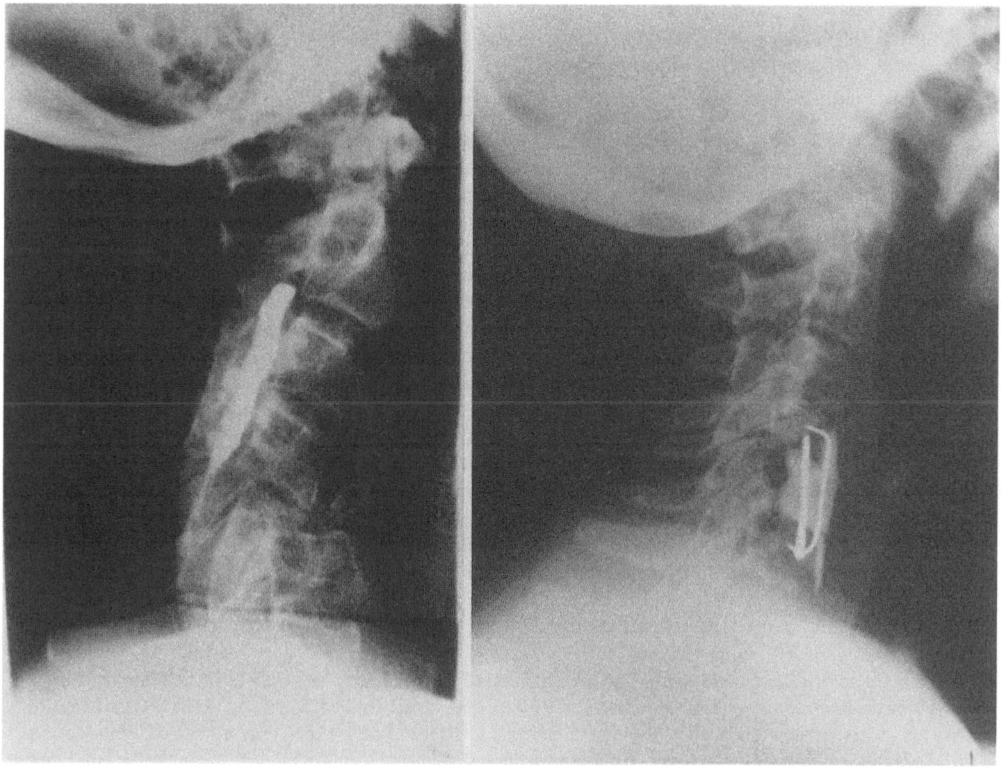

a b

Fig. 2. Anterior decompression and stabilization of cervical pathologic fracture. a) Lateral radiograph demonstrating pathologic crush fracture with complete block at C5. b) Lateral radiograph demonstrating decompression and realignment of C5 with PMMA and wires.

Vertebral fractures are the most common site of skeletal metastases (Jaffe, 1958). Therapy is based on the treatment of cord compression. Causes of cord compression include pressure secondary to enlarging extradural deposit, pressure due to pathologic fracture dislocation, severe spinal angulation following vertebral collapse and, rarely, due to intradural metastases (Barron, 1959; Boland et al, 1982). It is important to determine exactly the extent and location of the tumor. This can be readily performed by myelogram and CT scan. Most of the patients' lesions are located in the vertebral body with anterior extradural pressure. The Memorial Sloan-Kettering Cancer Center has had wide experience in treating such patients with posterior laminectomy, fasciectomy, and stabilization with Harrington rods and/or, more recently, by using anterior decompression with stabilization by methylmethacrylate, wires and bone grafting (Boland, 1982; Sundaresan et al, 1981)(Figs. 1 and 2). Posterior decompression (Fig. 1a, 1b) has been associated with complete pain relief but stabilization of neurologic function with only moderate improvement. However, the 30% complication rate is usually related to skin necrosis secondary to operating in an area that has been radiated. In cases where radiation has been withheld after surgery, the complication rate was markedly decreased.

The posterior approach has been most useful in the treatment of patients with multiple root involvement, posteriorly located tumor with an intact vertebral body, or combined with an anterior approach. The anterior vertebral decompression (Fig. 2a, 2b) leads to removal of the tumor where it lies with adequate exposure. The complication rate is extremely small and adequate decompression of the dura can be achieved. Several adjacent levels can be decompressed in one maneuver. Stabilization is easily achieved with methylmethacrylate and wire fixation. Pain relief has been dramatic with this procedure, and the anterior decompression has led to a higher yield of regaining function following surgery. Anterior decompression after laminectomy will not provide adequate long-term stabilization. Consequently, a second posterior fixation procedure is necessary. It is a practice at Memorial Sloan-Kettering Cancer Center to study anteriorly decompressed patients who survive six months to determine whether it is advisable to carry out a delayed posterior stabilization at a time when the effects of radiation have been partially mitigated. Our recent experience (Boland et al, 1982; Sundaresan et al, 1981) is in general agreement with Harrington (1981) and suggests that Gilbert's preference for conservative radiotherapy (Springfield, 1982) alone may be outmoded by current surgical techniques.

Humeral lesions can be easily stabilized with intramedullary rods and methylmethacrylate (Harrington, 1976; Sim and Pritchard, 1982). Destruction of the humeral head may not permit fixation with intramedullary rods and necessitates use of an endoprosthesis particularly of the Neer variety. Patients who have had prior radiation frequently cannot achieve abduction beyond 90° due to postradiation fibrosis. Recently, at our Center, we have been using cast bracing of nondominant arms in individuals with breast cancer and pathologic diaphyseal fractures with excellent symptomatic relief and slow but definite healing. Individuals, however, who use the upper extremities for weight bearing (spinal or lower extremity involvement) require surgery.

Hip fractures can occur as subcapital, intertrochanteric and subtrochanteric (Lane et al, 1980). Investigators have used hip screws with methylmethacrylate, particularly for limited intertrochanteric fractures (Harrington et al, 1976). Zickel has advocated the use of his device for subtrochanteric fractures and has a series of over 60 patients in whom cement was not necessary to obtain fixation (Zickel and Mouradean, 1976). At Memorial Sloan-Kettering Cancer Center, we have adopted the principle of using a long-stem endoprosthesis (Lane et al, 1980). The advantages are that the complete bone can be stabilized from the femoral head to the distant diaphysis with methylmethacrylate so that long-term survivors will not require a second operation in the same bone. We have noted a small but significant number of individuals previously treated with hip screws or Zickel devices in whom the tumor has outgrown the device leading to difficulty in performing a secondary reconstruction, especially when the device is cemented. Initially, Austin-Moore custom devices were used, but recently bipolar devices have been used (Fig. 3). These devices appear to provide better pain relief with a low incidence of secondary

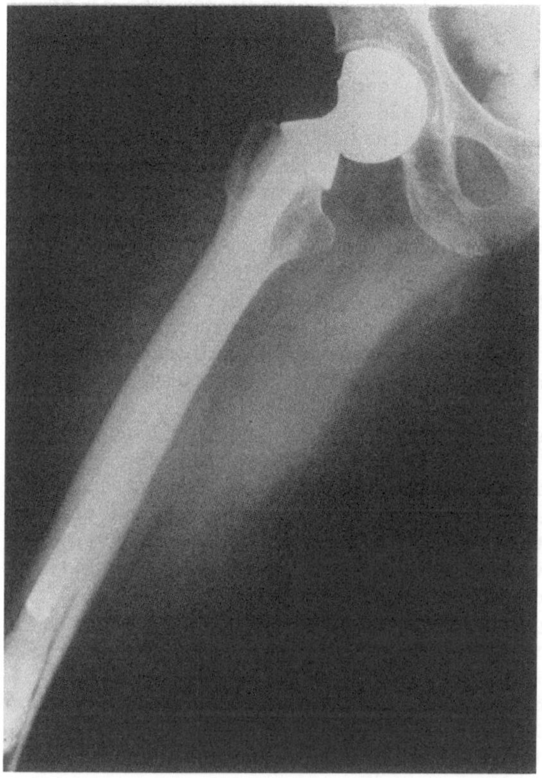

Fig. 3. Long-stem bipolar endoprosthetic replacement of a pathologic fracture at base of neck of the right hip from metastatic breast cancer.

dislocation (less than 2% in 100 consecutive cases). Total hip arthroplasty has been performed in selected patients in whom the principal area of deficiency was in the acetabular side. It is recognized, however, that the subchondral bone may be the strongest remaining bone and at this point none of the 100 bipolar devices has had to be revised to carry out a total hip operation inspite of the frequent occurrence of acetabular involvement. Frank fractures and destruction of the acetabular side can be stabilized using acetabular rings, extra cement, wiring, and occasionally bone grafting as advocated by Harrington (1981). The use of long-stem devices and pressurized cement in a more liquid state has been associated with significant hypoxemia in severely debilitated patients. These individuals should be carefully monitored throughout the course of surgery and endotrachial oxygen support should be administered during the postoperative course to prevent pulmonary failure.

Femoral shaft fractures are most easily handled with intramedullar rod devices augmented with cement (Harrington et al, 1976; Lane et al, 1980; Zickel and Mouradean, 1976). If the patient has a slowly progressive tumor and suggestion of early lesions in the proximal femur, it would be wise not to cement in the rod since it would be difficult to remove the device if a conversion to an endoprosthesis became necessary. Fluted intramedullary rods without cement have been quite suitable in providing support even in the presence of a tumor. It can be easily removed if an endoprosthesis has to be used later.

The treatment of pathologic fractures should lead to a 95% incidence of significant pain relief (Harrington et al, 1976; Lane et al, 1980). Function, however, may not be completely regained (Lane et al, 1980). In a careful study of patients treated with endoprosthesis for pathologic fractures of the hip, it has been noted that if patients were confined to bed for two months before surgery, only 50% of them could achieve home ambulation. Conversely, in patients who had been able to walk up to several weeks before the fracture, anywhere from 75 to 95% became at least ambulatory at home although they could not go outside. Upper humeral fractures usually lead to excellent pain relief and return to normal function. Range of motion, however, is frequently compromised, mostly due to radiation fibrosis and capsulitis about the shoulder.

While treating pathologic fractures, representative tissue samples should be sent for pathologic examination (Sherry et al, 1982). They may show clearer diagnosis of an unknown primary tumor, help the oncologist in planning subsequent chemotherapy, and help the radiologist to judge the efficacy of presurgical radiation treatment. During surgical intervention, as much tumor should be removed as possible. Cryosurgery has been performed by Marcove* and others to extend local control. In the presence of copious bleeding, the complete involved area should be surgically excised and reconstruction carried out with an endoprosthesis.

In hypernephromas, hemostasis is notoriously difficult to obtain. Ideally, in these tumors, the vascular supply to the lesion should be identified before attending to the fracture itself.

Finally, the rehabilitation of these patients is critical. Most patients have multiple lesions and it is wise to explain to them that subsequent fractures of other bones may occur. To hold back ambulation after repairing the hip because of fear that a fracture of the other leg may occur is defeating the major goal of intervention.

Proper patient selection and appropriate stabilization of the bone with a large long device and methylmethacrylate should result in major pain relief, restoration of function, and a return of the patient to his home and family.

----

* Personal communication

References

1. Barron, K.D., Herano, A., Araki, S. and Ferry, R.D.: Experiences with metastatic neoplasms involving the spinal cord. Neurology 9, 91, 1959.

2. Beals, R.K., Lawton, G.D. and Snell, W.E.: Prophylactic internal fixation of the femur in metastatic breast cancer. Cancer 28, 1350, 1971.

3. Boland, P.J., Lane, J.M. and Sundaresan, N.: Metastatic disease of the spine. Clin. Orthop. Rel. Res. 169, 95, 1982.

4. Galasko, C.B.: Skeletal metastases and mammary cancer. Ann. Royal Coll. Surg. (Eng.) 50, 3, 1972.

5. Galasko, C.B.: Pathological fracture secondary to metastatic cancer. J. Coll. Surg. (Edinb.) 19, 351, 1974.

6. Gilbert, R.N., Kim, J.M. and Posner, J.B.: Epidural spinal cord compression from metastatic tumor. Diagnosis and treatment. Ann. Neur. 3, 40, 1978.

7. Harrington, K.D.: The use of methylmethacrylate for vertebral body replacement and anterior stabilization of pathological fracture dislocation of the spine due to metastatic disease. J. Bone Joint Surg. 63A, 36, 1981.

8. Harrington, K.D.: The management of acetabular insufficiency secondary to metastatic malignant disease. J. Bone Joint Surg. 63A, 653, 1981.

9. Harrington, K.D., Sim, F.H., Enis, J.E., Johnston, J.C., Dick, H.M. and Gristina, A.G.: Methylmethacrylate as an adjunct in internal fixation of pathological fractures. J. Bone Joint Surg. 58A, 1047, 1976.

10. Jaffe, W.L.: Tumors and Tumeroid Condition of the Bone and Joints. Lila and Febeger, Philadelphia, 1958.

11. Lane, J.M., Sculco, T.P., and Zolan, S.: Treatment of pathological fractures of the hip by endoprosthetic replacement. J. Bone Joint Surg. 62A, 954, 1980.

12. Saville, P.D.: The syndrome of spinal osteoporosis. Clin. Endocrin. Metab. 2, 177, 1973.

13. Schocker, J.D and Brady, L.W.: Radiation therapy for bone metastasis. Clin. Orthop. Rel. Res. 169, 38, 1982.

14. Sherry, H.S., Levy, R.N. and Siffert, R.S.: Metastatic disease of bone in orthopaedic surgery. Clin. Orthop. Rel. Res. 169, 44, 1982.

15. Sim, F.H., and Pritchard, D.J.: Metastatic disease in the upper extremity. Clin. Orthop. Rel. Res. 169, 83, 1982.

16. Springfield, D.S.: Mechanisms of metastasis. Clin. Orthop. Rel. Res. 169, 15, 1982.

17. Sundaresan, N., Galicich, J.H., Lane, J.M., and Greenberg, M.S.: Treatment of odontoid fractures in cancer patients. J. Neurosurg. 54, 187, 1981.

18. Thompson, J.E., and Keiller, V.M.: Multiple skeletal metastasis from cancer of the breast. Surg. Gynecol. Obstet. 38, 369, 1924.

19. Zickel, R.F. and Mouradean, W.H.: Intramedullary fixation of the pathological fractures and lesions of the subtrochanteric region of the femur. J. Bone Joint Surg. 58A, 1061, 1976.

Fibrous Dysplasia

PART I: PATHOLOGY

M.E.A. Bell

The characteristic histopathologic findings of monostotic and polyostotic fibrous dysplasia are similar.

Gross Pathologic Findings

Lesions totally excised are well defined, but unencapsulated, delineated by a thin overlying cortex or only periosteum. Cut surface shows dense gray-white fibrous connective tissue with gritty foci. Numbers of cysts of different sizes vary, with serous fluid within the cavities. Some cases have occasional foci of cartilage. Following a fracture, hemorrhages or hemosiderin deposition may be observed, giving the lesion a brown color.

Microscopic Appearance

Lesions consist predominantly of spindle shaped fibroblasts, sometimes in a whorled or storiform pattern, in a dense collagenous matrix with variable vascularity. In some cases, this is the only type of tissue present. Generally, irregular trabeculae of immature woven bone and/or osteoid are seen, without osteoblastic activity (Fig. 1). Polarized light shows irregularly distributed birefringent fibers characteristic of woven bone (Fig. 2). Areas of cartilage are seen most commonly in cases of polyostotic fibrous dysplasia suggesting an association with enchondromatosis (Sanerkin and Watt, 1981). Cartilage may also be present in lesions with a fracture as part of the callus formation.

In areas of degeneration, the stroma may have a myxoid appearance. Cysts may be lined by stromal cells or by multinucleated giant cells suggesting a possible aneurysmal bone cyst. Some areas may contain foam cells which would require differentiation from an xanthofibroma, or numerous multinucleated giant cells requiring differentiation from a giant cell tumor (Fig. 3).

A variant of fibrous dysplasia, termed ossifying fibroma or osteofibrous dysplasia (Campanacci, 1976; Campbell and Hawk, 1982), usually involves the cortex, rather than the medulla, of the tibia or the fibula. These lesions resemble fibrous dysplasia except for the presence of osteoblasts rimming many of the bony trabeculae. The trabeculae near the centre of

Current Concepts of Diagnosis and Treatment
of Bone and Soft Tissue Tumors.

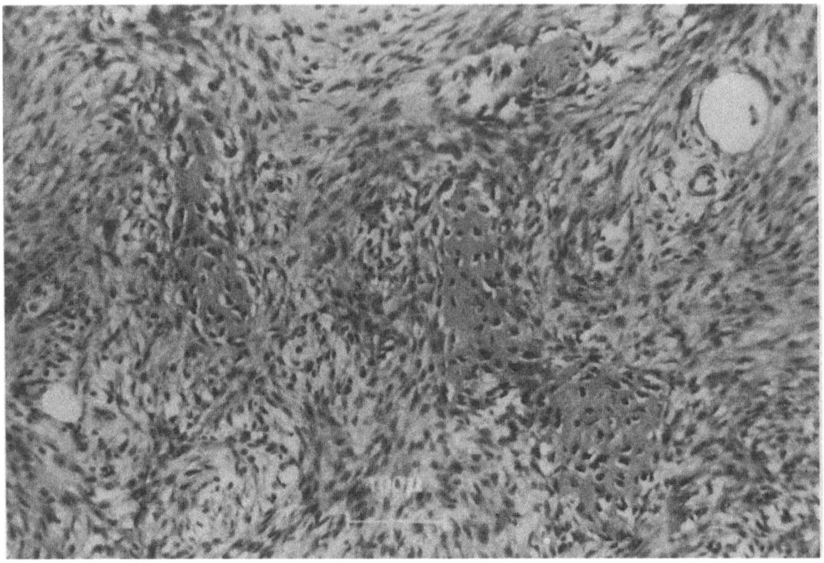

Fig. 1. This photomicrograph shows a typical fibrous dysplasia with irregular trabeculae of bone, without osteoblastic rimming, surrounded by fibrous connective tissue.

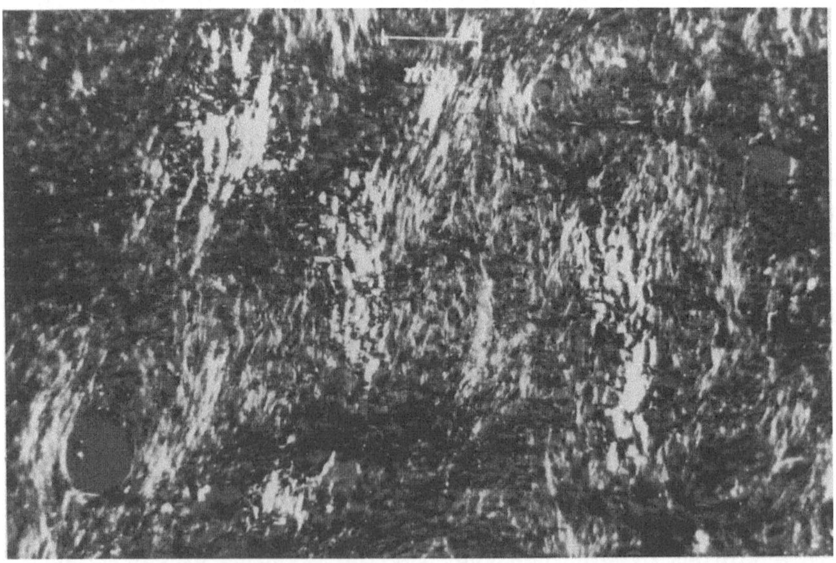

Fig. 2. This is the same field as in Fig. 1 under polarized light showing cross-hatch pattern of the woven bone trabeculae blending into the fibrous stroma.

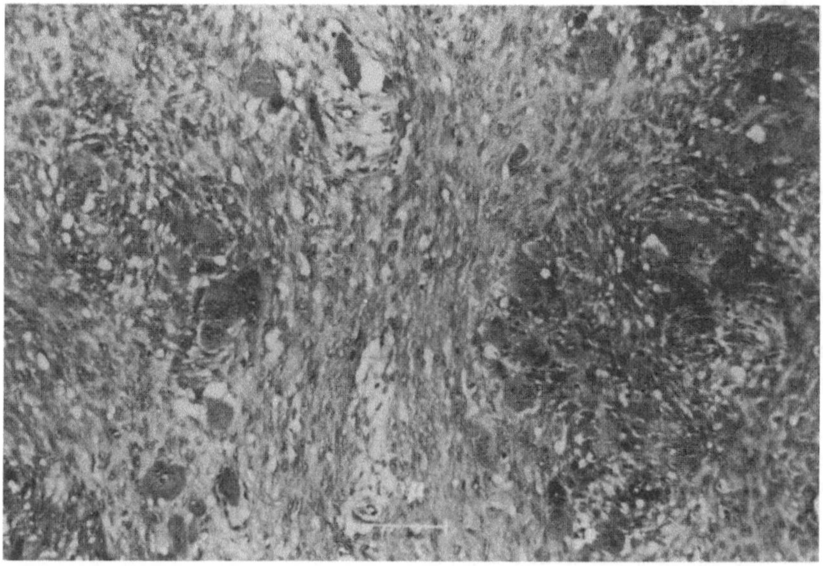

Fig. 3. An area with many multinucleated giant cells and hemorrhage, suggesting a possible giant cell tumor.

the lesion have a woven pattern and those near the cortex have a more mature lamellar structure.

Malignant transformation described in the literature has consisted of an associated osteogenic sarcoma, fibrosarcoma, giant cell sarcoma (Schwartz and Alpert, 1964), or chondrosarcoma (DeSmet et al, 1981; Huvos et al, 1972).

References

1. Campanacci, M.: Osteofibrous dysplasia of long bones: A new clinical entity. Ital. J. Orthop. Trauma 2, 221, 1976.

2. Campbell, C.J., and Hawk, T.: A variant of fibrous dysplasia (osteofibrous dysplasia). J. Bone Joint Surg. 64(A), 231, 1982.

3. DeSmet, A.A., Travers, H., and Neff, J.R.: Chondrosarcoma occurring in a patient with polyostotic fibrous dysplasia. Skel. Radiol. 7, 197, 1981.

4. Huvos, A.G., Higgenbothem, N.L., and Miller, T.R.: Bone sarcomas arising in fibrous dysplasia. J. Bone Joint Surg. 54(A), 1047, 1972.

5. Sanerkin, N.A., and Watt, I.: Enchondromata with annular calcification in association with fibrous dysplasia. Brit. J. Radiol. 54, 1027, 1981.

6. Schwartz, D.T., and Alpert, M.: The malignant transformation of fibrous dysplasia. Am. J. Med. Sc. 241, 35, 1964.

PART II: CLINICAL ASPECTS

E.R. Mindell

Symptomatology

Fibrous dysplasia is a relatively common abnormality of bone which may present problems in diagnosis and treatment. It involves one or several bones and is more common in women than in men. It usually becomes manifest in childhood or adolescence, but may, in fact, have developed on a congenital basis. Some observers feel that the lesions result from an anomaly in bone development (Dahlin, 1981). Brownish patches of skin pigmentation are often seen in these patients, particularly when widespread bony disease is present. The skin is not elevated and has normal texture. However, abnormal skin pigmentation is not always found even in patients with marked skeletal changes (Jaffe, 1958). When a combination of polyostotic fibrous dysplasia, cutaneous pigmentation, and signs of endocrine abnormalities (such as précocious puberty in girls) are observed, the term " Albright's Syndrome " is often applied.

When only one bone is affected, the patient's complaints are usually mild and may consist of pain and deformity. Not infrequently, however, the lesions are asymptomatic and are often detected radiographically by chance, sometimes as an incidental finding. A bone scan is usually positive. Occasionally, the radiographic appearance suggests a more serious lesion and biopsy is contemplated. Usually, however, the radiogram is characteristic enough to establish a diagnosis, obviating a biopsy. The lesion itself has a homogenous appearance, often described as "ground glass", and is sharply delineated from the adjacent normal bone. The medullary canal is distended and the cortices, although not completely penetrated, are eroded from within. When several lesions are present, bones in the same extremity are frequently involved. Radiographically, the lesions vary in size from a few centimeters to involving most of the bone.

Diagnostic and Clinical Features

When the lesions are widespread and in the lower extremities, serious deformity and crippling disability may occur. Affected bones may be expanded, bowed, shortened or otherwise deformed. Such bones are often susceptible to spontaneous fracture. Deformities such as coxa vara and femoral bowing are not infrequent. When the spine is involved, a significant kyphoscoliosis may develop. The facial bones are sometimes involved, occasionally leading to grotesque deformities and secondary changes such as exophthalmus. The biochemical findings in the blood and urine of patients with fibrous dysplasia are normal, although a slight elevation of the serum alkaline phosphatase level is sometimes noted in patients with widespread disease (Jaffe, 1958).

The differential diagnosis in fibrous dysplasia includes simple bone cyst from which it is easily distinguished radiologically. Nonossifying fibroma is sometimes confused radiologically with fibrous dysplasia, particularly when the nonossifying fibroma is large. Fibrous osteoma or ossifying fibroma of jaw bones in the skull are considered to be a manifestation of fibrous dysplasia.

The only condition that sometimes creates a problem in differential diagnosis from widespread fibrous dysplasia is enchondromatosis or Ollier's disease (Jaffe, 1958). Radiographs in Ollier's disease, however, demonstrate localized rarefactions in the shaft of the phalanges and in the metacarpal bones. The bulging contours of these bones suggest enchondromatosis rather than fibrous dysplasia. The cartilage found occasionally in fibrous dysplasia is deposited within fibrous tissue while, in contrast, the cartilage in enchondromatosis is located within bone.

Clinical course

The period of active progression of skeletal involvement usually ends at the onset of adult life. Bones not involved by then generally do not become involved later on. However, one of the involved bones may show progressive deformity even into adult life. Some lesions that were stable for years can subsequently take on the appearance of growth activity although the lesion itself is only undergoing degenerative changes.

Sarcomatous degeneration can occur in fibrous dysplasia. This has been reported in cases of fibrous dysplasia treated by radiation but may also occur spontaneously (Huvos et al, 1972). Malignant degeneration in a fibrous dysplasia lesion, although rare, should be considered when a patient develops increasing symptoms or increased swelling in a lesion that had been quiescent for many years.

Treatment

Many lesions of fibrous dysplasia are completely asymptomatic and are probably never discovered.

The prognosis is quite good for most patients with fibrous dysplasia with the exception of patients with endocrine abnormalities noted in early childhood. Most patients present only a limited or moderate expression of the disorder and usually have a normal life expectancy.

Fibrous dysplasia lesions should be considered for surgical treatment if they are painful, predispose to fracture, or have led to significant deformity and, as sometimes seen in facial bones, have led to disfigurement. Curettage and bone grafting of a lower extremity lesion is occasionally indicated where fracture and subsequent deformity is imminent. These procedures may be worthwhile in such locations as the femoral neck in a young child to prevent coxa vara. Corrective osteotomies of long bones are sometimes necessary in patients who have developed significant deformities, particularly in the femur. For instance, high femoral osteotomy for coxa vara often leads to significant functional improvement. Recurrence of the lesion, however, and sometimes deformity following surgical treatment may occur in the young child.

Malignant transformation of a fibrous dysplasia lesion occurs rarely (Huvos et al, 1972). This complication must be recognized early and treated by adequate surgical removal, usually amputation. Radiotherapy is contraindicated as it is of little value and introduces the hazard of inducing malignant change.

The clinical manifestations of fibrous dysplasia are varied. A thorough understanding of its natural history is necessary to plan appropriate treatment for each patient.

References

1. Dahlin, David C.: Bone Tumors. 3rd Ed. C.C. Thomas, Springfield, 1981.

2. Jaffe, Henry: Tumors and Tumorous Conditions of the Bones and Joints. Lea and Febiger, Philadelphia, 1958.

3. Huvos, A., Higgenbothem, N., and Miller, T.: Bone sarcomas arising in fibrous dysplasia. J. Bone Joint Surg. 54(A), 1047, 1972.

Surgical Management of Osteosarcoma at the Mayo Clinic

D. J. PRITCHARD

Introduction

During the past years, important new developments have occurred in the management of patients with osteosarcoma. These developments include a better understanding of the histologic criteria for diagnosis, better staging procedures, improved surgical techniques, and the use of adjuvant treatment.

Histologic Criteria for the Classification of Osteosarcoma

Different varieties of osteosarcoma must be recognized and classified separately from conventional osteosarcomas if meaningful therapeutic results are to be derived. These histologic variants are now generally recognized; however, clinical reports do not always indicate whether these variants are included with so-called classic osteosarcoma. Large differences are evident in the clinical behavior of some of these histologic variants. For example, osteosarcoma of the jaw, low-grade central osteosarcoma, and parosteal osteosarcoma have relatively good prognoses whereas the telangiectatic variety of osteosarcoma has a particularly poor prognosis. Hence, if a clinical series contains patients with the telangiectatic variety, the results might be worse. Conversely, if the clinical study contains many patients whose tumors consist of histologic variants with a better prognosis, an unwarranted better survival picture would be implied (Dahlin and Unni, 1977).

Staging Procedures

The clinical staging of osteosarcoma has been improved. In 1970, my colleagues and I began the routine use of whole lung tomograms in the preoperative assessment of patients with sarcoma and found that we could identify some patients with pulmonary metastatic lesions that were not previously recognized. Since October 1975, we have used computerized tomography (CT) for the detection of pulmonary nodules. CT has proved to be more accurate than either plain radiographs or whole lung tomograms. To evaluate the role of CT scan of the chest in staging patients with biopsy-proved classic osteosarcoma, we studied the results of 125 CT scans performed on 50 patients. In 124 studies in which plain chest radiographs and CT were performed, 83 revealed the same number of nodules --

Current Concepts of Diagnosis and Treatment
of Bone and Soft Tissue Tumors.

from no nodule to two nodules. However, in 41 patients, more nodules were detected by CT. Of the patients in whom nodules were detected by CT scan, 22 had normal plain chest radiographs. Of 35 patients studied by both CT and whole lung tomograms, 22 had the same number of nodules, 11 had more, and two had fewer on CT scan. Of the 11 patients with more nodules on CT, five showed no nodules on whole lung tomography (Muhm and Pritchard, 1980).

In approximately 15% of our patients, a solitary lesion was found by chest radiographs or whole lung tomograms (or both), but multiple lesions were demonstrated by CT. In addition, the CT scan may reveal bilateral disease when other studies suggest only unilateral involvement. Thus, CT scan of the chest has resulted in improvement in the clinical staging of patients with osteosarcoma; this allows a more accurate selection of appropriate therapy and increases the accuracy of the selection of patients for inclusion into various protocols. Moreover, we may expect to see some improvement in survival statistics on this basis alone, that is, on the basis of more accurate staging, and therefore, more accurate selection of appropriate therapy.

We also use CT to evaluate the primary lesion. The CT scan is superior to radiography for visualization and evaluation of the extent of soft tissue involvement. This information is especially helpful in the preoperative evaluation of patients with pelvic or sacral lesions. The CT scan often provides valuable information about the status of adjacent major blood vessels and nerves and, in our hands, has been more reliable than arteriography which we now seldom use. CT is also useful in evaluating the extent of a tumor within the medullary canal and in the detection of recurrence of previously resected bone tumors. In such a situation, routine radiography is often of little value because the recurrent lesion frequently is either predominantly or entirely within the soft tissues, and CT is far superior to the radiograph for the detection of soft tissue recurrence.

Surgical Procedures

A number of new developments have occurred in surgical treatment. Generally, amputation is the procedure of choice for most primary lesions. The level of amputation for a lesion located in the distal femur has been the subject of considerable controversy. We have routinely performed thigh amputation rather than hip disarticulation; and in recent years, we have had only two patients with recurrences in the femoral stump. The general consensus among orthopedists who have a primary interest in the treatment of bone tumors is that thigh amputation is the treatment of choice as long as the amputation level is well above the most proximal extent of the tumor. At the Mayo Clinic, we try to achieve a margin of approximately 8 cm (Sim and Pritchard, 1978); that is, the desired level of amputation is 8 cm above the most proximal extent of the tumor. Since the functional results with thigh amputation are so much better than the functional results with hip disarticulation, we believe that proponents of disarticulation must establish that the procedure improves survival to justify its use. We recognize that skip areas do occasionally occur (Enneking and Kagan, 1975); however, the presence of skip areas is rarely a clinical problem. With longer survival times, we may see more local recurrences.

Whereas amputation is usually the accepted treatment of choice, we have two aims with our surgical treatment: first, to totally remove the tumor; second, to preserve as much function as possible. Both aims can occasionally be accomplished with local resection. With the development of new surgical techniques, particularly the new developments in prosthetic design, it has been possible, in some patients, to resect tumors and to reconstruct the limb and thereby accomplish as much by local resection as with amputation. However, this is an experimental approach that has not yet become an established procedure.

For some lesions, the tumor may be resected without the need for elaborate reconstruction. For example, a lesion in the wing of the ilium may be resectable without requiring a reconstruction procedure. Most lesions, however, occur in the major long bones, with about half of these tumors occurring in the region of the knee. If lesions located in the major long bones are resected, some reconstructive surgery will be necessary. In most patients, reconstruction is accomplished by doing an arthrodesis (fusion of a joint) by means of an internal prosthesis or an allograft implant.

Four areas are of major current interest at the Mayo Clinic in the surgical treatment of osteosarcoma:

1) resection of a lesion of the proximal femur and replacement with a custom-type total hip arthroplasty (Fig. 1);
2) resection of a tumor in the distal femor or proximal tibia and replacement with a modified Walldius hinged-type of prosthesis (Fig. 2);
3) resection of a section of the proximal humerus and replacement with a ceramic or metal prosthesis (Fig. 3); and
4) occasional resection of a diaphyseal lesion with replacement by a fiber metal implant (Fig. 4).

For tumors about the hip and knee, a number of factors must be considered before attempting resection. The pathologic changes in each case are different. Thus, for large tumors that are histologically classified as high grade and that have extensive soft tissue involvement, resection would not be considered. Patients with osteosarcoma are generally in the lower age group and, hence, the prosthetic devices will be subjected to considerably more stress. The prosthetic device will be much larger than devices conventionally used for patients with arthritic conditions, and since soft tissues as well as bone have to be resected, the joint mechanics will be considerably altered. After a resection procedure we feel obligated to do follow-up studies of these patients in our biomechanics laboratory to learn more about functional results and prosthetic design. In our biomechanics laboratory we can follow numerous measurements in the functional evaluation of the patient and relate these clinical findings to the technical aspects of surgery and the design of the prosthesis. To date, we have examined 120 patients in the biomechanics laboratory who have had internal prostheses implanted for various bone tumors (Fig. 5).

Finally, attention must be given to the total surgical management of these patients. We have adopted an aggressive attitude toward the surgical treatment of pulmonary metastatic lesions. We follow up our patients closely with monthly chest roentgenograms, and we perform CT on them

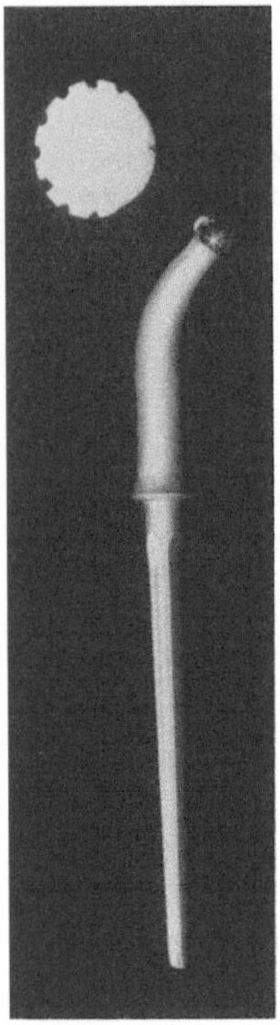

Fig. 1

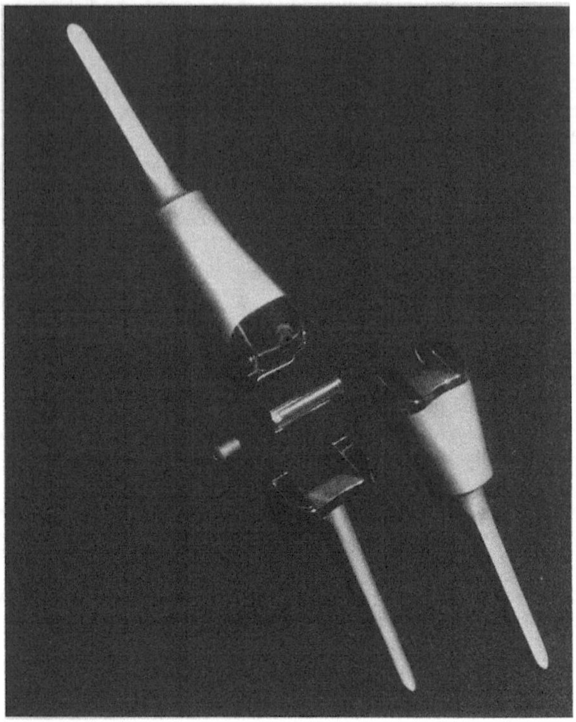

Fig. 2

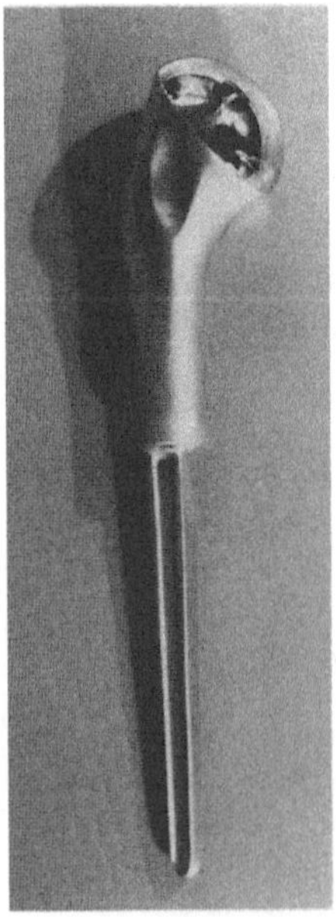

Fig. 3

Fig. 1. Custom-type total hip replacement.

Fig. 2. Custom-type Walldius total knee units for distal femur or proximal tibial replacements.

Fig. 3. Custom-type proximal humerus endoprosthesis.

Fig. 4

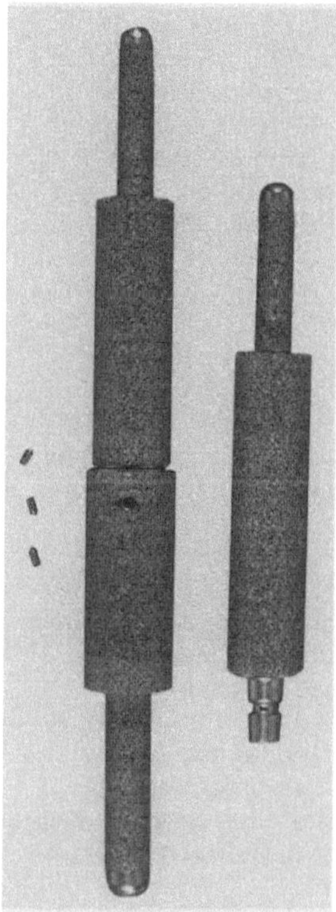

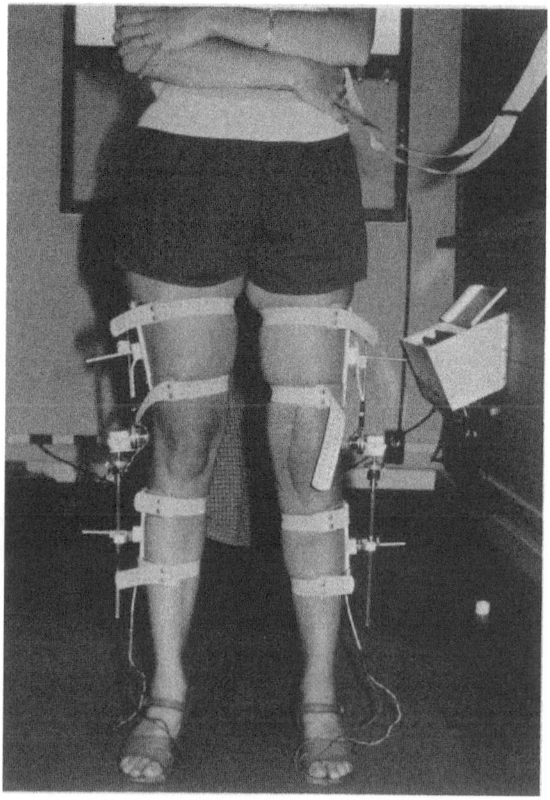

Fig. 5

Fig. 4. Fiber metal replacements for diaphyseal lesions.

Fig. 5. Patient undergoing evaluation in Biomechanics Gait Laboratory.

every three months. When a pulmonary metastatic lesion occurs, our thoracic surgery colleagues explore the chest and resect the tumor. Additional pulmonary resections are recommended should subsequent metastatic lesions occur. With this aggressive approach, about one-third of our patients with pulmonary disease have achieved long-term survival. (Spanos et al, 1976; Telander et al, 1978).

Adjuvant Treatment

Perhaps the greatest recent interest in the treatment of osteosarcoma has focused on the development of adjuvant treatment particularly chemotherapy. Numerous reports have indicated improvement in duration of survival associated with the use of various adjuvant modalities. Both disease-free survival and actual survival (that is, survival accompanied by relapse and subsequent control of the disease) are improved when comparisons are made with results achieved in the past. However, those reports are studies without concurrent untreated controls.

The problem of performing controlled studies is highlighted by our report of a study wherein patients, after amputation of the primary osteosarcoma, were randomized: 1) to receive prophylactic radiation to the lung fields or 2) to be simply observed for possible development of pulmonary metastasis. In that study, 26 patients were treated with a dose of 1,500 rad to both lung fields, and 27 patients were controls. The interval between surgical treatment and the development of pulmonary metastasis in the two groups was identical. Similarly, the duration of survival for the two groups was identical. It was of considerable importance, however, to note that at 24 months after surgery nearly 60% of the patients in this study were still alive (Rab et al, 1976). If we had used only the experimental treatment group, that is, the group given prophylactic pulmonary radiation, and had not included a control, these results would have appeared to be a considerable advance compared with the historical control curve of Marcove and coworkers (Marcove et al, 1970). This control curve is generally considered to be the standard reference against which clinical trials are compared. Hence, the value of concurrent controls is readily apparent.

In another study, we compared the effectiveness of chemotherapy and immunotherapy. We randomized patients to receive standard surgical treatment followed by either combination chemotherapy, including methotrexate, doxorubicin, and vincristine, or transfer factor as a form of immunotherapy. In this prospective randomized study, 18 patients were randomized to receive chemotherapy and 17 patients were to receive transfer factor. Although there was no significant difference between the two treatment groups, both disease-free survival and actual survival were better in both groups than in historical experience (Ritts et al, 1978).

We reviewed our more recently treated patients who did not receive any adjuvant treatment. Whereas historical survival experience has generally shown that about 20% of patients are alive at five years, we wanted to know whether our patients treated in the more recent past had an identical experience. We, therefore, set out on an extremely comprehensive and exhaustive clinical, pathologic, and statistical study. Numerous factors were evaluated to make our series comparable with those reported in the literature. We excluded all patients who did not have classic osteosarcoma on histologic examination; we excluded all patients with osteosarcoma in sites other than the extremities; and we excluded all patients who were more than 21 years of age. Unexpectedly, we found considerable improvement in both disease-free survival and actual survival of these patients who received no adjuvant therapy as compared with the survival reported in the literature (Taylor et al, 1978). Thus, these patients were treated in the same manner as patients were treated before chemotherapy was available; yet, their survival experience was better than that of patients reported in the past.

Actual survival has increased steadily in recent years, and now more than 50% of our patients treated without adjuvant treatment are alive at five years. There are probably many reasons for this improvement but we have not been able to identify statistically the responsible factors. We do not know whether patients are being referred for treatment sooner than they were in the past. However, the sizes of the tumors were the same; the age, sex, and site distributions were also similar; and the treatments were generally similar. Better staging techniques and more aggressive pulmonary resections may account for some of this improvement in survival. The importance of all this is that we now recognize that survival has improved even though adjuvant treatment (ie, chemotherapy) has not been used. Therefore, one should be cautious in interpreting clinical results of studies without concurrent controls.

In 1980, we gave a preliminary report of a randomized study in which patients with osteosarcoma after surgical excision of their tumor either were given high-dose methotrexate or no adjuvant therapy (Edmonson, 1980). Within one month after definitive surgery, 37 patients with osteosarcoma were randomized either to receive no adjuvant therapy but with observation every six weeks or to receive a regimen of high-dose methotrexate every three weeks. After total surgical removal of their osteosarcomas, with a mean follow-up of approximately 18 months, 56% of the patients remained disease free after one year, and 52% at two years. The estimated rate of survival was 93% at one year and 74% after two years. No significant differences in survival rates were noted between the patients who had received the high-dose methotrexate regimen and those who did not. The survival rates in this study were similar to the overall institutional survival rates, which included all patients with nonmetastatic osteosarcoma who received definitive surgical treatment at the Mayo Clinic during the time of this study. Further observations have revealed no significant change in the appearance of these survival curves, and there has been no apparent advantage to the use of this regimen of high-dose methotrexate over the use of surgery alone.

We remain firmly committed to the concept of adjuvant chemotherapy and believe that eventually one or more chemotherapeutic regimens will be shown to have clinical value. Perhaps some of the more recently reported clinical trials will demonstrate improved survival due to adjuvant chemotherapy, with more patients followed for longer periods. Even when a particular regimen appears to have therapeutic value, the extent of this value is not readily apparent without the use of concomitant controls. The magnitude of improvement due to adjuvant chemotherapy may be so great that, even without concomitant controls, the value must be recognized. This, however, has not yet been established.

Summary

At the Mayo Clinic, in recent years, patients with osteosarcoma who have been surgically treated have experienced improved disease-free intervals and increased survival. All of the factors responsible for these improvements [in survival] are not apparent, but they are not dependent on any form of postsurgical adjuvant treatment (Ivins et al, 1979).

It is important that the tumors of patients used for clinical trials be classified according to the known histologic variants of osteosarcoma and that this histologic classification be applied consistently. Clinical staging of tumors has been improved considerably by the use of computerized tomography of the chest.

There is much interest in the use of limb salvage procedures as an alternative to amputation, and many procedures aimed at limb salvage are now available. The precise value of these innovative forms of surgical treatment is not yet established. However, aggressive surgical removal of pulmonary metastatic lesions has therapeutic value and should be employed whenever feasible (Ivins et al, 1979).

The precise value of adjuvant chemotherapy has not yet been established, but we hope that this issue will soon be resolved.

References

1. Dahlin, D.C., and Unni, K.K.: Osteosarcoma of bone and its important recognizable varieties. Am. J. Surg. Pathol. 1, 61, 1977.

2. Edmonson, J.H., Green, S.J., Ivins, J.C., et al: Post-surgical treatment of primary osteosarcoma of bone--comparison of high dose methotrexate vs observation: preliminary report (Abstract). Proc. Ann. Meet. Am. Assoc. Cancer Res. Am. Soc. Clin. Oncol. 21, 476, 1980.

3. Enneking, W.F., and Kagan, A.: "Skip" metastases in osteosarcoma. Cancer 36, 2192, 1975.

4. Ivins, J.C., Edmonson, J.H., Gilchrist, G.S., et al: Osteosarcoma--current status of the search for effective surgical adjuvant treatment. Recent Adv Orthop 3, 19, 1979.

5. Marcove, R.C., Miké, V., Hajek, J.V., et al: Osteogenic sarcoma under the age of 21: a review of 145 operative cases. J. Bone Joint Surg (A) 52, 411, 1970

6. Muhm, J.R., and Pritchard, D.J.: Computer tomography for the detection of pulmonary metastasis in patients with osteogenic sarcoma (Abstract). Proc. Ann. Meet. Am. Assoc. Cancer Res. Am. Soc. Clin. Oncol. 21, 148, 1980.

7. Rab, G.T., Ivins, J.C., Childs, D.S. Jr., et al: Elective whole lung irradiation in the treatment of osteogenic sarcoma. Cancer 38, 939, 1976.

8. Ritts, R.E., Jr., Pritchard, D.J., Gilchrist, G.S., et al: Transfer factor versus combination chemotherapy: an interim report of a randomized postsurgical adjuvant study in osteogenic sarcoma. Prog. Cancer Res. Ther. 6, 293, 1978.

9. Sim, F.H., Ivins, J.C., and Pritchard, D.J.: Surgical treatment of osteogenic sarcoma at the Mayo Clinic. Cancer Treat. Rep. 62, 205, 1978.

10. Spanos, P.K., Payne, W.S., Ivins, J.C., et al: Pulmonary resection for metastatic osteogenic sarcoma. J. Bone Joint Surg (A) 58, 624, 1976.

11. Taylor, W.F., Ivins, J.C., Dahlin, D.C., et al: Trends and variability in survival from osteosarcoma. Mayo Clin. Proc. 53, 695, 1978.

12. Telander, R.L., Pairolero, P.C., Pritchard, D.J. et al: Resection of pulmonary metastatic osteogenic sarcoma in children. Surgery 84, 335, 1978.

Limb Preservation in Primary Bone Disease

J. M. LANE, P. BOLAND, K. ABOU ZAHR, B. HURSON, G. ROSEN,
B. CAPARROS, A. HUVOS, and J. OTIS

Introduction

There has been a dramatic improvement in the survival and surgical management of osteogenic sarcoma during the last decade. In the past, radical amputation and disarticulation were the only surgical methods available for the treatment of malignant osteogenic sarcoma (Dahlin and Coventry, 1967; Lewis and Lotz, 1974; Sweetman, 1973). Recent accomplishments in chemotherapy, progress in the methodology of en bloc resection, and advances in joint replacement have all resulted in limb salvage procedures for this highly malignant bone tumor (Huvos et al, 1977; Huvos, 1979; Marcove et al, 1970; Marcove, 1977; 1979; 1980; Pade et al, 1942). To highlight the advances in limb preservation, we present in this paper a comparison of the survival and function between ablation and limb preservation in patients with osteogenic sarcoma of the lower limb.

Materials and Methods

Between December 1977 and December 1981, 74 consecutive patients, 48 men and 26 women, with primary malignant osteogenic sarcoma of the lower extremity were treated at Memorial Sloan-Kettering Cancer Center by the same orthopedic surgeon. Locations of the tumors included proximal femur (2), femoral shaft (6), distal femur (38), proximal tibia (22), distal tibia (2), proximal fibula (2), distal fibula (1) and calcaneous (1). The ages ranged from 5 to 62 years. The patients were classified according to the surgical staging system of the Musculoskeletal Tumor Society (Enneking et al, 1980). Two patients were in Stage IIA, 53 patients in Stage IIB, and 19 patients in Stage III. Patients were surgically treated by either ablative surgery (51) or en bloc resection (23). The ablative surgical procedures included hemipelvectomy (1), hip disarticulation (11), above-knee amputation (34), knee disarticulation (2) and below-knee amputation (3).

En bloc resections included custom titanium semiconstrained total knee replacement (14) (Fig. 1), total femoral replacement (2), reduction resection Van Ness turniplasty (Van Ness, 1950)(3) (Fig. 2), soft tissue reconstruction (2), bipolar proximal femoral replacement (1), and diaphyseal allotransplant (1). All patients underwent a routine history, physical examination, and laboratory evaluation (alkaline phosphatase). The extent of the disease was determined by means of anteroposterior and lateral roentgenograms of the involved bone, bone scan, gallium scan, computerized axial tomography of the involved area with perfusion, chest tomography and

Current Concepts of Diagnosis and Treatment
of Bone and Soft Tissue Tumors.

biplane arteriography of the involved area. Criteria for en bloc resection in our group were absence of tumor invasion of neurovascular bundle and the probability of preserving a cuff of normal tissue around the entire tumor. Patients undergoing en bloc or ablative procedures received preoperative chemotherapy (Rosen et al, 1978; 1979; 1982) consisting of high dose methotrexate with a citrovorum rescue, combined with either vincristine and doxorubicin or bleomycin, cyclophosphamide and dactinomycin. Six to 12 weeks following preoperative chemotherapy, the patient underwent the operative procedure and the specimen was carefully evaluated postoperatively for the effect of adjuvant chemotherapy. In specimens with less than 90% necrosis (Huvos et al, 1977; Huvos, 1979; Rosen et al, 1982) cisplatin was added to the postoperative chemotherapy program. Patients with over 90% necrosis were continued with the same therapeutic plan as before surgery.

Results

Patients with Stage IIA (intracompartmental), Stage IIB (extracompartmental) and Stage III (metastatic) high-grade osteogenic sarcoma of the lower extremity were evaluated for survival and function following limb preservation by en bloc resection or amputation. The majority of the patients in all stages had lesions located about the knee (62 out of 74, or 84%). The overall survival of osteogenic sarcoma depended on the stages of the tumor. Stage II disease was significantly superior to Stage III ($p < .005$). Forty-seven out of 55 (85.5%) patients with Stage II tumors (no distant metastases) had survived compared with 6 of 19 (31.6%) patients with Stage III disease (lung metastases).

Within Stage II disease (median follow-up of 27 months), both patients with intraosseous disease (IIA) had survived. Stage IIB disease consisting of extraosseous extension had an overall survival of 45 out of 53 patients (84.9%) and did not differ significantly from the limited number of patients with IIA disease. Combining both IIA and IIB disease, 19 patients had limb preservation (35%) and 36 had amputations (65%). There was no significant difference in the survival among the various surgical treatments, either the distal femur (76.9% vs 66.6%) or proximal tibia (100 vs 100%). The slightly improved survival of the tibial lesions was statistically significant ($p < .02$, chi-square analysis). None of the Stage IIA or IIB en bloc resections was associated with local recurrences. The methods of construction in Stage II disease following en bloc resection included 11 patients using a semiconstrained type titanium total knee that allowed 5° of varus valgus and rotational motion (Fig. 1). It was designed with a cementless fluted fixation stem proximally and a cementable tibial component distally. Two patients underwent total femoral replacement, two patients had a Van-Ness turniplasty (Fig. 2), two patients only required soft tissue reconstruction, one patient had a bipolar proximal femoral endoprosthetic hip replacement for a proximal femoral lesion, and one had diaphyseal allograft lesion of the shaft of the femur (freeze dried).

In terms of function, results in patients with limb preservation were statistically superior as to overall gait and specifically velocity, stride length, oxygen consumption and percent use of maximum aerobic capacity both at normal speed and 20% augmented velocity ($p < .05$). The limb-preserved patient used less oxygen and less maximum aerobic capacity compared with the amputated patients at each level of tumor presentation.

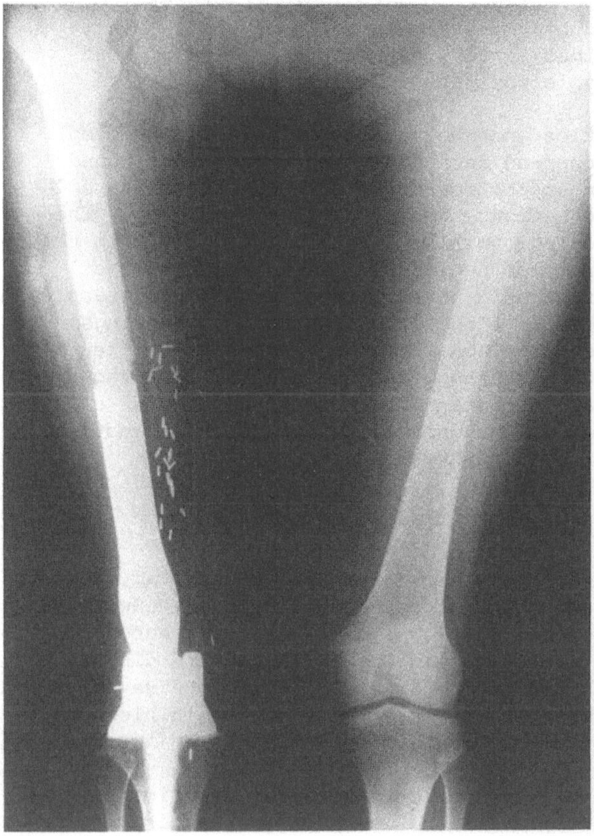

Fig. 1

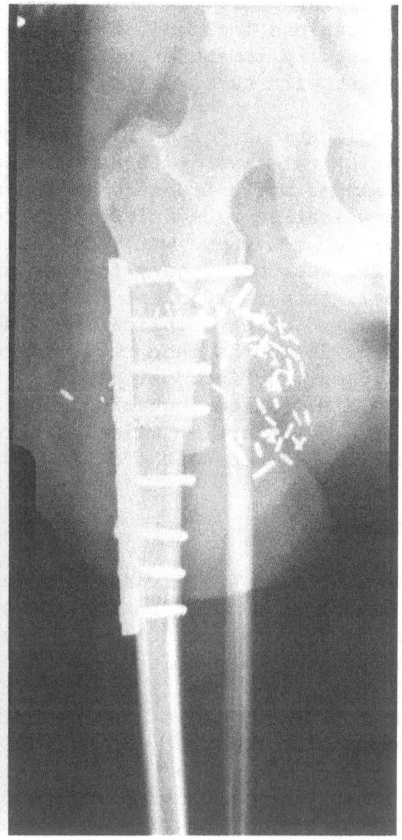

Fig. 2

Fig. 1. AP radiograph of a patient following en bloc extra-articular resection of a distal femoral osteogenic sarcoma and the insertion of a B-L semiconstrained titanium custom total knee replacement.

Fig. 2. AP radiograph of a patient following en bloc extra-articular resection of a distal femoral osteogenic sarcoma and reconstruction with a Van Ness turniplasty (recent). Union was achieved in 8 weeks despite chemotherapy.

Patients presenting with Stage III osteogenic sarcoma (distant metastases) could be saved with combined local surgery and thoracotomies in a setting of appropriate chemotherapy. Six out of 19 patients (32%) with a mean follow-up of 22 months remained NED. Two additional patients were alive with the disease with little likelihood of ultimate survival. There appeared to be little difference whether these patients were treated with amputation or en bloc resection in terms of their ultimate survival as long as there was an even chance of curing their lungs with thoractomy.

In 51 patients in both Stage II and III who underwent amputation, the indications in descending order of importance included neurovascular involvement, large size of tumor, and occurrence of pathologic fractures. Other indications included an infected biopsy site, location of tumor in a skeletally immature child, previous surgery with contamination of soft tissue with tumor cells, unreliable patients and palliation.

Complications for en bloc resection and amputations differed. The complication rate was 6% for amputations and included stump overgrowth, wound hematoma and a fracture in the near postoperative period. There was a 48% complication rate for en bloc resections, usually at a minor level. The most common cause was skin necrosis requiring a wound revision within the first week of surgery for any areas of necrosis. It was believed that small eschars adjacent to the prosthesis boded ill and required aggressive repair. Other complications included three late prosthetic loosenings after 24 months that required revision, one death secondary to chicken pox pneumonia during postoperative chemotherapy, and one patient with recurrent hip dislocation requiring a Chiari osteotomy to increase acetabular capacity. An infection which occurred in the only allograft during chemotherapy ultimately necessitated a revision to an AK amputation.

Discussion

En bloc limb preserving procedures can achieve comparable survival rates to that of amputation for the same stage of osteogenic sarcoma. Survival rates approaching 90% can be expected for lower limb lesions treated by amputation or en bloc resection procedures in Stage IIA and IIB disease. On the other hand, patients with distant metastases can only be saved in approximately 1/3 of the cases whether they are treated with en bloc or amputation for the primary tumor site.

Limb-preserved en bloc resections leave patients with far superior gait function in terms of gait characteristics and lower oxygen requirement for ambulation. Initially, a significantly higher rate of complications was observed following en bloc resection requiring secondary procedures such as plastic wound revision for skin necrosis. This has decreased recently as experience by the operating team has increased. Local recurrence was not seen in patients with Stage IIA or IIB lesions, probably partly due to better staging of the tumor with perfusion CT scan, arteriography, bone scan, gallium scan, etc. Assuming the tumor is of a reasonable size, neurovascular bundle is free, skin contamination is not involved, adequate coverage can be achieved at the site and the patient has approached maturity, then limb preservation is a viable alternative to amputation. If margins are close, however, one runs the risk of local recurrence. (Comparable indications and results have been achieved at our institution with upper extremity limb preservation.) Prior experience has shown in our center as well as others that in cases of local recurrences, few of these patients can be saved by a second procedure. Future solutions include earlier recognition of the disease to enable more patients to benefit from en bloc resection, improved reconstruction techniques to decrease the rate of device failure or loosening, continued improvement in preoperative and postoperative chemotherapy to improve even more the survival rate of 90% in Stage II disease and, finally, recognition of the disease sufficiently early to avoid distant metastases.

References

1. Dahlin, D.C., and Coventry, M.B.: Osteosarcoma: A study of 600 cases. JBJS 49A, 101, 1967.

2. Enneking, W.F., Spanier, S.S., and Goodman, M.A.: A system of the surgical staging of musculoskeletal sarcoma. Clin. Orthopaed. 153, 106, 1980.

3. Huvos, A.G., Rosen, G., and Marcove, R.C.: Primary osteogenic sarcoma. Pathologic aspects in 20 patients after treatment with chemotherapy, en bloc resection and prosthetic bone replacement. Arch. Pathol. Lab. Med. 101, 14, 1977.

4. Huvos, A.G.: Bone Tumors: Diagnosis, Treatment and Prognosis. W.B. Saunders, 47-93, 1979.

5. Lewis, R.J., and Lotz, M.J.: Medullary extension of osteosarcoma. Cancer 33, 371, 1974.

6. Marcove, R.C., Mike, V., Hajek, J.V., et al: Osteogenic sarcoma under the age of 21. A review of 145 pre-operative cases. JBJS (Am) 502, 411, 1970.

7. Marcove, R.C.: En bloc resection for osteogenic sarcoma. Cancer J. Surg. 20, 521, 1977.

8. Marcove, R.C. and Rosen, G.: En bloc resections for osteogenic sarcoma. Cancer 45, 3040, 1980.

9. Marcove, R.C.: En bloc resection for osteogenic sarcoma. Bull. N.Y. Acad.Med. 55, 744, 1979.

10. Pade, G.T., McNeer, G., and Coley, B.L.: Interscapulothoracic amputation for malignant tumors of the upper extremity. A report of 31 consecutive cases. Surg. Gynecol. Obstet. 74, 1610, 1942.

11. Rosen, G., Marcove, R.C., Caparros, R., et al: Primary osteogenic sarcoma. The rationales for pre-operative chemotherapy and delayed surgery. Cancer 43, 2163, 1979.

12. Rosen, G., Huvos, A.G., Mosenda, C., et al: Chemotherapy and thoracotomy for metastatic osteogenic sarcoma. A model for chemotherapy and rationale: The timing of the thoracic surgery. Cancer 41, 841, 1978.

13. Rosen, G., Caparros, B., Huvos, A.G., et al: Pre-operative chemotherapy for osteogenic sarcoma: Selection of post-operative adjuvant chemotherapy based on the response of the primary tumor to pre-operative chemotherapy. Cancer 49, 1221, 1982.

14. Sweetman, R.: Amputation in osteosarcoma. JBJS 55B, 189, 1973.

15. Van Ness, C.P.: Rotationplasty in the treatment of congenital defects of the femur. JBJS (Br) 32B, 12, 1950.

Surgical Treatment of Giant Cell Tumors, Chondrosarcomas, and Chordomas of the Spine

B. STENER

Introduction

This communication reports a series of 19 primary tumors of the spine -- giant cell tumors, chondrosarcomas, and chordomas -- which, even though aggressive, appear to have been amenable to curative surgical treatment. Five illustrative cases will be described.

Giant Cell Tumors

Six tumors of this kind have been treated, three located above the sacrum and three in the sacrum. In 1968, three entire vertebrae (T11, T12, L1) were removed in a 26-year old woman (Stener and Johnsen, 1971). She is well after more than 14 years. In 1975, a 33-year old woman underwent total removal of T11, including parts of the adjacent vertebrae (Stener, 1977). She is well after more than seven years. Her reconstructed spine has successfully stood the test of a completed pregnancy (May 1979). In 1971, the left half of the sacrum, except the uppermost part of the ala, was removed in a 34-year old woman (Stener and Gunterberg, 1978); the adjacent part of the left ilium was included in the specimen as were all the ipsilateral sacral nerves. The patient is free of symptoms after more than 11 years, and has normal urogenital and anorectal function. More recently, in December 1982, a 23-year old woman underwent an almost identical operation on the right side (the S1 nerve could be preserved). The tumor was removed in one piece. The postoperative course was uneventful. Below, one suprasacral and one sacral case of giant cell tumor in the spine will be described in more detail.

Case 1. In September 1980, a 23-year old woman was admitted with a giant cell tumor which had destroyed most of the body of L4 (Fig. 1A); a remaining anterior part had been displaced forward, behind the bifurcation of the aorta and the inferior end of the vena cava. Laminectomy and biopsy had been done elsewhere before referral. The patient suffered severe pain and was unable to walk. Angiography showed that the highly vascular tumor extended outside the affected vertebra on both sides. The operation was performed in two stages, first through an anterior approach and then, four weeks later, a posterior approach. For the anterior approach, a large V-shaped incision was made through all layers of the abdominal wall except the peritoneum (see postoperative scar in Fig. 1B). The large myocutaneous flap, with the underlying viscera enclosed in the peritoneal sac, could be

Current Concepts of Diagnosis and Treatment of Bone and Soft Tissue Tumors.

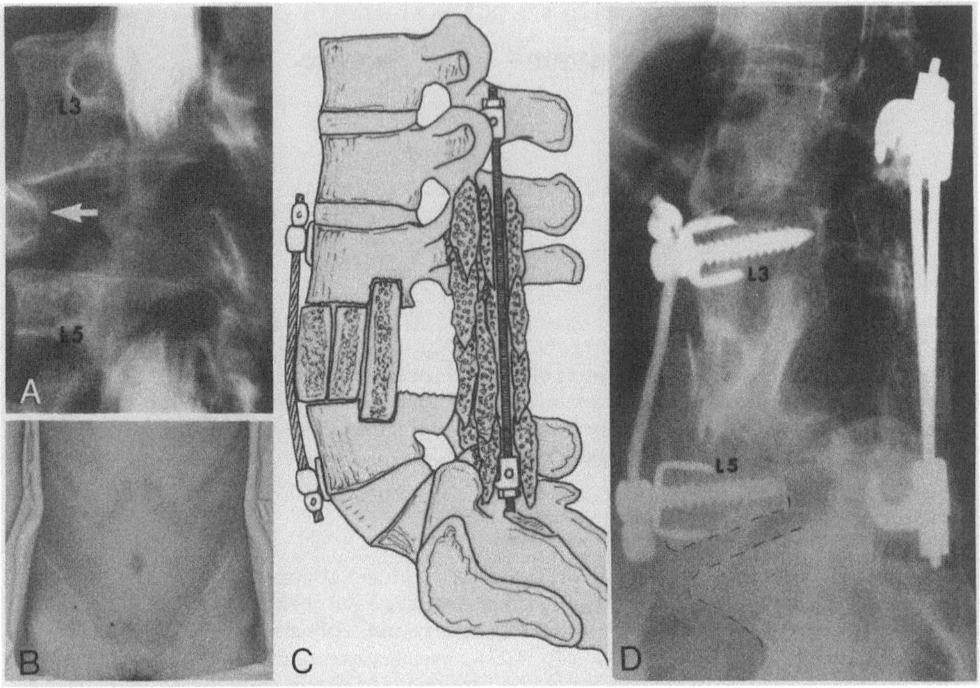

Fig. 1. *Case 1*. A: the lumen of the dural sac has been obliterated by a giant cell tumor in L4 whose body has disappeared almost completely; a small remaining part (arrow) is displaced anteriorly. B: Postoperative scar indicates anterior extraperitoneal approach. C: After removal of the whole of L4 and parts of L3 and L5, the spine was reconstructed using iliac bone grafts and a Dwyer tension wire anteriorly, and two Harrington compression rods posteriorly. D: Radiography two years after operation. Fusion has occurred between the bone grafts and the remaining parts of L3 and L5.

retracted to the right or left side as required for suitable exposure. All that remained of the body of L4, including the adjacent discs, was removed along with all tumor tissue that could be reached by the anterior approach. Corticocancellous blocks of iliac bone were inserted between the bodies of L3 and L5 (Fig. 1C); a Dwyer tension wire was used for axial compression. After the operation the patient was placed in a Stryker bed. During the posterior approach, all remaining parts of L4 and some posterior parts of L3 and L5 were removed along with all residual tumor tissue. Strips of cancellous iliac bone were placed on each side along a Harrington compression rod joining L2 to L5 (Fig. 1C). Four months after the posterior operation, the patient was allowed to start walking. For the next eight months she used a corset when out of bed. As shown radiographically (Fig. 1D), a solid block of bone has been formed by the anterior and posterior grafts and the remaining parts of L3 and L5. Two years and four months after the last operation, the patient is free of symptoms and is working a full day as a dentist's nurse. She has recently become pregnant.

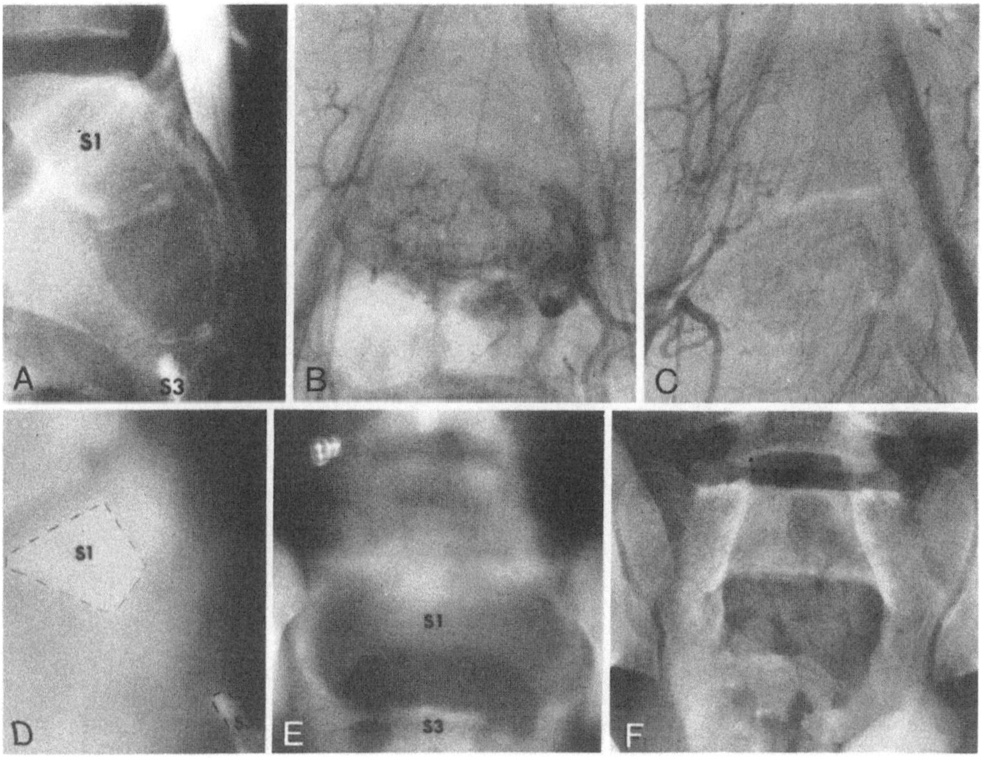

Fig. 2. *Case 2.* A and B: Highly vascular giant cell tumor in S2. C: Angiography 26 months after operation. No evidence of recurrence of the tumor. D and E: Lateral and anteroposterior postoperative tomograms. The intervertebral discs adjacent to S2 have been removed (cp A and D). E: Radiography 26 months after operation. Removed sacral bone partly restored.

Case 2. In January 1979, a 15-year old boy was admitted with a giant cell tumor which had largely destroyed the second sacral vertebra (Fig. 2A); the tumor was highly vascular (Fig. 2B). At operation, the tumor was removed piecemeal through a combined anterior and posterior approach; the S1 through S4 sacral nerves were preserved bilaterally. Most of what remained of the second sacral vertebra was removed, as were the adjacent intervertebral discs and the anteriorly-bulging periosteum (cp A and D in Fig. 2). Laterally, the curettement came close to the cartilage of the sacroiliac joints (Fig. 2E). Angiography (Fig. 2C) and plain radiography (Fig. 2F) 26 months after operation revealed no signs of recurrence of the tumor. He is still well after four years.

Chondrosarcomas of the Spine

Four patients with spinal chondrosarcoma have undergone surgical treatment which may have resulted in a definite cure. In two patients, a substantial

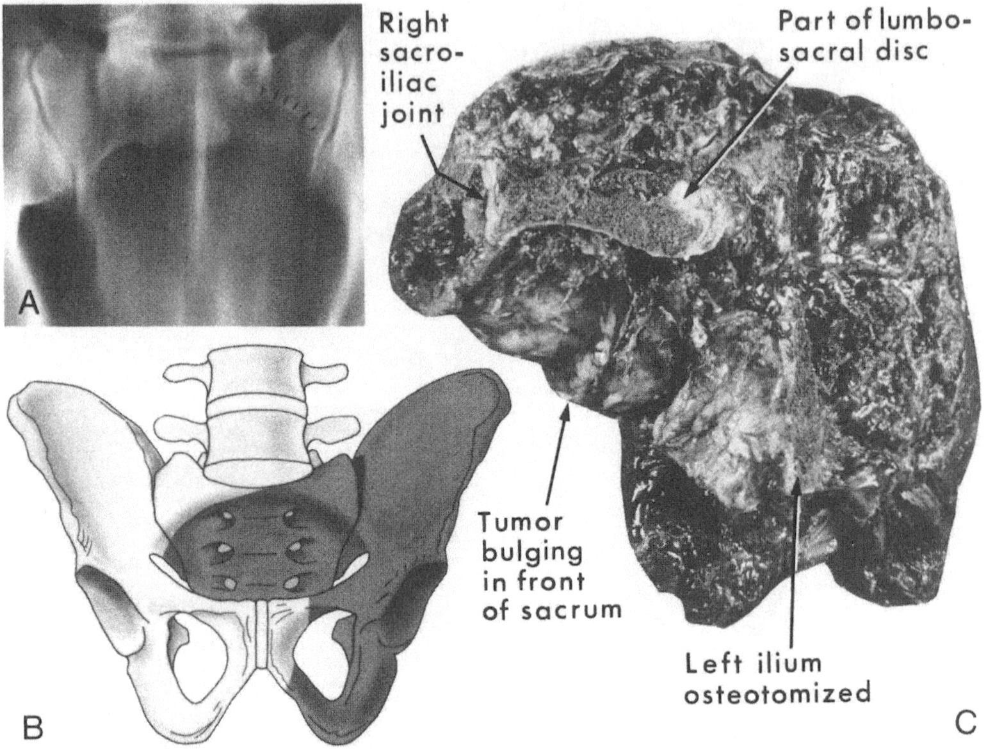

Fig. 3. *Case 3.* A: Sacrum destroyed by chondrosarcoma up to the roof of the first sacral hole on the right side, and beyond this level on the left side; left ilium also affected. B: A left hemipelvectomy, somewhat reduced anteriorly, has been combined posteriorly with a subtotal amputation of the sacrum. C: Surgical specimen seen from above. The iliac wing was removed during operation to facilitate the osteotomy through the sacrum.

portion of the dura had to be included in the surgical specimen; since the follow-up time is still rather short in these cases, they will not be further described here. In one of the other patients, reported previously (Stener, 1971), all of T7 and half of T6 and T8 were removed because of a chondrosarcoma which protruded both into the spinal canal and the mediastinum. The spine was successfully reconstructed using iliac bone grafts. The patient is without evidence of disease after 14 years. The fourth patient (Case 3) underwent a hemipelvectomy combined with a subtotal amputation of the sacrum.

Case 3. In December 1979, a 46-year old man presented with a chondrosarcoma which had destroyed the sacrum up to the roof of the first sacral hole on the right side and beyond this level on the left side (Fig. 3A). Also, the left ilium was affected by the tumor. The patient had severe pain and a useless left leg due to nerve involvement. After incisional biopsy from

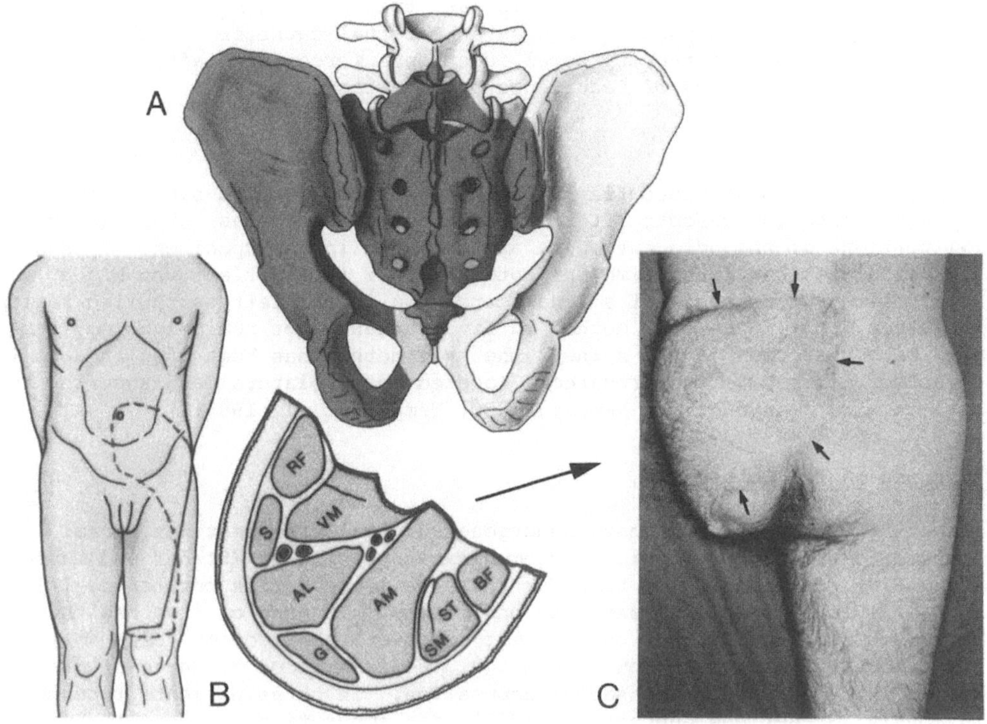

Fig. 4. *Case 3*. A: Posterior aspect of skeletal loss (dark area). B: Thick myocutaneous flap spared from the whole length of the left thigh (RF = rectus femoris, S = sartorius, VM = vastus medialis, AL = adductor longus, AM = adductor magnus, G = gracilis; SM = semimembranosus, ST = semitendinosus, BF = biceps femoris). C: Posterior aspect of patient after operation. Myocutaneous flap indicated by arrows.

behind, a subtotal amputation of the sacrum was performed in addition to a left hemipelvectomy (Fig. 3B). Only a thin bridge of bone was spared between the lumbar spine and the right ilium, above the first sacral hole. All the sacral nerves were included in the specimen. Figure 3C shows the surgical specimen from above. To substitute the large posterior defect (Fig. 4A), a long, thick, myocutaneous flap from the left thigh was used (Fig. 4B); with preserved circulation, it contained all muscles of the thigh except the vastus lateralis and the lateral half of the vastus intermedius. As the femoral nerve was preserved, the flap retained perfect sensibility within most of its area. The flap, sutured into the defect, healed successfully (Fig. 4C). It now serves as a living cushion between the inferior end of the lumbar spine and the modified Canadian prosthesis with which the patient was fitted. Despite the amputation at the lumbosacral level, he almost looks as if he had undergone a hip joint disarticulation. When he is resting in the standing position, he loads the prosthesis more than the right leg; when he is sitting, he distributes the load more equally. He is able to drive his car, transporting mail over long distances. Being incontinent, he uses an external catheter. With a suitable diet, he has satisfactory bowel function. Three years after operation, there are no signs of recurrence or metastasis.

Comment

A long, thick, myocutaneous flap from the ipsilateral thigh substitutes well for large posterior defects after extended hemipelvectomies where substantial portions of the adjacent spine are included in the specimen. I used this method for the first time in September 1977 in a patient who had a large solitary metastasis of a renal carcinoma in the left sacroiliac region. Altogether I have used this method in five patients for hemipelvectomy combined with resection of the spine. The same method has been used in a hemipelvectomy for a huge chondrosarcoma located posterolateral and superior to the ilac wing (Mnaymneh and Temple, 1980; Temple et al, 1982).

Chordomas of the Spine

Since 1971, seven patients have undergone surgical treatment for sacral chordoma according to principles communicated previously (Stener and Gunterberg, 1978). The follow-up time for these patients varies between 11-1/2 and 2-1/2 years (mean six years and one month). In none of them has the tumor recurred locally, but one patient underwent thoracotomy for a solitary pulmonary metastasis four years ago. He is without evidence of disease 11-1/2 years after the sacral amputation. In three patients reported previously (Stener and Gunterberg, 1978), the amputation was done through both first sacral holes, with sacrifice of all the sacral nerves except the first. One of these patients was free of disease when she died of a heart attack two years and four months later. The other two patients have been continuously free of disease for 11 years and one month, and seven years and six months.

In Case 4, an even higher sacral amputation necessitated the sacrifice of all the sacral nerves; in Case 5, all of L3 and parts of L2 and L4 were removed because of chordoma.

Case 4. In June 1980, a 55-year old man was admitted with a sacral tumor which extended high up in the sacral canal, as shown by myelography (Fig.5A). A posterior incisional biopsy confirmed the suspicion of chordoma. Neurologic symptoms and signs indicated that even the first sacral nerves were involved in the tumor. The operation was performed first through an anterior approach and then a posterior approach. As the biopsy had been done from behind, there was no need for including the rectum in the surgical specimen. Therefore, the anterior dissection was done extraperitoneally through a semicircular incision in the abdominal muscular wall. After division and ligation of the internal iliac artery and vein on both sides, a track was made with a chisel through the anterior cortex of the first sacral vertebra (Fig. 5B); bilaterally, it passed above the first sacral hole, beneath the lumbosacral nerve trunk, and through the sacroiliac joint (cp right side in Fig. 3B). After closure of the anterior wound and with the patient in the prone position, the actual amputation was performed from behind with an osteotome directed against the anterior transcortical

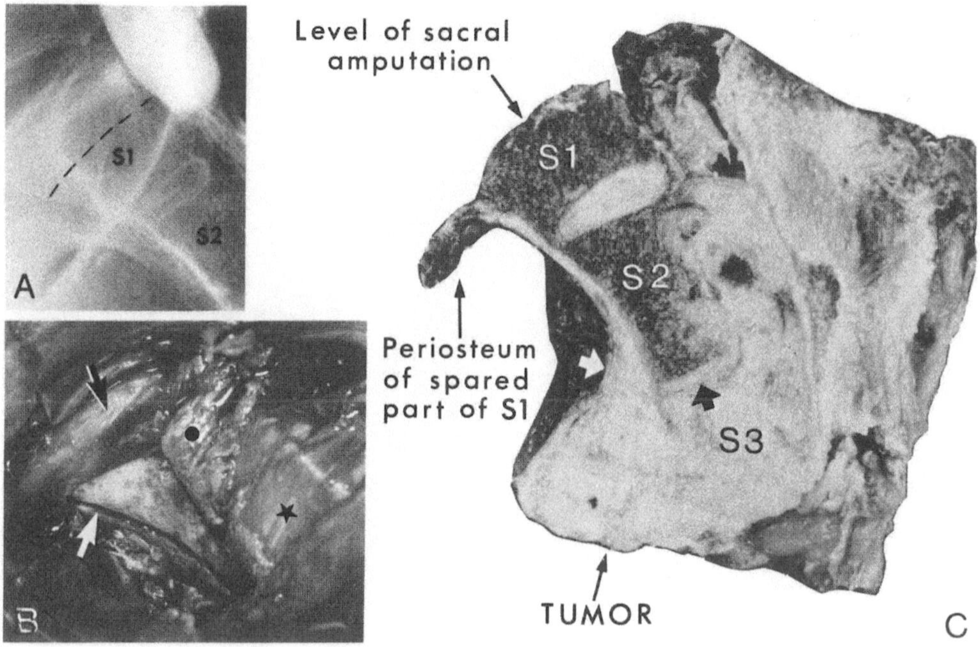

Fig. 5. *Case 4.* A: Dural sac deformed by sacral chordoma extending cephalad in the sacral canal. Interrupted line indicates level of sacral amputation. B: Surgical field after extraperitoneal anterior approach. White arrow indicates track made by chisel in the right sacral wing, above the first sacral hole (cp Fig. 3B). Black arrow indicates the right lumbosacral nerve trunk. Round dot indicates periosteum reflected from the right sacral wing. Asterisk indicates the lumbosacral disc. C: Surgical specimen (superior part) divided in the middle sagittal plane. Having destroyed the body of S3, the tumor protrudes anteriorly elevating the periosteum and presacral fascia (white arrow); extending cephalad in the sacral canal it also affects the body of S2. The disc between S2 and S3 (black arrow) serves as a temporary barrier against tumor growth.

track, which on each side was palpable through the greater sciatic notch. The specimen included the posterior part of each iliac wing and a muscular cuff consisting of parts of all muscles attached to the sacrum: the sacrospinalis, gluteus maximus, and piriformis. At 2-1/2 years after operation, the patient is without evidence of disease. Cortical bone has formed at the site of the osteotomy through S1, illustrating Wolff's law. Being incontinent, the patient uses an external catheter. Despite the loss of all the sacral nerves, he is able to walk even on tiptoe.

Comment

Figure 5C shows a sagittal section through the specimen; it demonstrates how a sacral chordoma usually grows. Knowledge of this is important for

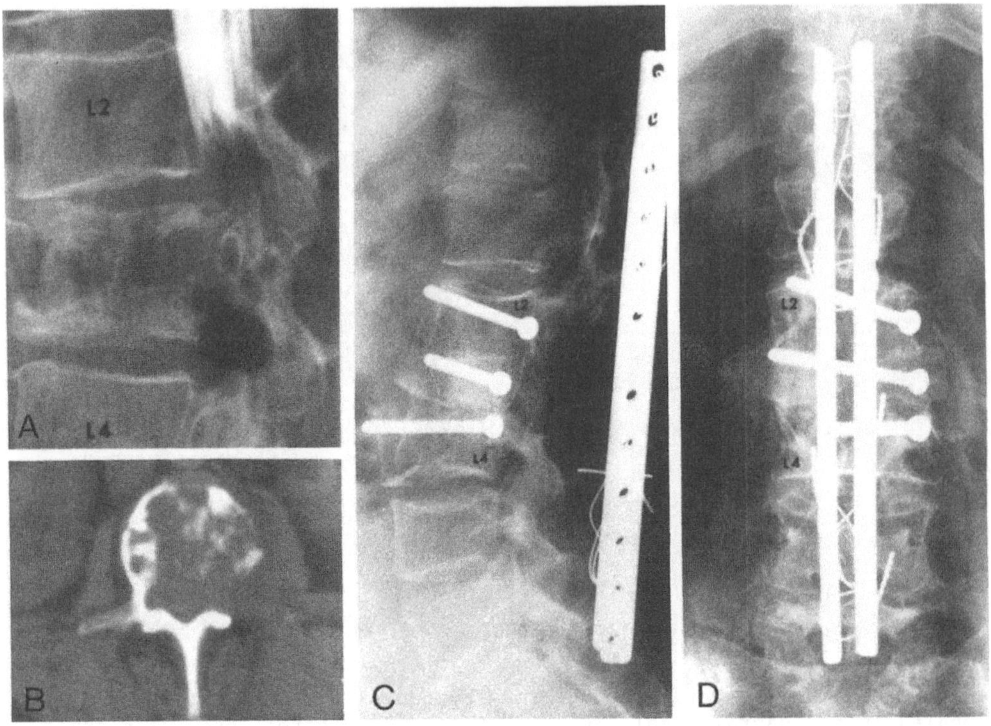

Fig. 6. Case 5. A and B: Chordoma of L3. The tumor has grown outside the body of the affected vertebra as revealed by myelography (A) and computed tomography (B). After removal of the whole of L3 and parts of L2 and L4, the spine was reconstructed by inserting iliac bone grafts anteriorly, and by wiring ASIF plates to spinous processes posteriorly. C and D: Radiography 15 months after operation. A block vertebra has been created by the grafts and the remaining parts of L2 and L4.

the planning of adequate surgical treatment. Arising from the body of a sacral vertebra (S3 in Case 4), the tumor often grows out far in an anterior direction, but for a long time it respects the anterior periosteum and the presacral fascia. A transrectal biopsy is therefore contraindicated because it entails a risk of spreading the tumor in the retrorectal space; if it has been done, the rectum must be included in the surgical specimen. The tumor also grows in a posterior direction and may penetrate the posterior sacral holes and the hiatus of the sacral canal. A posterior muscle cuff should therefore be included in the specimen. The rudimentary sacral discs may for a long time serve as a barrier (the S2-S3 disc in Fig. 5C). Growing cephalad in the sacral canal, though, the tumor may pass around this barrier and affect another vertebra (S2 in Fig. 5C). Myelography should therefore be done to find out how far the tumor extends cephalad in the sacral canal.

Case 5. In December 1980, a 51-year old woman was admitted with a chordoma in L3 (Fig. 6A and B), diagnosed cytologically by fine needle aspiration biopsy. Myelography showed a block of the dye at the inferior part of L2 (Fig. 6A). The patient had continuously suffered pain in the back for more than two years, and had gradually become unable to walk. For extirpation of the tumor it was necessary to remove the entire affected vertebra. The whole operation was performed from behind, with the patient in the prone position. A large skin flap, including the underlying latissimus dorsi muscle, was raised on both sides, uncovering the erector spinae muscles. After release from the lumbar spine, these muscles were retracted laterally to allow work on the posterior elements of the vertebra, and medially for exposure on both sides of the bodies of L2, L3 and L4. After complete laminectomy of L3 and partial of L2 and L4, the affected vertebra was removed, without opening the tumor, by a transcorporal osteotomy above the L2-L3 disc and below the L3-L4 disc. The spine was then reconstructed with iliac bone grafts inserted according to a method described previously (Stener, 1977). ASIF plates were wired to spinous processes above and below the inserted grafts. The stability of the reconstructed spine allowed dispensing with all external support until, after about four months, the patient was allowed to start walking; she was then fitted with a corset for security. Figures 6C and D show that the inserted iliac grafts have fused with the remaining parts of the bodies of L2 and L4. Now, two years and one month after operation, the patient is free of symptoms. She has no pain, does not need a corset, and is able to walk without support.

Comment

The follow-up time is, of course, too short for stating that the patient has been definitely cured. There may be grounds, however, for a more optimistic view than that expressed in a recent review on the subject of chordoma (Mindell, 1981), which states that a chordoma affecting a vertebra above the sacrum is not amenable to curative treatment.

References

1. Mindell, Eugene: Current concepts review. Chordoma. J. Bone Jt. Surg. 63A, 501, 1981.

2. Mnaymneh, Walid and Temple, Wally: Modified hemipelvectomy utilizing a long vascular myocutaneous thigh flap. Case report. J. Bone Jt. Surg. 62A, 1013, 1980.

3. Stener, Bertil: Total spondylectomy in chondrosarcoma arising from the seventh thoracic vertebra. J. Bone Jt.Surg. 53B, 288, 1971.

4. Stener, Bertil: Total spondylectomy for removal of a giant-cell tumor in the eleventh thoracic vertebra. Spine 2, 197, 1977.

5. Stener, Bertil and Gunterberg, Björn: High amputation of the sacrum for extirpation of tumors. Principles and technique. Spine 3, 351, 1978.

6. Stener, Bertil and Johnsen, Odd: Complete removal of three vertebrae for giant-cell tumor. J. Bone Jt. Surg. 53B, 278, 1971.

7. Temple, Wally, Mnaymneh, Walid, and Ketcham, Alfred: The total thigh and rectus abdominis myocutaneous flap for closure of extensive hemipelvectomy defects. Cancer 50, 2524, 1982.

Free Vascularized Bone Transfers: Their Use in the Surgery of Tumors and Tumor-like Conditions

W. RENNIE and R. K. DANIEL

Introduction

During the past decade, a number of reports have appeared in the literature concerning radical en bloc resection of malignant and aggressive benign bone tumors and reconstruction of the resulting skeletal defect with a variety of materials (Brown and Cruess, 1982; Enneking et al, 1980; Mankin et al, 1976; Parrish, 1973). Traditional solutions for such massive bone defects have included:

1) massive autogenous grafts from the ilium or tibia;
2) transferance of the fibula for a tibial defect or fibular bypass by tibiofibular synostosis;
3) custom prosthetic replacement;
4) massive allograft;
5) hemicylindrical sliding graft.

Solutions Nos. 1 and 5 have resulted in a high rate of failure when the defect to be bridged is greater than 6 cm. Custom prosthetic replacement is prone to loosening and fibular transference is obviously only applicable to the tibia. While massive prepared allografts have been reported to be successful in a number of series (Enneking et al, 1980; Mankin et al, 1976; Parrish, 1973), their use has not been without complications. Nonunion, fracture of the allograft, and infection have all been reported. Despite these complications, this technique remains useful, especially for metaphyseal and epiphyseal tumors where the entire end of the long bone must be replaced.

In 1975, for the first time in clinical practice, a free vascularized fibular transplant with microvascular anastomoses was reported for the reconstruction of large tibial defects by Taylor et al (1975). While such grafts have been used in a variety of clinical situations, their use in reconstructive tumor surgery has not been extensive (Daniel and Weiland, 1982; Weiland et al, 1977).

Advantages and Indications

The fibula is the bone most likely to be transferred for tumor reconstruction. It is transferred on a vascular pedicle consisting of the peroneal

Current Concepts of Diagnosis and Treatment of Bone and Soft Tissue Tumors.

artery, supplying a vessel to the endosteum and musculoperiosteal branches to the periosteum (Taylor, 1977). The fibula is a long straight bone whose triangular cross section resists both angular and rotational stresses. In the adult, between 22 and 26 cm of the bone are available for transfer. Its size allows it to be fitted snugly into the medullary canal of femur, tibia or humerus. In addition, morbidity in the donor leg has been found to be minimal, although one is sacrificing one major vessel in the limb and, theoretically, can place the lateral popliteal nerve in jeopardy.

Other advantages of free fibular transfer include:

1) more rapid bone healing and earlier mobilization than with a nonvascularized fibular graft;
2) a higher incorporation rate than cancellous grafts or massive allografts;
3) more rapid bone hypertrophy;
4) a decreased risk of infection;
5) the potential transfer of associated muscle (peroneus longus) and/or an epiphyseal plate.

It is difficult to give absolute indications for the use of free fibular transfer in tumor reconstructive surgery. While ideal for diaphyseal lesions, use of the technique may also be considered for benign metaphyseal lesions where wide margins of normal tissue are not required. Such aggressive benign lesions as giant cell tumor or large unicameral cyst may be resected and reconstructed in this manner. Its use in malignant lesions requires that a radical en bloc resection can be performed. Stage I-A tumors are perhaps ideal, although some Stage II-A tumors may also be considered.

Tumor cell types previously reported in the literature (Enneking et al, 1980; Mankin et al, 1976; Parrish, 1973) as candidates for radical resection and allograft reconstruction include chondrosarcoma, giant cell tumor, adamantinoma, osteogenic sarcoma, parosteal osteogenic sarcoma, fibrosarcoma and malignant fibrocystiocytoma. With reference to the preceding criteria as to site within the bone and tumor staging, patients with any of these tumor cell types could be candidates for a free fibular transfer.

Disadvantages and Contraindications

One of the stated disadvantages for this type of procedure has been noted to be the prolonged operating time. In his report of 1977, Weiland et al noted the operating time to be between 14 and 18 hours. Currently, with two teams, one for the donor fibula and another for excision of the tumor and preparation of the recipient bed, our operating time has been decreased to 5-6 hours. The major disadvantage of this procedure is that as yet no reliable noninvasive test exists for anastomic patency in the immediate postoperative period. Where free bone transfers can be harvested with overlying skin, such as from the iliac crest, one can monitor skin blood flow as an index of anastomic patency. Such information is not available with free fibular transfer. As noted above, donor site morbidity has been minimal. Finally, unlike reconstruction with a massive allograft, no

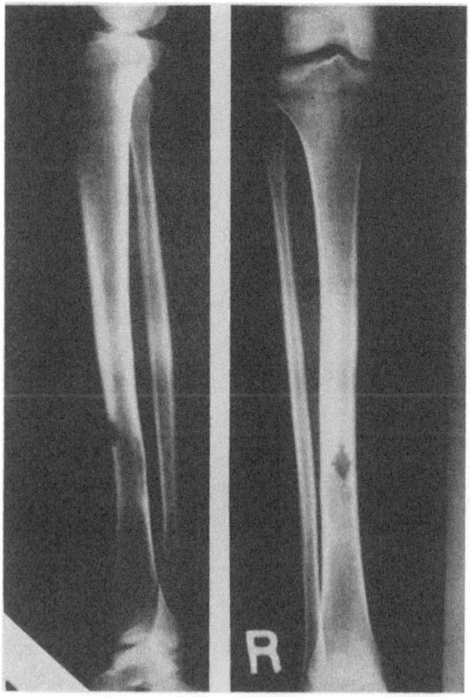

Fig. 1. Case C.P.

a) A.P. and lateral views of (R) tibia showing central chondrosarcoma (postbiopsy).

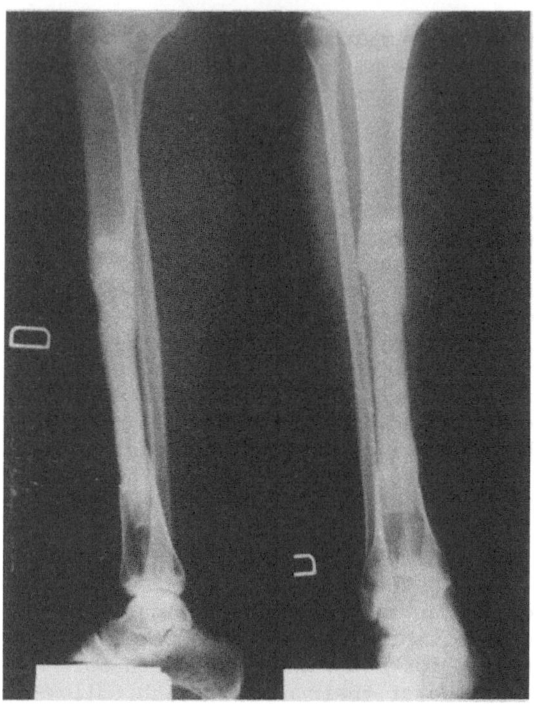

b) A.P. and lateral views of (R) tibia at 6 years showing good incorporation and hypertrophy of free fibular graft.

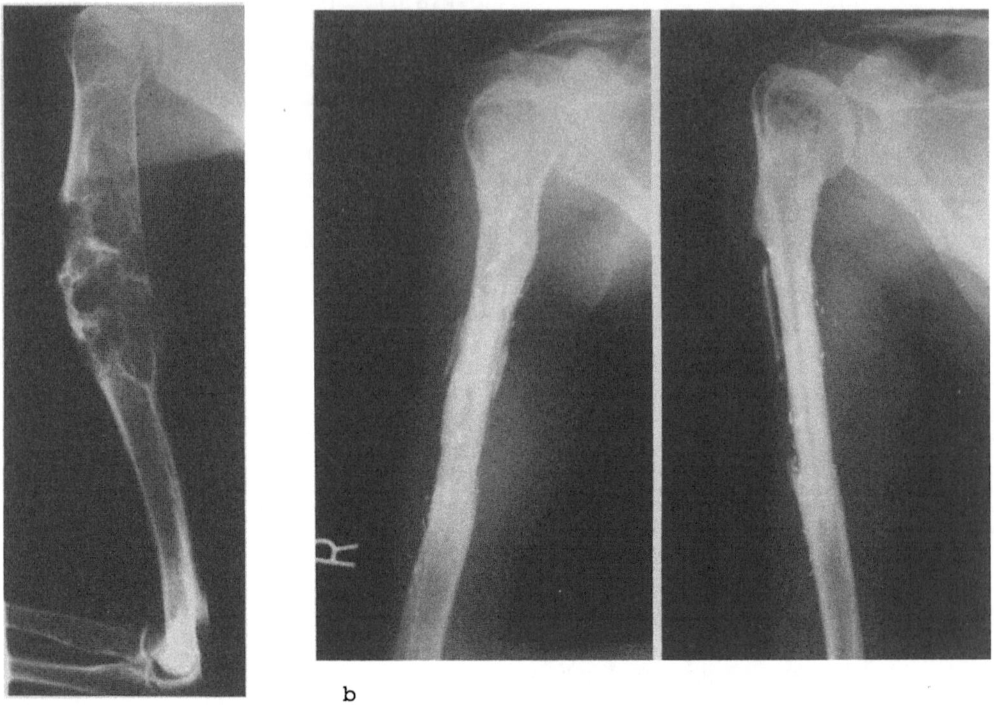

b

a

Fig. 2. Case G.R.: a) Lateral view of (R) humerus showing large recurrent unicameral bone cyst. b) A.P. and lateral views of (R) humerus at 3 years. Graft has hypertrophied. Full function of the arm.

articular surfaces are available, with the exception of the use of the proximal fibula for distal radial substitution.

Contraindications to the use of this technique are primarily related to both the grade and stage of the tumor and to its site within bone. Stage II-B lesions and probably all Stage III lesions would likely not be considered candidates for this type of reconstruction. In addition, distal metaphyseal or epiphyseal lesions are unsuitable for free fibular transfer unless one is willing to accept concomitant joint arthrodesis as a part of the procedure, as with many of the patients in Parrish's series (Parrish, 1973).

Results

To date, we have performed four free fibular transfers in four patients, one for a low grade chondrosarcoma of the tibia, one for a recurrent unicameral bone cyst in the humerus, and two for congenital pseudarthrosis of the tibia. All patients are alive, free of their disease with follow-

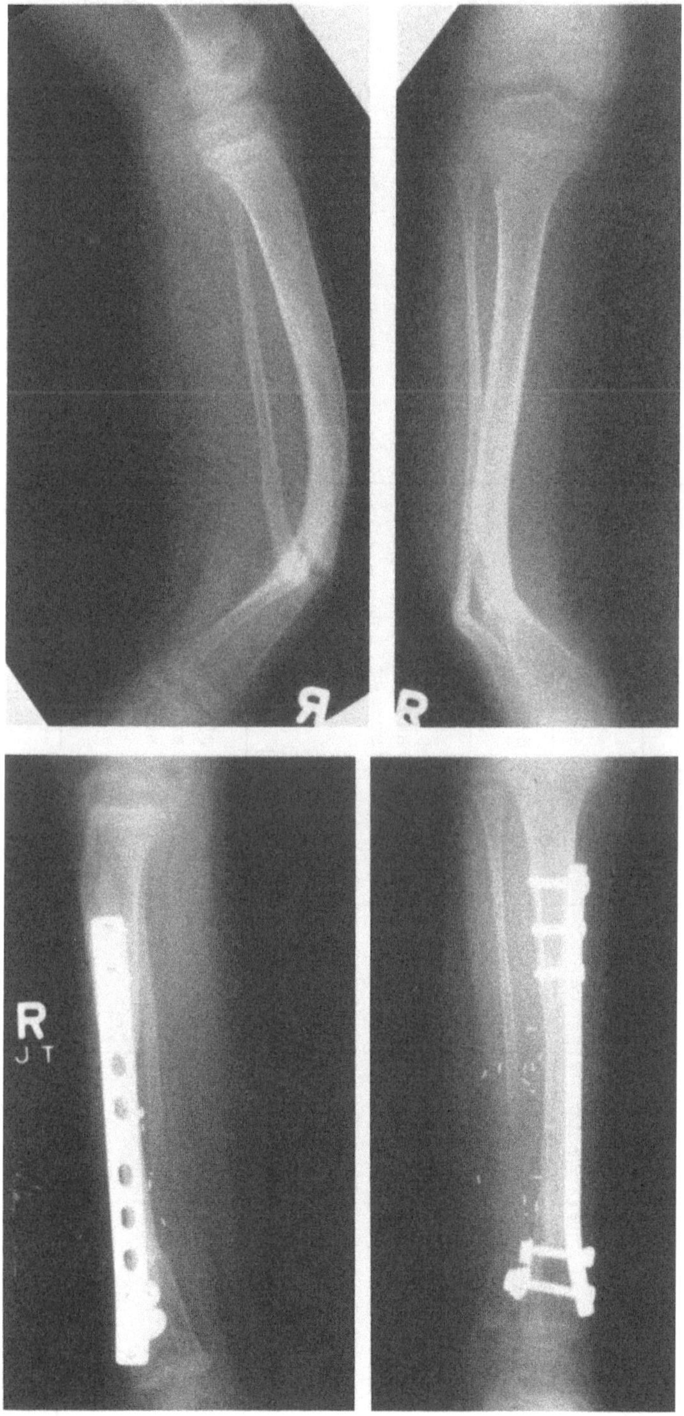

Fig. 3. Case T.C.:

a) A.P. and lateral views of (R) tibia with congenital pseudarthrosis.

b) A.P. and lateral views of (R) tibia at 1 year. Note the rapid graft hypertrophy and the absence of deformity.

Table I. Summary of free fibular transfers performed in four patients

Patient Age/Sex	Pathologic findings	Previous surgery	Ischemia time h:min	Surgery time h:min	Fixation	Time to union (mos)	Defect/ fibula length (cm)	Compli-cations	Follow-up and comments
C.P. 22/F (Fig.1)	Low-grade chondrosarcoma (R) tibia	Open biopsy	3:30	11:30	Hoffman and 2 screws proximal	3	11/18	None	6 years. Full function; no recurrence; good graft hypertrophy
G.R. 16/M (Fig.2)	Recurrent unicameral bone cyst (R) humerus	Bone graft x 2; resorbed	4:30	10:[illegible]0	Wagner and 2 screws distal	3	11.5/19	None	5 years. Full function; no recurrence; good graft; hypertrophy
N.W. 5/M	Congenital pseudarth-rosis (L) tibia; suggestive of neuro-fibromatosis	None	3:30	10:10	Plate and screws	12	10/12	Proximal nonunion. Bone graft at 8 mo; successful	3.5 years. Walking with brace; LLD 4.5 cm; good graft; hypertrophy
T.C. 8/F (Fig.3)	Congenital pseudarth-rosis (R) tibia; non-diagnostic	None	2:20	6:40	Plate and screws	3	8/11	None	1 year. Walking with brace; good graft; hypertrophy

ups ranging from one to six years, and all have demonstrated at least some of the above-mentioned advantages of this technique (Table I).

Of particular interest is the fact that complications have been few and all grafts have shown excellent hypertrophy to provide a functional limb. The duration of surgery can be significantly reduced with increasing experience (case T.C.). It should also be noted that donor leg morbidity has not been a problem in our experience.

Conclusion

Free bone transfer with microvascular anastomoses represents, in our opinion, a useful clinical procedure for the reconstruction of large skeletal defects created by the resection of certain bone tumors and tumor-like conditions. While obviously not applicable to all problems, the use of these grafts have the advantages of being living autogenous bone, capable of rapid union and functional hypertrophy to provide a limb of normal or near normal proportions. Although our experience has been limited, complications have been rare and the end results highly satisfying. We are encouraged to continue.

References

1. Brown, K.L.B. and Cruess, R.L.: Bone and cartilage transplantation in orthopaedic surgery. A review. J. Bone Joint Surg. 64A, 270, 1982.

2. Daniel, R.K. and Weiland, A.J.: Free tissue transfers for upper extremity reconstruction. J. Hand Surg. 7, 66, 1982.

3. Enneking, W.F., Eady, J.L., and Burchardt, H.: Autogenous cortical bone grafts in the reconstruction of segmental skeletal defects. J. Bone Joint Surg. 62A, 1039, 1980.

4. Mankin, H.J., Fogelson, F.S., Thrasher, A.Z., and Jaffer, F.: Massive resection and allograft transplantation in the treatment of malignant bone tumors. N. Engl. J. Med. 294, 1247, 1976.

5. Parrish, F.F.: Allograft replacement of all or part of the end of a long bone following excision of a tumor. J. Bone Joint Surg. 55A, 1, 1973.

6. Taylor, G.I.: Microvascular free bone transfer. A clinical technique. Orthop. Clin. North Am. 8, 425, 1977.

7. Taylor, G.I., Miller, G.D.H., and Ham, F.J.: The free vascularized bone graft. Plast. Reconstr. Surg. 5, 533, 1975.

8. Weiland, A.J., Daniel, R.K. and Rily, L.H. Jr.: Application of the free vascularized bone graft in the treatment of malignant or aggressive bone tumors. Johns Hopkins Med. J. 140, 85, 1977.

Giant Cell Tumors of Bone: Clinical Aspects and Staging

E. R. MINDELL

Incidence

Giant cell tumors of bone are uncommon lesions in general orthopedic practice. However, in large series of patients, such as those of the Mayo Clinic (Dahlin, 1981) they constitute 5% of the total number of bone tumors and if grouped with the benign lesions, constitute 20%.

Location

Practically all giant cell tumors are epiphyseal in location, being eccentrically placed with some involvement of the metaphysis. Most observers feel that the tumor develops in the epiphysis with extension into the metaphysis. Others, however, state the opposite, namely that the lesion begins in the metaphysis. In the unusual circumstance in which the patient is a child, the metaphysis may be the only site of the tumor.

The bone involved is usually a long bone and commonly either the upper tibia or lower femur. About half of all these tumors occur around the knee joint. The distal radius is also a common location and, with the distal femur and proximal tibia, constitutes 60-70% of all giant cell tumor locations. Giant cell tumors may occur in many other bones and have been reported in practically every enchondrally-formed bone (Dahlin, 1981; Goldenberg et al, 1970; Mirra, 1980).

Occasionally, patients are seen who have multiple lesions which histologically suggest the diagnosis of giant cell tumor, especially in patients with hyperparathyroidism. Furthermore, a rare condition of multicentric giant cell tumor of bone is found mainly in small bones of the hand (Peimer et al, 1980).

Age

Most patients are young adults. Three-fourths of all cases occur between 20 and 40 years of age. Giant cell tumors of bone are infrequent under the age of 20 and are very rare under the age of ten. Although the tumor may be found in the elderly, it is rarely seen after 50 years of age.

Current Concepts of Diagnosis and Treatment
of Bone and Soft Tissue Tumors.

Clinical Features

The initial symptoms are usually mild and may even be absent until a pathologic fracture occurs. Often, however, dull aching pain with slight swelling is present. A firm tender expansion of the involved bone is usually palpable. The adjacent joint may be swollen. Occasionally, the swelling and disability are marked, particularly if a pathologic fracture has occurred. On rare occasion, the vascularity in a giant cell tumor in a superficial location is so marked that a thrill may be detected and a primary vascular tumor suspected. The clinical features will vary considerably depending on the location of the tumor.

Apparently 15% of gian cell tumors act as malignant lesions from the onset, leading to pulmonary metastases and even death (Jaffe, 1958).

Radiographic Findings

The lesions are osteolytic, eccentrically located, usually in the epiphyseal-metaphyseal end of a long bone. The adjacent cortex is thinned and may be slightly expanded. The osteolytic process may be fairly well localized or may have destroyed a large segment of bone. Portions of the adjacent subchondral cortex are usually directly involved by the tumor although there is generally no evidence of articular cartilage having been completely penetrated with direct extension into the joint.

Periosteal new bone is usually not seen unless a pathologic fracture has occurred. Soft tissue involvement in most cases cannot be demonstrated on routine roentgenogram or CAT scans.

The bone scan is almost always positive and usually for some distance about the tumor because of the vascularity which is almost always present.

Gross appearance: Grossly, the contour of the bone is generally expanded but the lesion is nevertheless contained by a thin shell of bone which may not even be seen radiographically. The tumor is soft, friable, reddish brown, with firm greyish-yellow areas where fibrosis has occurred. Regions of hemorrhage, necrosis and cystic areas may also be found. The medullary canal, metaphyseal border of the tumor, often does not have a reactive tissue margin but is fairly regular in its outline.

Staging Giant Cell Tumors

Giant cell tumors of bone are serious, somewhat unpredictable lesions. Some act in a benign manner and can be cured following simple curettage while others have a high recurrence rate and may even metastasize to the lungs and, on rare occasion, cause death. Investigators have attempted to correlate the histologic appearance with the clinical course and degree of aggressiveness but, except for the frankly malignant lesions, have been unable to do so (Dahlin, 1981; Goldenberg et al, 1970).

The biopsy must be carefully done to avoid hemorrhagic or necrotic areas and should provide viable tissue to the pathologist for evaluation. Histologically, a giant cell tumor consists of two main types of cells: stromal and giant cells. Most pathologists feel that the significant cell is the stromal cell (Dahlin, 1981; Goldenberg et al, 1970; Jaffe, 1958; Mirra, 1980). In the usual giant cell tumor, the stromal cells are mononuclear, plump, spindle shaped or ovoid in various proportions. The nucleus is elongated or round. Occasionally, mitotic figures are seen among the stromal cells. The multinucleated giant cells are distributed between the stromal cells. Their nuclei tend to be located towards the middle of the cell and some cells may have several dozen nuclei. On the whole, the nuclei in the giant cells do not differ much from those of the stromal cells and may be indistinguishable from them. Some investigators feel the giant cells represent a conglomeration of stromal cells. Occasionally, a giant cell may show pyknosis of its nuclei apparently reflecting degeneration of the cell rather than aggressive behavior. The amount of cytoplasm of the giant cell is usually considerable, frequently granular and sometimes vacuolated.

High concentrations of acid phosphatase may be found within the giant cell. However, acid phosphatase is also found in normal bone osteoclasts, in foreign body giant cells, and in Langhans' giant cells of tuberculosis, and is therefore not pathognomic.

In some areas of the giant cell tumor, healing may be evident. For instance, abundant intercellular collagen may be present. Giant cells may become scarce and foam cells appear, reflecting spontaneous regression and healing occurring within a portion of the lesion. Occasionally, this healing reaction may predominate over the characteristic giant cell tumor tissue.

Dahlin (1981) has recently stated that grading of giant cell tumors has not been of help in predicting which lesions will recur or which will become sarcomatous.

Occasionally, one does encounter a giant cell tumor which is frankly sarcomatous at the outset. In these circumstances the stromal cells appear sarcomatous. Some authors prefer to call these tumors fibrosarcoma with giant cells while others use the term malignant giant cell tumor (Dahlin, 1981).

It is disturbing to the clinician that the giant cell tumor can recur, even metastasize, and still maintain histologically a seemingly benign giant cell tumor pattern and benign cellular detail. In addition, the pulmonary metastases may appear histologically to be perfectly benign (Dahlin, 1981).

Although a giant cell tumor can produce metastasis, although cytologically benign, it does so rarely. Only two such cases of 264 giant cell tumors were found by the Mayo Clinic. Giant cell tumors of the vertebrae or sacrum have a better prognosis than one would expect from the experience of such tumors at conventional sites and this is helpful in predicting their course (Dahlin, 1981).

Observations have been recorded of histologic findings of plugs of tumor cells within a vein adjacent to a giant cell tumor. These tumor cells within a vessel show no histologic evidence of malignancy. Although this histologic observation is not often followed by the appearance of metastasis, it is a worrisome finding. Thus, the significance of vascular invasion with giant cell tumor tissue remains unknown.

It is also disturbing to the clinician to realize that, when studying large microscopic sections of this lesion, marked variation within the same lesion is seen from aggressive to benign-appearing tissue. This suggests the importance of adequately sampling the lesion and examining several portions of the tumor.

Occasionally, the giant cell tumor appears to have changed its histologic appearance, and its metastases have the appearance of a fibrosarcoma. Careful review is necessary to determine whether this lesion was a sarcoma from the beginning. It is well known that radiation may induce malignant degeneration to fibrosarcoma from a previously benign giant cell tumor, and this accounts for the reluctance of most physicians to treat these lesions with radiation therapy.

The rapid onset of the lesion, the appearance on the roentgenogram, the rate of progression, and the overall clinical picture is more helpful to the clinician in gauging the future behavior of the so-called benign giant cell tumor than is the histologic appearance. Aggressive lesions may penetrate the cortex and invade the surrounding fascia and muscle and rarely develop a capsule or pseudocapsule. On the rare occasions when a giant cell tumor does invade through articular cartilage or extends far up in the medullary canal, aggressive behavior of the tumor is suggested.

Instead of grading giant cell tumors, it is probably more appropriate today to stage giant cell tumors according to the Enneking classification (Enneking et al, 1980) for malignant neoplasms. Most giant cell tumors could probably be staged as IA.

Thus, grading of the typical giant cell tumor on the basis of histologic appearance is not considered helpful. However, there is hope that in the future proper staging including many variables may be of value in establishing a prognosis and treatment.

References

1. Dahlin, D.C.: Bone Tumors. Charles C. Thomas, Springfield, 1981.

2. Enneking, W.F., Spanier, S.S., and Goodman, M.A.: Surgical staging of musculoskeletal sarcomas. Clin. Orthopaed. Rel. Res. 153, 106, 1980.

3. Goldenberg, R.R., Campbell, C.J., and Bonfiglio, M.: Giant cell tumor of bone, J. Bone Joint Surg. 52(A), 619, 1970.

4. Jaffe, H.L.: Tumors and Tumorous Conditions of the Bones and Joints. Lea and Febriger, Philadelphia, 1958.

5. Mirra, J.M.: Bone Tumors: Diagnosis and Treatment. Eds: R.H. Gold and R.C. Marcove. Lippincott, Philadelphia, 1980.

6. Peimer, C.A., Schiller, A.L., Mankin, H.J. and Smith, R.J.: Multicentric giant-cell tumor of bone. J. Bone Joint Surg. 62(A), 652, 1980.

Aneurysmal Bone Cyst

F. H. SIM

Introduction

Aneurysmal bone cyst may present difficult diagnostic and therapeutic problems. Before 1942, when aneurysmal bone cyst was first described by Jaffe and Lichtenstein (1942), this lesion was frequently confused with and diagnosed as a giant cell tumor or a sarcoma. The most important consideration in the diagnosis of aneurysmal bone cyst remains its accurate histologic differentiation from other lesions. One must recognize that many entities have aneurysmal bone cyst-like areas. The lesion exists as a component of giant cell tumor, chondroblastoma, chondromyxoid fibroma, fibrous dysplasia, and telangiectatic osteosarcoma. For this reason an adequate biopsy specimen must be obtained if sampling errors are to be avoided.

It is generally accepted that the aneurysmal bone cyst is not a true neoplasm but that it is a reactive non-neoplastic process. The relationship to trauma is obscure; although some believe that an aneurysmal bone cyst arises from a precursor lesion, such as fibrous dysplasia or arteriovenous fistula, most believe that the cyst arises *de novo* (Dahlin, 1978). Little is known about its pathogenesis. The most widely discussed theories involve either increased vascular pressure from alteration of the vascular mechanics or possibly a local release of a fibrinolytic material that destroys the bone and results in a reactive lesion.

Clinical Findings

In the Mayo Clinic series (Dahlin, 1978), 134 patients underwent treatment for aneurysmal bone cyst (Fig. 1). The series consisted of slightly more women than men. In striking contrast to the age distribution in giant cell tumor, 85% of which occur in paients who are 20 years old or older, 78% of the patients with aneurysmal bone cyst are less than 20 years old. The long bones and the vertebral column are the usual sites for aneurysmal bone cyst, but it may occur in almost any bone. Although the lesion generally is metaphyseal, after epiphyseal closure it may extend to the end of the bone. Multiple vertebral involvement is not uncommon. Pain and swelling are the usual presenting symptoms (Nobler et al, 1968; Tillman et al, 1968), and these may be present from several weeks to several years, although the average is approximately three months. When the lesion is near a joint, there may be swelling or limitation of range of motion. When the lesion is located in the spinal region, stiffness and radicular pain or occasionally paralysis may be the symptoms. Surprisingly, orthopedic fractures are uncommon.

Current Concepts of Diagnosis and Treatment
of Bone and Soft Tissue Tumors.

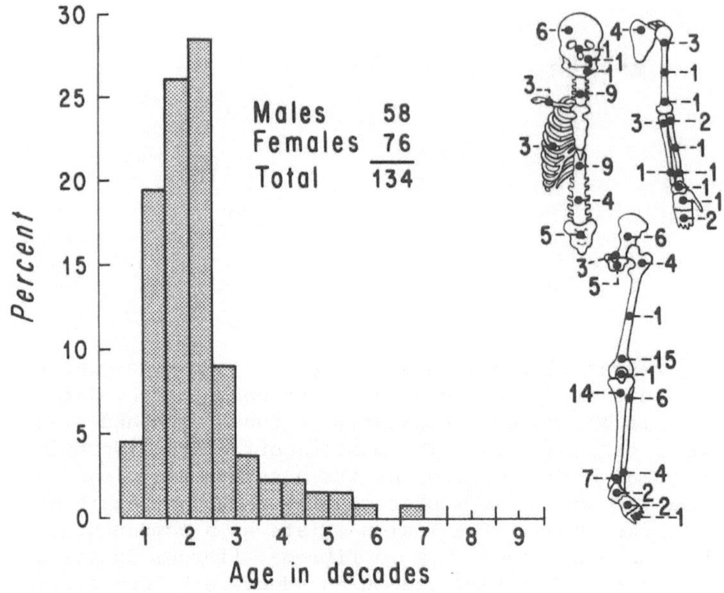

Fig. 1. Age and sex incidence of patients with aneurysmal bone cyst in Mayo Clinic series. (From Dahlin, D.C.: Bone Tumors: General Aspects and Data on 6,221 Cases. Third Ed. Chas. C. Thomas, Springfield, Ill, 1978. By permission.)

Radiographic Findings

The classification of aneurysmal bone cysts presented by Dabska and Buraczewski (1969) has been helpful in the management of this condition. The initial phase is osteolysis with periosteal elevation. The second phase is the growth phase characterized by rapid increase in the size of the cyst, with extensive bone destruction and poor margination. This is followed by the third phase, which is a stabilization phase, that is, one characterized by the development of a bony shell and osseous septa. The fourth phase is the healing phase, with progressive ossification of the lesion. Aneurysmal bone cyst is probably the only lesion that derives its name from its radiographic appearance. Classically, it presents with a ballooned or aneurysmal cystic expansion. Typically, there is a surrounding rim of sclerosis and periosteal new bone formation (Fig. 2). The lesion may be very aggressive and result in cortical destruction, soft tissue extension, and the formation of Codman's triangles. Thus, the differential diagnosis must include sarcoma. Approximately 20% of aneurysmal bone cysts occur in the spinal region, where a lytic expansile lesion of the dorsal elements is typical. Aneurysmal bone cysts have the capability of involving multiple adjacent vertebrae.

The differential diagnosis of aneurysmal bone cyst based on roentgenographic appearance alone includes many entities. Metabolic conditions such as hyper-

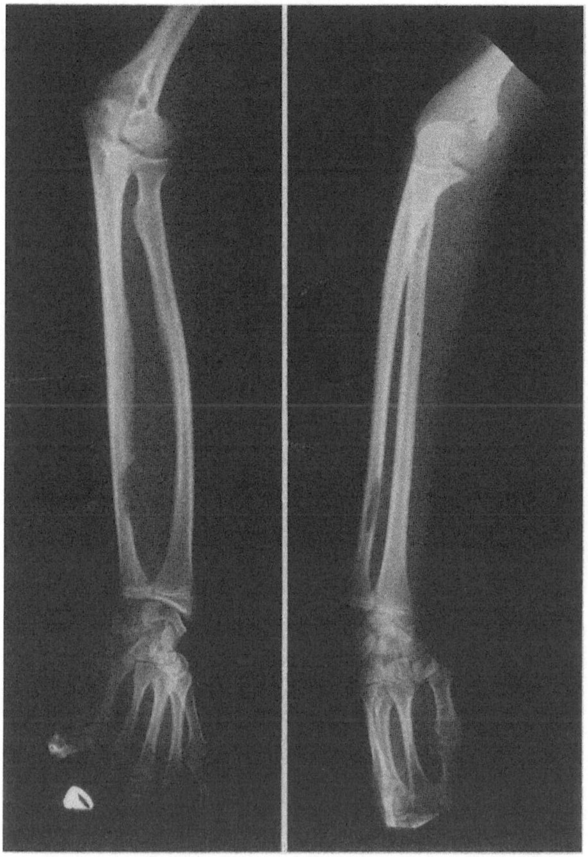

Fig. 2. Oblique views of right forearm showing typical aneurysmal bone cyst in metaphysis of distal ulna.

parathyroidism must be ruled out. There is a need to differentiate an aneurysmal bone cyst from a telangiectatic osteosarcoma; a benign giant cell tumor may be confused with an aneurysmal bone cyst. Other conditions such as chondromyxoid fibroma, osteoblastoma, histiocytosis X, and, rarely, a unicameral bone cyst may mimic this lesion.

Pathologic Findings

An aneurysmal bone cyst contains anastomosing cavernous spaces usually filled with unclotted blood (Jaffe and Lichtenstein, 1942). This space is surrounded by an eggshell-thick layer of subperiosteal new bone formation. Histologically, the essential feature is the presence of cavernous spaces, whose walls lack the normal features of blood vessels (Fig. 3). The septa are thin and contain bony trabeculae which have been called "fiber-osteoids." Although mitotic activity may be seen, the stroma of the spindle cell lacks atypia, and confusion with malignancy should not occur. The presence of benign giant cell tumors, which may be frequent, can cause confusion with a giant cell tumor. Mineralization may occur within an aneurysmal bone cyst, and this may cause confusion with osteosarcoma.

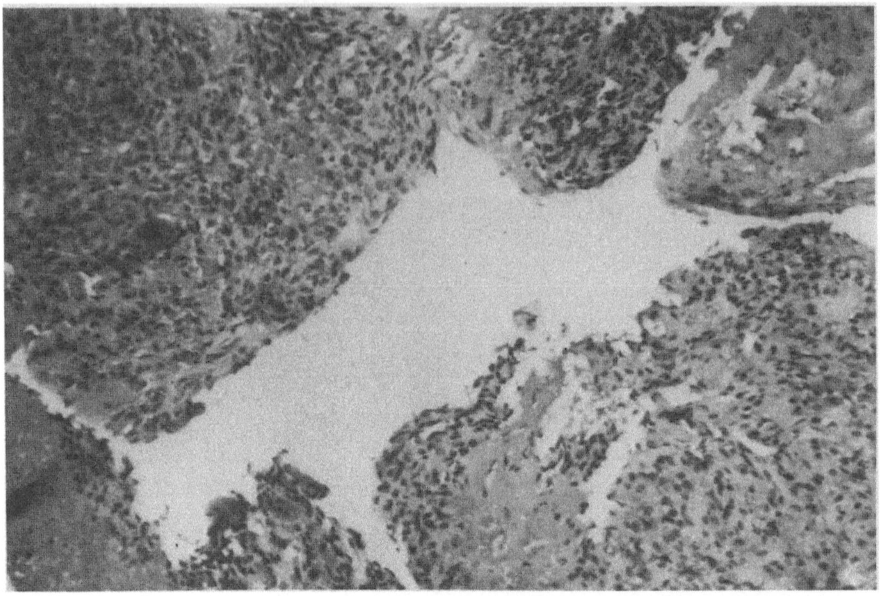

Fig. 3. Cavernous spaces between septa composed of loosely arranged spindle cells and benign giant cells. (Hematoxylin and eosin; x 100)

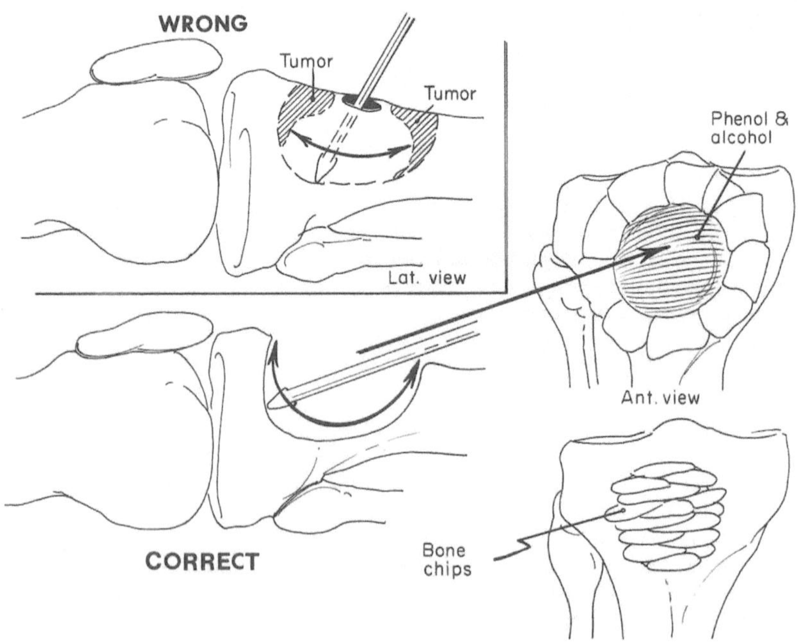

Fig. 4. Diagram of technique of curettage. Wide exteriorization is essential.

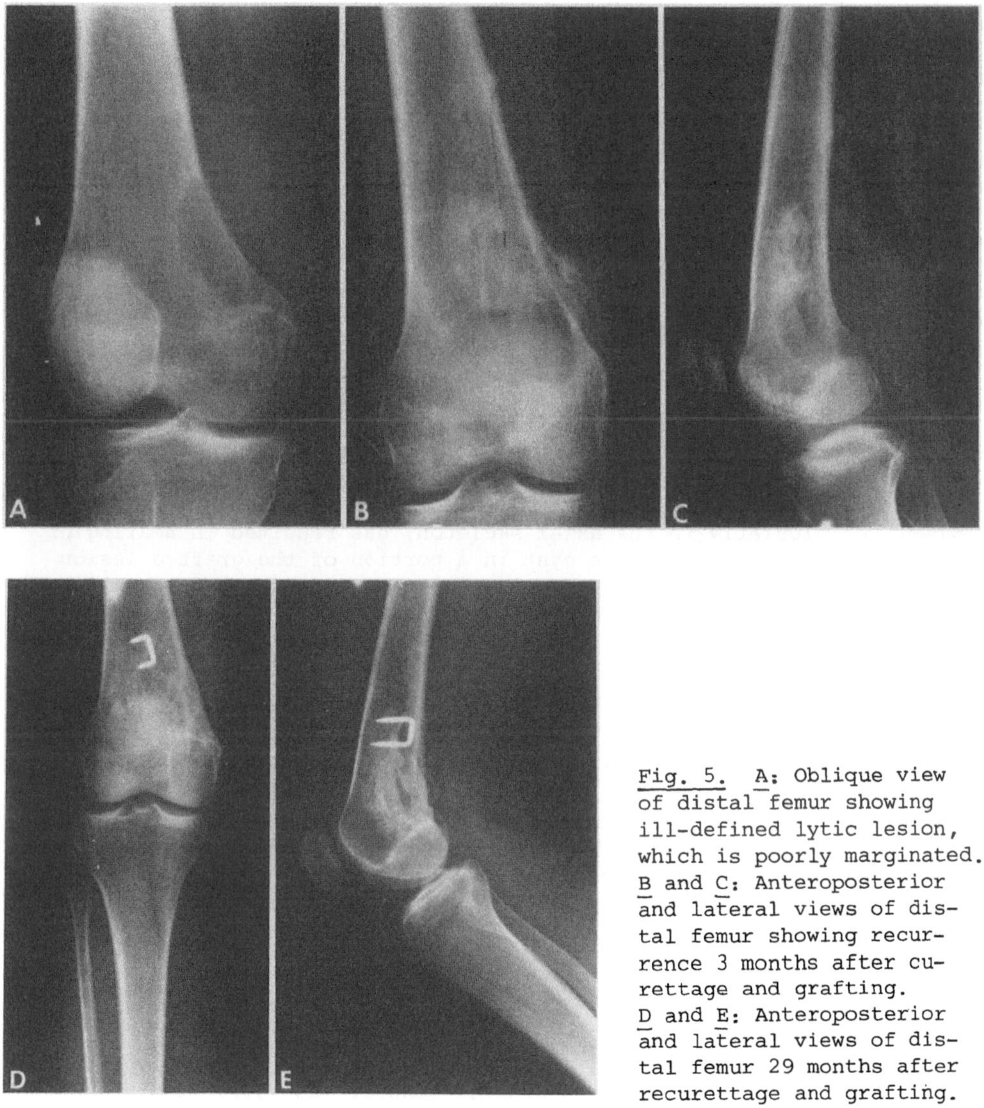

Fig. 5. A: Oblique view of distal femur showing ill-defined lytic lesion, which is poorly marginated. B and C: Anteroposterior and lateral views of distal femur showing recurrence 3 months after curettage and grafting. D and E: Anteroposterior and lateral views of distal femur 29 months after recurettage and grafting.

Treatment

Aneurysmal bone cysts are amenable to various treatment modalities. The treatment of choice is en bloc resection when the lesion is located in expendable bone such as the fibula, rib, patella, or portions of the scapula. When en bloc resection cannot be done because the location of the lesion would cause too much functional loss, curettage and bone grafting are preferred. Amputation is rarely necessary. While cryosurgery has been reported to be helpful, we have not used this modality. We do not recommend radiation therapy in the management of aneurysmal bone cyst because of the potential for malignant transformation. The only three incidences of malignant transformation in our experience occurred after radiation therapy (Tillman et al, 1968). When curettage and grafting are used, wide exteriorization is essential to the success of this treatment (Fig. 4). We use chemical cautery with phenol and alcohol and prefer autogenous bone grafts.

Our overall recurrence rate has been 21%, and for patients treated by curettage and grafting, it has been 29%. Recurrence usually develops within two years and may develop very rapidly after the initial treatment. This is heralded by recurrence of pain and swelling and roentgenographic evidence of dissolution of the bone grafts. The treatment for a recurrence is generally the same as for a primary lesion (Fig. 5), but it may involve a more radical procedure, such as en bloc resection and reconstruction. A number of factors have been reported to influence recurrence including youth, sex, lesion location, mitotic index, lesion size, and completeness of the initial excision. Of these factors, probably the most significant is incompleteness of the initial excision. However, incomplete excision of aneurysmal bone cyst, particularly in the axial skeleton, has resulted in healing of these lesions. Persistence of a cyst in a portion of the grafted lesion that is not expanding, as observed on serial roentgenograms, in a patient who is asymptomatic does not require further treatment. Close observation may be all that is needed.

Summary

Aneurysmal bone cysts are reactive non-neoplastic lesions that are found predominantly in the metaphyseal region of long bones in patients less than 20 years old. Treatment of choice is en bloc resection for expendable bones. When en bloc resection is not feasible, wide exteriorization with careful curettage and autogenous bone grafting is recommended. Although aneurysmal bone cysts may present diagnostic and therapeutic problems, close collaboration among the clinician, radiologist, and pathologist will help avoid pitfalls in management.

References

1. Dabska, M., and Buraczewski, J.: Aneurysmal bone cyst: Pathology, clinical course and radiologic appearance. Cancer 23, 371, 1969.

2. Dahlin, D.C.: Bone Tumors: General Aspects and Data on 6,221 Cases. Third Ed., Charles C. Thomas, Springfield, Ill. 1978.

3. Jaffe, H.L., and Lichtenstein, L.: Solitary unicameral bone cyst: With emphasis on the roentgen picture, the pathologic appearance and the pathogenesis. Arch. Surg. 44, 1004, 1942.

4. Nobler, M.P., Higinbotham, N.L., and Phillips, R.F.: The cure of aneurysmal bone cyst: Irradiation superior to surgery in an analysis of 33 cases. Radiology 90, 1184, 1968.

5. Tillman, B.P., Dahlin, D.C., Lipscomb, P.R., and Stewart, J.R.: Aneurysmal bone cyst: An analysis of ninety-five cases. Mayo Clin. Proc. 43, 478, 1968.

Presentation and Pathology of Soft Tissue Tumors

T. D. BROWER

If one were to analyze all the soft tissue tumors of the extremities, including ganglia, seen in a surgeon's office, the following generalizations could be made.

Seven out of ten such tumors would obviously be benign. There are ganglia, subcutaneous lipomas, neurofibromas and other benign tumors. One would be frankly malignant (Enneking: personal communication). This is usually determined by a rapidly increasing, large, subfascial tumor. Two would fall into the undeterminable category.

Soft tissue tumors are not always painful and, therefore, may be discovered by the patient as a lump. Some tumors cause neurologic symptoms but rarely vascular occlusion. Some physical observations may help in differentiating a benign from a malignant soft tissue tumor.

1) Deep tumors, subfascial, are more likely to be malignant than the superficial ones.

2) Tumors in young persons are less likely to be malignant than in the older population.

3) Rapidly growing tumors are obviously more likely to be malignant than the static ones.

4) Adherence to skin or to the bone implies a malignant tumor.

5) Hard tumors are more likely to be malignant than soft.

6) The rare tumor that is tender on palpation implies a rapid growth and, therefore, is likely to be malignant.

It can be confidently stated that soft tissue tumors are the most mismanaged neoplasms in the U.S.A. All surgeons treating breast tumors have a clearly defined plan of attacking the problem and are cognizant of the available alternatives. Not so with tumors of the soft tissues of the extremities. The usual surgical procedure is to do a biopsy of the apparently innocuous mass and then frantically seek help should the report show a malignancy. Centers treating soft tissue sarcomas receive most of their patients after a surprise biopsy or the patient and surgeon recognize a recurrence after inadequate therapy.

Current Concepts of Diagnosis and Treatment
of Bone and Soft Tissue Tumors.

Before discussing the specific pathologic pattern of soft tissue tumors, I would like to describe their gross pathologic appearance in general terms. Malignant soft tissue tumors often arise *do novo* within muscle compartments. The tumor enlarges rather concentrically, pushing and compressing normal tissue ahead of it. This tissue becomes fibrous and is referred to as the "capsule". The more malignant tumors send pseudopods through the capsule and the tumor tends to spread preferentially proximally within the muscle or muscle planes as satellite nodules or intramuscular metastases. Only after the tumor has become extremely large does it break through intermuscular septa. Malignant soft tissue tumors find the loose areolar tissue about nerves and blood vessels advantageous for expansion. Very late in the natural history of a deep soft tissue tumor, it bursts through the overlying deep fascia and ulcerates the skin.

The classic muscle compartment in the lower extremity is that of the quadriceps femoris. Here, the tumor is confined for some time to the medial and lateral intermuscular septa and the tendinous muscle origins and insertions. Posteriorly, the thigh muscle compartment is violated by the sciatic nerve and the loose tissue about the nerve allows ready escape by the aggressive tumor. The gluteal muscles, however, are deep and broad and tend to confine tumors for a long time. Other compartments of the extremities can become tumor sites in the lower and upper extremities.

Should the soft tissue tumor arise outside a muscle close to a major nerve or vessel, no anatomic barrier blocks its proximal spread. Tumors arising in the hand, foot, antecubital fossa, or popliteal fossa have no well defined anatomic structures to limit the extension of the lesion.

With these general principles in mind, we shall now discuss specifically the microscopic pathologic aspects of some of the soft tissue malignancies. The most complete listing of the histologically identified soft tissue tumors, I have found, is in Enzinger's excellent text (1983). He tabulates almost 200 benign and malignant soft tissue tumors. The purpose of this comprehensive classification is to offer us some ability to predict the tumor's behavior from past experience. In theory, this would seem helpful and, ultimately, I think it will be. Over the past 30 years, pathologists have shown their humanness by their tendency to diagnose by whim. The 1950's was the decade of the synovioma. Pathologists seemed to vie for the frequency with which this diagnosis could be made. The 1960's was the decade for liposarcoma and, of course, in the 1970's fibrous histiocytoma came to the fore. Indeed, the identification of histiocytoma by Enzinger has seemed to calm the waters. This lesion has a varied histologic presentation and, before its identification, provoked multiple diagnoses from a single slide. The use of electron microscopic and special staining techniques, I hope, will stabilize this previously faddish approach to soft tissue tumor diagnosis.

A definite histologic classification of the soft tissue tumor offers the clinician some confidence in the predictable course of the disease and, therefore, more insight into the proper therapy. A myxoid liposarcoma is of low grade malignancy. Local recurrence occurs almost always before distant metastasis. A pleomorphic liposarcoma is usually much more aggressive and, therefore, may metastasize to the lungs early. The peculiarities of the malignant fibrous histiocytoma are discussed by Dr. Bouchard.

Grading the degree of malignancy of sarcomas has been a challenge to pathologists. These tumors have not lent themselves to categorizing as well as epithelial tumors. Enzinger proposes three grades but readily admits that grading does not have the same predictive value in all sarcomas. Some tumors have high mitotic activity and these are predictably aggressive. In others, such as the malignant fibrous histiocytoma, the number of mitoses does not predict its behavior.

The clinician should obtain from the pathologist the histologic classification of the tumor for predictive information about the course of the lesion and appropriate therapy. If the pathologist can give some estimate of the grade of malignancy of the tumor, this can be added to the equation. Two grades should be sufficient -- high or low.

In summary, the clinical presentation of a soft tissue tumor can provide significant information about its gross pathologic feature, its location, position, and rapidity of growth. Knowledge of the behavior of soft tissue tumors along with the histologic classification and possibly a grade, should provide the clinician with the information needed to develop a proper therapeutic plan.

References

1. Enneking, William F.: Personal communication.

2. Enneking, William F., Spanier, S. S., and Malwar, M.M.: The effect of the anatomic setting on the results of surgical procedures for soft part sarcoma of the thigh. Cancer 47, 1005, 1981.

3. Enzinger, Franz M., and Weiss, Sharon: Soft Tissue Tumors. C.V. Mosby, St. Louis, 1983.

Preoperative Assessment of Soft Tissue Sarcomas of the Extremities

P. KINNARD

Introduction

In evaluating a soft tissue mass, the physician must always bear in mind the possibility of a soft tissue sarcoma and, to determine precisely the diagnosis, it is imperative to collect all the pertinent data by a careful history and physical examination supplemented by special imaging modalities.

Local recurrence, metastases, and survival of patients with sarcomas are mostly affected (Simon and Enneking, 1976) by the location of the tumor and by the adequacy of the surgical procedure; therefore, the evaluation of the patient must be done in a stepwise fashion to perform the correct treatment consisting of, whenever possible, a functional limb-saving wide excision (Hajdu, 1979).

Prebiopsy Planning

History and physical findings

Previous surgical excision of a benign looking mass, changes in tumoral size and pain pattern carry a high index of suspicion for malignancy but tumoral behavior often may be not striking.

Roentgenograms

Well performed conventional radiography (Levine et al, 1979) is mandatory to differentiate tumors from other pathologic conditions and to separate a primary or metastatic bone tumor invading soft tissues from a soft tissue sarcoma whether or not invading adjacent bony structures.

Good quality films are as good as xeroradiography (Madewell and Sweet, 1981), allowing definition of the radiographic density of the tumor, the tumoral interface with adjacent structures, tumoral growth rate and associated calcifications.

Radiographic density may be unreliable because, beside lipoma, all soft tissue masses are of water density and distinction between benign and

Current Concepts of Diagnosis and Treatment
of Bone and Soft Tissue Tumors.

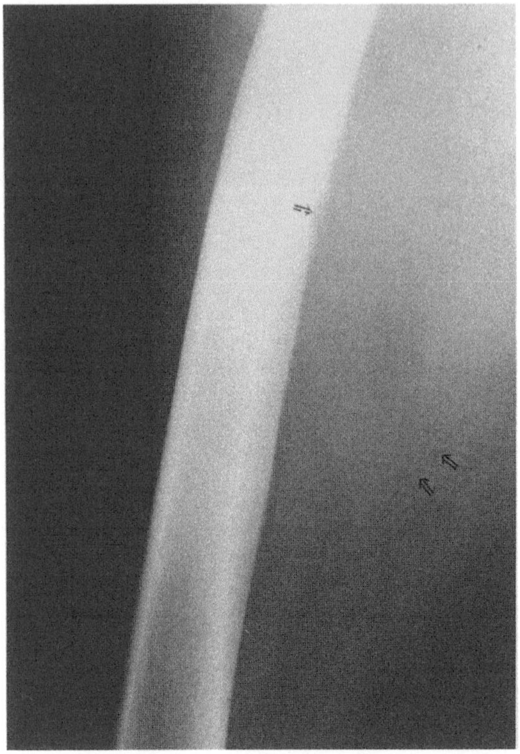

Fig. 1. Fibrosarcoma of the thigh with bone erosion (arrow) and faint calcifications (double arrow).

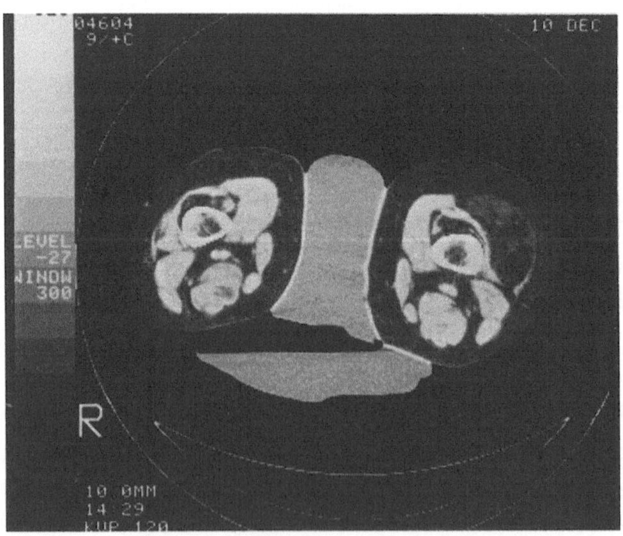

Fig. 2. Liposarcoma of the left thigh not invading bone.

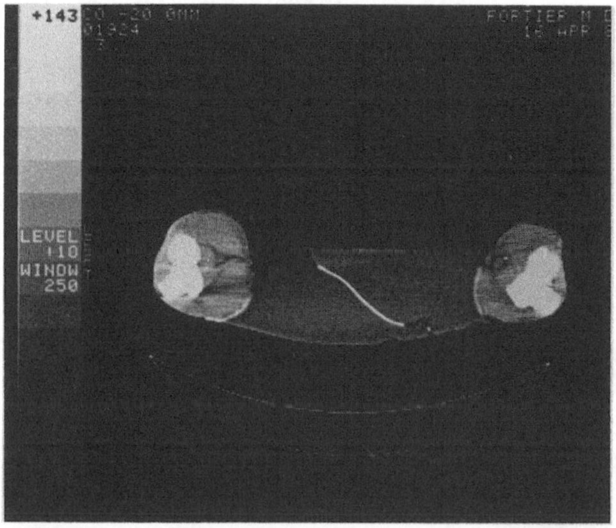

Fig. 3. Hemangiopericytoma of the left forearm demonstrated as a diffuse undefined swelling.

malignant tumor will not be possible (Forrester and Moore, 1977). Associated calcifications, if they are faint and irregular (Fig. 1) are more suggestive of a malignant process (Forrester and Moore, 1977).

Computerized axial tomography

Computerized images of the tumoral site performed by a well trained radiologist is now the next logical step in the assessment of soft tissue tumor, replacing angiography for most of the cases, providing an accurate and detailed cross sectional image of these structures (ie, extent, size, density and vascularity of the mass) (Schumaker et al, 1978).

By demonstrating possible encroachment on vital structures and bone involvement, it helps to establish the feasibility of surgical treatment (Fig. 2). Moreover, the content of the tumor may be precisely evaluated by specific roentgenogram absorption (Schumaker et al, 1978), coefficient values, and its vascularization studied by contrast enhancement using meglumine diatrizoate (Gilula, 1980).

Contrast enhancement may be particularly helpful in poorly defined or isodense lesions by outlining either the tumor or its pseudocapsule (Levine et al, 1979).

The actual shortcoming of this technique is the differentiation of a mass from the adjacent soft tissues in areas such as the leg and forearm where the fat is sparse (Fig. 3) (Gilula, 1980).

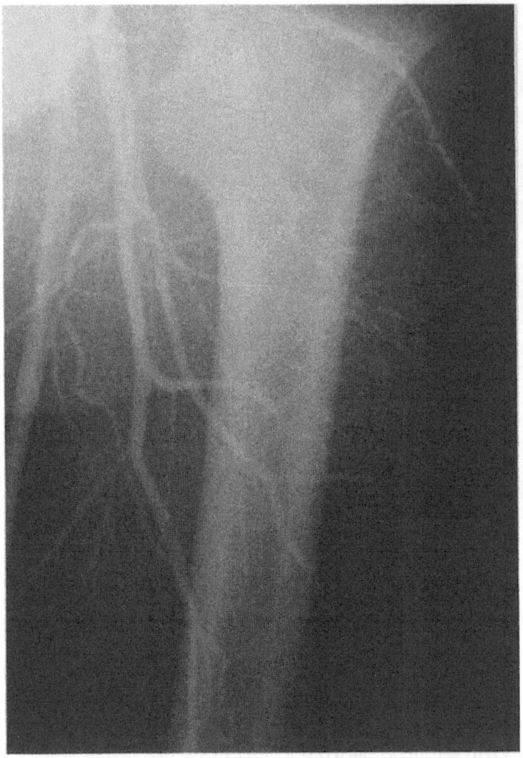

Fig. 4. Cirsoid vascular **hemartoma** of the lateral aspect of the thigh with neovascularization suggesting malignancy **by** arteriography.

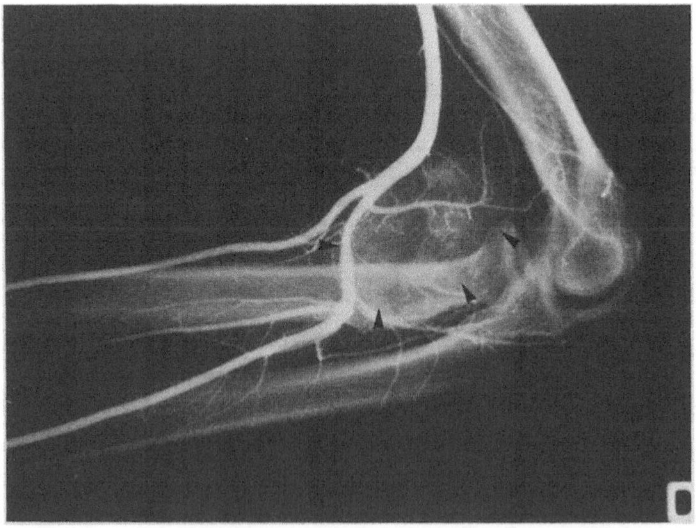

Fig. 5. Same as Fig. 3, well delineated by arterio graphy.

Angiography

Several angiographic findings such as neovascularization, increased blood flow and venous pooling, although nonspecific, may be suggestive of malignancy (Finck and Moore, 1977; Viamonte et al, 1973), but may also be demonstrated in angiodysplasias, hemartomas and inflammatory conditions (Finck and Moore, 1977; Martel and Abell, 1973). Angiography remains superior to computed tomography in tumors with possible direct involvement of an artery or vein, in judging the adequacy of the blood supply to the lower extremity of an elderly patient undergoing en bloc resection, and in areas where fat is sparse (Fig. 4 and 5) (Hudson et al, 1975; Levine et al, 1979; Viamonte et al, 1973).

Sonography

Sonography is not particularly helpful in the assessment of a tumor but may be of help in judging the response of a mass to nonsurgical treatment (Madewell and Sweet, 1981).

Radionuclide scanning

Radionuclide scanning is invaluable in detecting distant metastasis, influencing, therefore, management and prognosis (Alazraki, 1981; Levine et al, 1979; Madewell and Sweet, 1981; Resnick, 1981; Telfer, 1977).

Locally, a hot scan cannot differentiate bone reaction from invasion, but invariably means there is only, in the best circumstances, a narrow margin of reactive tissue between the lesion and bone; therefore, dissection of the tumor away from bone will likely be followed by recurrence and is not consistent with an adequate wide excision (Simon, 1982).

Other diagnostic procedures

The other diagnostic procedures such as computed tomography of the lungs (Sagel and Stanley, 1980) or skull, radionuclide imaging of the liver, lungs or spleen, lymphography and intravenous pyelography belong to the postbiopsy stage and will be indicated according to the histologic type and behavior of the tumoral specimens.

Biopsy Planning

Biopsy should be regarded as a final diagnostic procedure and should not be lightly undertaken.

Tumoral specimens may be obtained by needle aspiration or open incisional or excisional biopsy.

Although needle biopsy may be the method of choice for difficult sites, an open incisional biopsy, when feasible, is clearly superior provided proper samplings are taken and controlled by frozen section (Schajowicz, 1981).

Indeed, the major source of error in diagnosis is inadequate tissue due to defective technique (Schajowicz, 1981); either the specimen is taken from peripheral healthy tissue or from necrotic tissue inside the tumor.

Excisional biopsy must be condemned because, invariably, it will leave behind tumoral cells in the surgical cleavage plane which remain a potential source of spreading.

Although in many instances definitive treatment may be carried out after frozen section, there is no harm in waiting a few days for the final conclusions before undertaking the definitive treatment provided the procedure will be able to remove the biopsy site and tract with a good margin of safety (Schajowicz, 1981; Simon, 1982; Watts, 1980).

Conclusions

Although all the modalities outlined above assist the surgeon in obtaining morphologic information about the soft tissue mass, the preoperative diagnosis remains speculative and almost invariably a biopsy is required to establish the diagnosis; however, the surgeon will have a thorough knowledge of some of the three dimensional aspects of the tumor, enabling him to undertake the best treatment for his patient.

Acknowledgement

I wish to acknowledge the help of Dr. Antoni Trias in the preparation of this manuscript.

References

1. Alazraki, N.: In, Diagnosis of Bone and Joint Disorders, Eds: Resnick and Niwayama. Saunders, 1981, pp 639-676.

2. Finck, E. and Moore, T.: Angiography for mass lesions of bone, joint and soft tissue. Orthop. Clin. North Amer. 8, 999, 1977.

3. Forrester, D. and Moore, T.: The radiology of bone and soft tissue sarcomas. Orthop. Clin. North Amer. 8, 973, 1977.

4. Gilula, L.: In, Computed Tomography, Ultrasound and X-Rays: An Integrated Approach, Eds: Moss, A., and Goldberg, H. Academic Press, 1980. pp 335-370.

5. Hajdu, S.: In, Pathology of Soft Tissue Tumors. Ed: S. Hajdu. Lea Febiger, 1979, pp 537-541.

6. Hudson, T., Haas, G., Enneking, W., and Hawkins, I.: Angiography in the management of musculoskeletal tumors. Surg. Gynec. Obstet. 141, 11, 1975.

7. Levine, E., Kyorakhee, J., Neff, N., Maklad, R., Robinson, R. and Preston, D.: Comparison of computed tomography and other imaging modalities in the evaluation of musculoskeletal tumors. Radiology 131, 431, 1979.

8. Madewell, J. and Sweet, D.: In, Diagnosis of Bone and Joint Disorders. Eds: Resnick and Niwayama, Saunders, 1981. pp 2691-2692.

9. Martel, W. and Abell, M.: Radiologic evaluation of soft tissue tumors, a retrospective study. Cancer 32, 352, 1973.

10. Resnick, D.: In, Diagnosis of Bone and Joint Disorders, Eds: Resnick and Niwayama. Saunders, 1981, pp 254-409.

11. Sagel, S. and Stanley, R.: In, Computed Tomo graphy, Ultrasound and X-Rays: An Integrated Approach, Eds: Moss, A. and Goldberg, H. Academic Press, 1980, pp 154-156.

12. Schajowicz, F.: In, Tumors and Tumorlike Lesions of Bone and Joints, Ed: Schajowicz. Springer-Verlag, 1981, pp 1-17.

13. Schumaker, T., Genant, H., Korobkin, M., and Bovill, E.: Computed tomography: Its use in space occupying lesions of the musculoskeletal system. J. Bone Joint Surg. 60A, 600, 1978

14. Simon, M. and Enneking, W.: The management of soft tissue sarcomas of the extremities. J. Bone Joint Surg. 58A, 317, 1976.

15. Simon, M.A.: Current concepts review: Biopsy of musculoskeletal tumors. J. Bone Joint Surg. 64A, 1253, 1982.

16. Telfer, N.: Nuclear medicine in the management of musculoskeletal tumors. Orthop. Clin. North Amer. 8, 1011, 1977.

17. Viamonte, M., Roen, S., and Lepage, J.: Nonspecificity of abnormal vascularity in the angiographic diagnosis of malignant neoplasms. Radiology 106, 59, 1973.

18. Watts, H.: Introduction to resection of musculoskeletal sarcoma. Clin. Orthop. Relat. Res. 153, 31, 1980.

Malignant Fibrous Histiocytoma of Soft Tissues

H.-L. BOUCHARD, R. LAGACÉ, and C. DELAGE

Introduction

Sarcomas of soft tissues, particularly those from the deep sites of the extremities, raise a problem of diagnosis and treatment. Histiocytic sarcomas, including malignant fibrous histiocytoma (MFH), represent a group of neoplasms with an unpredictable course and for which treatment varies widely. Kauffman and Stout (1961) reviewed 15 histiocytic tumors in children, of which three appeared to represent the malignant counterpart of the benign histiocytoma. O'Brien and Stout (1964) postulated that the histiocyte could behave as a phagocyte but was also able to produce connective tissue fibers. They selected 15 cases of "malignant fibrous xanthomas" and concluded that there existed a malignant counterpart of fibrous histiocytoma or xanthoma. Kempson and Kyriakos (1972) reported 30 cases of fibroxanthosarcoma which exhibited the histologic criteria of MFH, and proposed a new classification of fibrous histiocytomas. Yao-Shi Fu et al (1975) confirmed the histiocytic and fibroblastic component of MFH by electron microscope and tissue cultures. They suggested that those two major cellular components originated from a single pluripotential and undifferentiated stem cell. Enzinger (1979) described 41 cases of the angiomatoid variety of MFH which occurred principally in the lower extremities of young patients (5-25 years). A large excision was recommended as the appropriate treatment.

Definition and Clinical Features of MFH

MFH is a tumor characterized by a proliferation of pleomorphic sarcomatous cellular elements including histiocytes, fibroblasts, giant cells, and xanthomatous cells, more often dispersed in a typical storiform pattern (Weiss and Enzinger, 1978). The lower extremities are the most frequent anatomic site (±50%) and a rapid tumorous growth is the major symptom. This tumor may occur at any age but has a predominance for the 6th and 7th decades and occurs more often in men. Site, depth, volume, and number of muscles involved and the integrity of the neurovascular structures represent the most important clinical prognostic features. The inflammatory component, mitotic index, cellular polymorphism and paraneoplastic syndromes are other features of prognosis.

Current Concepts of Diagnosis and Treatment
of Bone and Soft Tissue Tumors.

Material and Methods

One hundred and forty-one soft tissue sarcomas from the files of the Department of Pathology, l'Hotel-Dieu de Quebec Hospital, between 1972-1982, were reviewed; 31 of these fulfilled the morphologic criteria of MFH (Lattes, 1982; Weiss, 1982) and were retained for the present study. This paper is a retrospective study of the clinical features, treatment and follow-up of 31 patients in whom a histopathologic diagnosis of MFH was made. The study group consisted of 18 men and 13 women, with a mean age of 58.5 years.

Results

The tumor was located in the inferior extremities in 67.8% and in the superior extremities in 19.4% of the patients. Localization in the trunk and head and neck was less frequent, 9.6% and 3.2% respectively. Size of the tumor varied from 2 to 15 cm in most of the patients (26). Involvement of deeper sites of soft tissues was the anatomic localization in 28 patients. Superficial tumors were uncommon (3). The chief complaint was the discovery of a painless mass in 17 patients, a painful mass in 12, a recurrence after local excision in seven, and in one patient a mass was found during a routine clinical examination. Time interval between the first symptom and initial treatment varied from 3 weeks to 24 months, for an average of 6.6 months. Alkaline phosphatases and CEA values proved of little value in this type of tumor whereas the erythrocyte sedimentation rate was significantly increased in 14 of the 31 patients (45.1%).

There was much variation in the therapeutic approach as summarized in Table I.

Table I. Treatment

Incomplete excision	4
Incomplete excision with radiotherapy	10
Large complete excision	10
Complete excision with radiotherapy	4
Amputation	3
Total	31

(Associated lymphadenectomy in 9)

No preoperative radiotherapy or chemotherapy was given. The patients' courses are shown in Table II.

The rate of local recurrence was 29.1%. The rate of local recurrence with metastases was 35.6%. Thirteen patients are alive with recurrence (41.9%) after 35.5 months, and nine are alive without recurrence after 25.6 months (29%). Seven patients died after an average of 21.7 months (26.8%).

Table II. Recurrences and metastases

Isolated local recurrence	9
Local recurrence with lymph node metastases	2
Local recurrence with pulmonary metastases	5
Systemic metastases	4
Total	20

Conclusion

MFH is a highly malignant tumor. Weiss and Enzinger (1978) reported a recurrence rate of 44% with a two-year survival of 60%. The best treatment remains a large excision. Local radiotherapy should be added in patients in whom large resection of the tumor appears either unsafe or insufficient. Depth, size, and the inflammatory component of the tumor seem to be of significant prognostic value in this tumor. Twenty-four of the 31 patients reported here had the usual histiologic type of MFH, six had a mixed pattern, and only one had the inflammatory variant.

References

1. Enzinger, Franz M.: Angiomatoid malignant fibrous histiocytoma. Cancer 44, 2147, 1979.

2. Fu, Yao-Shi, Gabbiani, Giulio; Kaye, Gordon I., and Lattes, Raffaele: Malignant soft tissue tumors of probable histiocytic origin (malignant fibrous histiocytomas): General considerations and electron microscopic and tissue culture studies. Cancer 35, 176, 1975.

3. Kauffman, Shirley L., and Stout, Arthur Purdy: Histiocytic tumors (fibrous xanthoma and histiocytoma) in children. Cancer 14, 469, 1961.

4. Kempson, Richard L., and Kyriakos, Michael: Fibroxanthosarcoma of the soft tissues. A type of malignant fibrous histiocytoma. Cancer 29, 961, 1972.

5. Lattes, Raffaele: Malignant fibrous histiocytoma: A review article. Am.J.Surg.Pathol. 6, 761, 1982.

6. O'Brien, Joseph E., and Stout, Arthur Purdy: Malignant fibrous xanthomas. Cancer 17, 1445, 1964.

7. Weiss, Sharon W., and Enzinger, Franz M.: Malignant fibrous histiocytoma: An analysis of 200 cases. Cancer 41, 2250, 1978.

8. Weiss, Sharon W.: Malignant fibrous histiocytoma: A reaffirmation. Am.J.Surg.Pathol. 6, 773, 1982.

Management of Soft Tissue Sarcomas

C. E. BROOKS

Introduction

Soft tissue sarcomas are difficult to manage, having a high incidence of recurrence and systemic spread following indiscriminate local resection. Recent trends in the management of soft tissue sarcomas reflect a better understanding of the biologic behavior of these neoplasms plus a more structured and aggressive surgical approach.

Until recently, radical surgery was felt essential to achieve local tumor control. Inspite of the most meticulous surgical technique, however, a large percentage of patients developed systemic metastases, thus rendering useless any planned surgical intervention. Extensive and often mutilating surgical procedures yielded poor results due to systemic dissemination of the disease.

Recently, however, evidence has been accumulating that limited surgery combined with radiotherapy and adjuvant chemotherapy yield better results than previously achieved with surgery alone (Antman et al, 1983; Morton et al, 1976; Rosenburg, 1981).

For this reason, precise surgical planning is crucial and preoperative assessment should include adequate investigations, a well planned and executed biopsy, a pathologic staging and, finally, based on the above variables, adequate multimodality therapy.

Preoperative assessment is of utmost importance in designing a suitable therapeutic approach to the problem. Diagnostic procedures should include routine hematologic evaluation and more specific diagnostic procedures related directly to the tumor under consideration.

Pulmonary assessment should consist of computerized axial tomography of both lungs.

Peripheral arteriography is useful in assessing the size of the tumor, its vascularity, and its relationship to adjacent major vessels. Computerized axial tomography of the tumor mass itself is helpful in assessing its compartmental status as well as its relationship to adjacent bone, vessels and nerves.

Current Concepts of Diagnosis and Treatment
of Bone and Soft Tissue Tumors.

SARCOMAS	STAGE GROUPING	
	Stage	Sarcoma
G1 - low grade G2 - moderate grade G3 - high grade	IA IB	G1, T1, N0, M0 G1, T2, N0, M0
N0 - no regional nodes N1 - regional nodes	IIA IIB	G2, T1, N0, M0 G2, T2, N0, M0
M0 - metastases absent M1 - metastases present	III	G3, T1-3, N0, M0
Soft Tissue	IVA IVB	G(any), T3, N(any), M0 G(any), T(any), N(any), M1
T1 - less than 5 cm T2 - greater than 5 cm T3 - invasion of bone, vessel, nerve		

Fig. 1. Soft tissue sarcomas. Staging - TNMG

Radionuclide bone scanning will assist in defining the relationship of the tumor to the adjacent bone.

Establishment of a correct tissue diagnosis is of paramount importance and must precede definitive therapy. Open biopsy is preferable and should be carefully planned in anticipation of further surgical intervention.

The biopsy may be excisional or incisional, depending on the size of the tumor. The incision should be placed so that it can be excised along with the pathologic specimen at the time of definitive surgery. Extensive tissue dissection should be avoided.

A task force on soft tissue sarcomas reported their findings in 1977. They suggested that a uniform staging septum be adopted as an aid in the definitive management of soft tissue sarcomas (Fig. 1). This has proven most helpful both in the management of those sarcomas and in assessing the prognosis of individual tumors (Enneking et al, 1980; Karakousis, et al, 1979; Manual for Staging of Cancer, 1977; Russell, et al, 1977).

Principles of Management

Following preoperative assessment, patients with localized soft tissue sarcomas are staged according to the principles outlined above.

Patients with Stage I and Stage II sarcomas are managed by local therapy alone consisting of surgery, with or without radiotherapy.

For patients with Stage IIB, Stage III and localized Stage IVA, the Montreal General Hospital Musculoskeletal Tumor Group has adopted a multimodality treatment protocol employing chemotherapy, surgery and radiotherapy.

Chemotherapy

Various chemotherapeutic regimens are used at the present time. It would appear that doxorubicin*is the most effective agent for adult soft tissue sarcomas (Rosenburg, 1981; Sutow and Maurer, 1981).

Recently, the intra-arterial delivery of cisplatin directly into the tumor bed has had modest success in the management of osteosarcoma (Benjamin, 1982). Based on the above evidence, preoperative chemotherapy has now been introduced into our protocol.

Following biopsy and staging, patients are started on doxorubicin 70 mg/m^2, continuous intravenous infusion over four days via a central venous catheter starting on day 1. Cisplatin, 120 mg/m^2, is given over a 2-hour intra-arterial infusion on day 6.

This course is repeated two weeks later followed by the appropriate surgical procedure two weeks after completion of the second cycle.

If clinical response is good or tumor necrosis has occurred with the above therapy, adjuvant treatment with doxorubicin and cisplatin will be given postoperatively. If no tumor progression or tumor necrosis is seen, no adjuvant therapy is given and the patients are closely followed.

Surgery

Surgical excision is the single most important factor in achieving local tumor control. This may be achieved either through amputation or local excision. The latter procedure must remove all sarcomatous tissue to be successful.

A better understanding of the pathologic and biologic behavior of soft tissue sarcomas has led to a more structured approach in surgical planning. A variety of surgical procedures has been advocated in attempts to achieve limb salvage.

Intralesional or marginal surgery is unsatisfactory for the tumors under discussion. Wide surgical excision consisting of an intracompartmental en bloc removal of the tumor with a cuff of normal tissue is believed to be satisfactory for small intracompartmentally contained sarcomas. Radical local resection is achieved by removing the entire muscle mass involved from origin to insertion and, transversely, the dissection passes at least one uninvolved anatomic structure away from the tumor. This may, of necessity, require sacrifice of neurovascular structures and any adjacent bone or joint if contained within the affected compartment (Simon, 1976).

* Adriamycin

Careful preoperative planning is essential when attempting a limb-salvaging procedure. If this approach is not feasible due to the extent or site of the tumor, then amputation is the obvious alternative.

Employing the principles outlined above, any amputation must be carried out above the level of origin of the muscles involved in the sarcomatous process. This may necessitate disarticulation or amputation above an uninvolved joint.

It is currently felt that en bloc surgical resection with at least one muscle or fascial plane between the tumor and the surgical margin is the preferred surgical approach. The disposition of the neurovascular bundle in relation to the tumor is obviously of crucial importance and must be accurately assessed if limb salvage surgery is undertaken (Karakousis et al, 1979).

Radiotherapy

In the current state of our knowledge, after inadequate surgery it appears unwise to rely on radiotherapy to kill viable tumor cells left in the wound. It is thus imperative that adequate surgical resection be carried out based on the principles outlined above (Carabell, 1981; Rosenburg, 1981).

Recent experience with megavoltage equipment and careful planning have shown that radiotherapy plays an important role in preventing recurrence. Postoperative radiation therapy has considerably reduced the incidence of local recurrence.

Radiotherapy is started approximately 3 to 4 weeks after surgery unless a full thickness skin graft was required in which case a 5- to 8-week lag period is necessary. The radiation field must encompass the entire surgical scar and a margin of at least 4 cm in all variables. For sarcomas involving muscle, the entire compartment is included in the radiation field although every effort must be made not to include the entire circumference of the extremities. If this is not feasible, a longitudinal strip of skin must be spared.

For Stage I lesions, a total tumor dose of 5,000 to 5,400 rads over 5-1/2 weeks is believed to be adequate. For Stage II or higher lesions, or gross residual tumor, a total dose of 6,000 to 7,000 rads is required.

It should be noted that the combination of surgical resection and radiotherapy alone does not achieve control over distant and pulmonary metastases.

Discussion

Although adequate surgical excision is the first line of defense in sarcomas of the extremities, it is recognized that 30% of patients may develop and succumb to distant metastases despite adequate local control. Even the most meticulous surgical procedure may fail unless this fact is recognized. The encouraging results obtained from a multidisciplinary

Patient eligibility : Proven soft tissue sarcoma I-IIA

Surgery : Wide excision
- intracompartmental en bloc excision

Radical resection
- extracompartmental en bloc excision

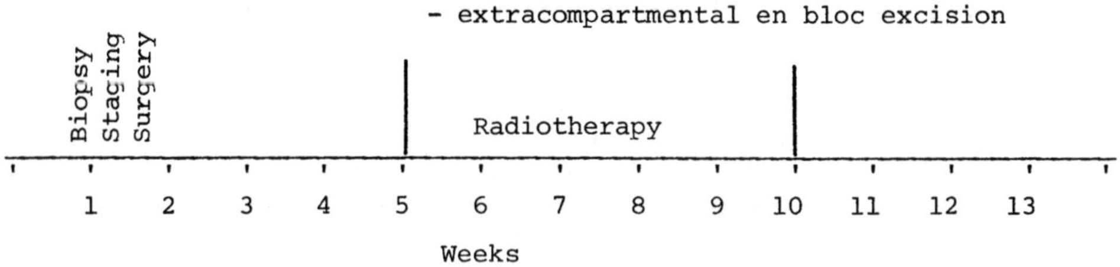

Fig. 2. Soft tissue sarcoma protocol la - Montreal General Hospital Musculoskeletal Tumor Group

Patient eligibility: Proven soft tissue sarcoma IIB-IVA
No significant cardiac history
WBC >4,000
Granulocyte count >1,500
Platelets >125,000
Creatinine clearance 765 mL/min

Surgery: Wide excision
- intracompartmental en bloc excision

Radical resection
- extracompartmental en bloc resection
- amputation

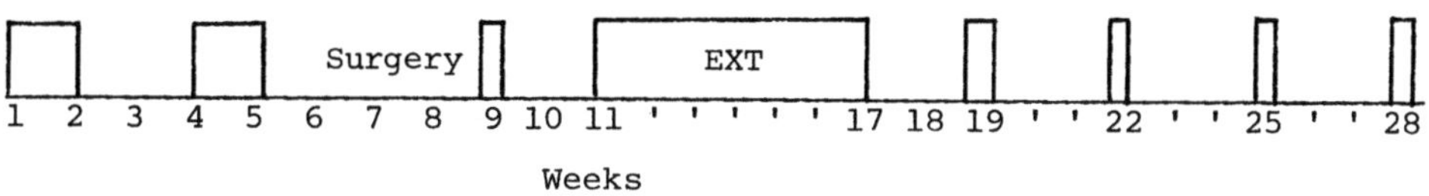

[box] doxorubicin 70 mg/m^2 i.v., continuous infusion over 4 days
cis-platin 120 mg/m^2 i.a., infusion over 2 hours on day 5 or 6

[narrow box] doxorubicin 60 mg/m^2 i.v.
cis platin 60 mg/m^2 i.v.

Fig. 3. Soft tissue sarcoma protocol lb - Montreal General Hospital Musculoskeletal Tumor Group.

approach in the treatment of osteogenic sarcoma has led to a similar approach in the management of soft tissue sarcomas.

At the present time, the Montreal General Hospital's Musculoskeletal Tumor Group has adopted two protocols based on tumor staging after biopsy.

Patients presenting with Stage I or Stage IIA sarcomas are managed by local surgical excision followed by radiotherapy when deemed necessary (Fig. 2).

Patients presenting with Stage II - IVA sarcomas are managed with pre-operative systemic chemotherapy as well as localized intra-arterial infusion chemotherapy. This treatment modality is followed by either radical resection or wide intracompartmental excision depending on the exact pathologic staging. Adjuvant chemotherapy is added to this regimen if there is evidence of a good clinical response and significant tumor necrosis. Postoperative radiotherapy is incorporated into the protocol to control local recurrence (Fig. 3).

It is recognized that these protocols must be subject to clinical review and that they are not without problems. The prospect of limb salvage without local recurrence is certainly an objective to be strived for. Prolonged and careful follow-up is necessary to resolve many of the questions which remain unanswered at the present time.

The Montreal General Hospital's Musculoskeletal Tumor Group, at present, believes that a multidisciplinary approach is warranted to improve the survival figures for patients with soft tissue sarcoma.

References

1. Antman, K.H., Blum, R.H., Wilson, R.E., Corson, J.M., Greenberger, J.S., Amato, D.A., Canellos, G.P. and Frei, E.: Survival of patients with localized high-grade soft tissue sarcoma with multimodality therapy. Cancer 51, 396, 1983.

2. Benjamin, R.S.: Preoperative chemotherapy for osteosarcoma. Proc. ASCO 1, 174, 1982.

3. Carabell, S.C., and Goodman, R.L.: Radiation therapy for soft tissue sarcoma. Seminars in Oncology 8, 201, 1981.

4. Enneking, W.F., Spanier, S.S. and Goodman, M.D.: A septum for the surgical staging of musculoskeletal sarcoma. Clin. Orthop. Rel. Res. 153, 106, 1980.

5. Karakousis, C.P., Lopez, R.E., Rao, U. and Holyoke, E.D.: Soft tissue sarcoma - Trends in treatment. N.Y. State J. Med. 79, 328, 1979.

6. Manual for Staging of Cancer, 1977, Chicago. American Joint Committee, p. 1.

7. Morton, D.L., Eilber, F.R., Townsand, C.M., Grant, T.T., Mirra, J., and Wisenburger, T.H.: Limb salvage surgery from a multidisciplinary treatment approach for skeletal and soft tissue sarcomas of the extremity. Ann. Surg. 184, 268, 1976.

8. Rosenburg, S.A.: Adjuvant chemotherapy for patients with soft tissue sarcoma. Surg. Clin. N.A. 61, 1415, 1981.

9. Rosenburg, S.A., and Glatstein, E.J.: Perspectives on the role of surgery and radiation therapy in the treatment of soft tissue sarcomas of the extremities. Seminars in Oncology 8, 190, 1981.

10. Russell, W.O., Cohen, J., Enzinger, F., Hajdu, S.I., Heise, H., Martin, R.G., Meissner, W., Miller, W.T., Schmitz, R.L. and Sult, H.D.: A clinical and soft tissue staging system for soft tissue sarcomas. Cancer 40, 1562, 1977.

11. Simon, M.A., and Enneking, W.F.: The management of soft tissue tumors of the extremities. J. Bone Joint Surg. 58A, 317, 1976.

12. Sutow, W.W., and Maurer, H.M.: Chemotherapy of sarcomas - A perspective. Seminars in Oncology 8, 207, 1981.

The Pathology of Giant Cell Tumor of Bone

V. L. FORNASIER

Introduction

Giant cell tumor of bone is a distinctive neoplasm formed by a combination of multinucleated giant cells and cells of a mononuclear stromal background. Over three-quarters of the giant cell tumors reported occur when the skeleton has reached maturity in the third decade of life and later. Those that are observed before the third decade generally occur close to 20 years of age. Rarely are tumors reported in children under 14 years of age, and one wonders whether these could represent variants of other tumors with the preponderance of giant cells as a histologic feature. Giant cell tumors of bone represent anywhere from 4-10% of all bone tumors in reported series and are said to have an overall incidence of approximately 0.39 cases per million of population.

Giant cell tumors occur on the metaphyseal side of the epiphyseal plate of long bones. The distal femoral metaphysis and the proximal tibial metaphyses are the most common sites. In the axial skeleton giant cell tumors occur most commonly in the sacrum. Another common site is the distal metaphysis of the radius. Because giant cell tumors of bone occur at a time when the epiphyseal plate is either closing or has closed, there is a direct bone marrow continuity across the epiphyseal plate or scar, and thus the tumor can quickly extend into the epiphysis proper. For this reason, many tumors at the time of presentation extend already to the subarticular region of the adjacent joint. This is an important distinction and it must be remembered for it is often incorrectly stated that giant cell tumor is an "epiphyseal" lesion. This is not strictly speaking a correct localization of the site of origin of the tumor, but represents extension into the epiphysis as part of the progression of the lesion across a closed epiphysis which is no longer a block to tumor growth since the bone marrow is now in continuity across the epiphyseal plate or scar from the epiphysis to the metaphysis.

Gross Findings

The tumor appears as a variegated mass that usually shows extensive hemorrhage. The viable tumor tends to have an off-white to cream color and is often quite gelatinous (Fig. 1). These areas tend to be firm and occasionally form distinct fibrous bands. Multifocal hemorrhage and associated necrosis accentuate the variegated appearance with areas of red and yellow.

Current Concepts of Diagnosis and Treatment
of Bone and Soft Tissue Tumors.

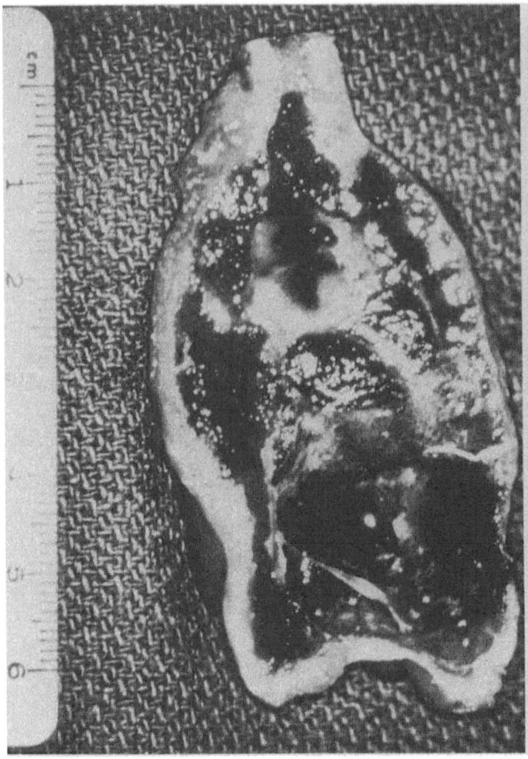

Fig. 1

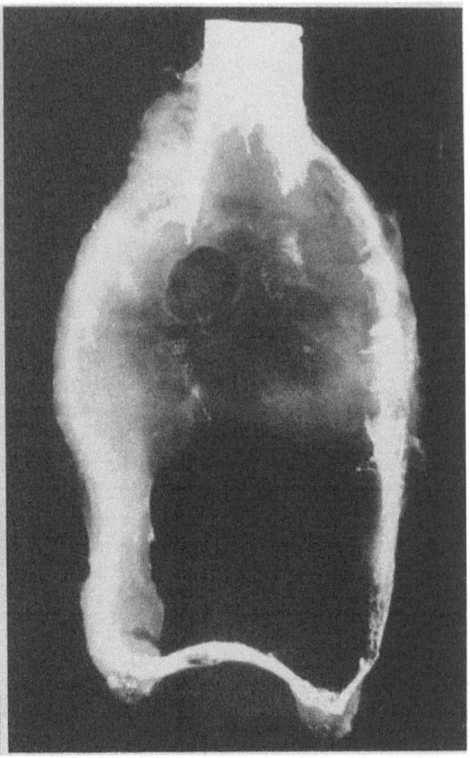

Fig. 2

Fig. 1. Gross appearance of a giant cell tumor of the distal end of the radius. The dark areas represent hemorrhage. The light areas represent fleshy, cellular viable tumor. The articular cartilage at the lower extreme of the specimen is closely apporixmated by the tumor. The extension of the tumor into the medullary cavity of the radius proximally contrasts with the expansion of the bone by the tumor. The lower third of the specimen is gelatinous to the point of translucency.

Fig. 2. A fine detail radiograph of the specimen seen in Fig. 1. The preservation of the subarticular bony plate is well denoted. The erosion and scalloping produced by the tumor on both the original radial bone and in the periosteal new bone are well illustrated in the proximal margin of the tumor. The lower half of the tumor is entirely radiolucent in keeping with the extensive degeneration evident in the gross appearance, and is an indication of impending cystic formation.

Liquefaction of these areas of tissue degeneration result in cysts that contain turbid yellow-brown fluid. Occasionally, this cystic degeneration may become a dominant feature and impart to the tumor the features that resemble aneurysmal bone cyst. The advancing margin of the tumor has a lobulated character that produces the scalloped erosions into the sur-

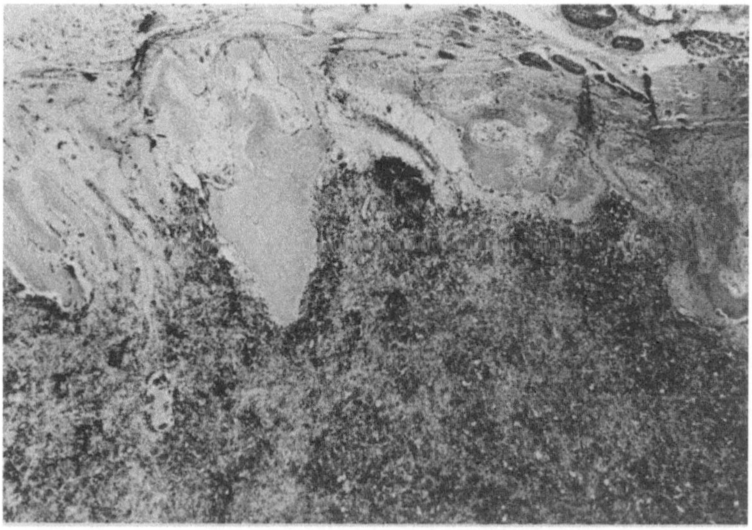

Fig. 3. Periosteal new bone and the advancing tumor. The tumor is eroding all of the newly formed bone by the displaced periosteal layer. The bone is made up of trabeculae which occasionally become thickened, but no semblance of a cortical outline is present. This sparsity of bone and the decreased mineralization of this newly formed bone result in the apparent loss of cortex visualized by a higher energy clinical x-ray. (H&E section x 100)

rounding bone. This is best seen on fine detail radiography of areas of the cortex that are being eroded by the tumor (Fig. 2). The tumor does affect both cortices and the periosteal sheath, which bulges out ahead of the advancing tumor (Fig. 3), but only fracture or surgical trauma tend to provide the opportunity for the tumor to extend out beyond the fibrous periosteal sheath into the adjacent soft tissues. Often, the periosteum consists only of the cellular cambium layer and its fibrous support with a total absence of any bony shell. Once the tumor has destroyed the cortical bony shell, all restriction to expansile movement of the lesion is removed; this will favor a transverse expansion across the bone leading to the soft tissue bulge rather than the longitudinal progression along the length of the bone. The fibrous septations that outline the areas of hemorrhage, necrosis and cystic degeneration may contain bony trabeculae, and thus appear gritty when cut. If the cysts are of long standing, the inner surfaces appear smooth and glistening.

Histologic Findings

The microscopic appearance of a typical giant cell tumor shows characteristic osteoclastic-type multinucleated giant cells in varying numbers

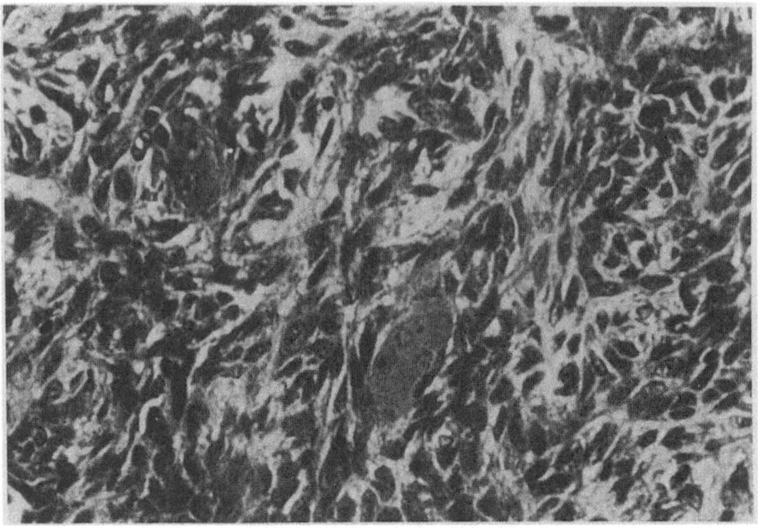

Fig. 4. A cellular part of a giant cell tumor with only two multinucleated giant cells easily distinguished. The range of nuclear size and shape within any one giant cell tumor can be seen in this illustration, and this represents the area of difficulty in histologic interpretation for the identification of a giant cell tumor, and for the categorization of the giant cell tumor as to its biologic potential. (H&E x 400)

scattered in a cellular mononuclear mesenchymal tissue (Fig. 4). The mononuclear cells tend to be quite uniform with poorly defined cell borders, moderate amount of cell cytoplasm, and a closely woven but filiform reticulin framework about each cell. As collagen-forming, spindle-shaped cells become more dominant the tissue becomes more hyaline and less cellular. Usually, these fibrous areas are part of the persisting septations after the tumor has undergone cystic degeneration. At the margins, and in particular in the fibrous septations, an abundance of vascular channels may be observed that may be associated with ossification (Fig. 5). The ossification that occurs is characterized by small cuffs of bone matrix that progress to form trabeculae covered by a well organized osteoblastic lining. This bone tends to blend with the bone in the surrounding tissue and is subject to resorption by osteoclasts and maturation to lamellar bone.

Discussion

Areas that are not distinguishable from giant cell tumor can be identified in many bone lesions including chondroblastoma, chondromyxoid fibroma, giant cell reparative granuloma, aneurysmal bone cyst, hyperparathyroidism, and malignant neoplasms such as telangiectatic osteosarcoma. Jaffe and

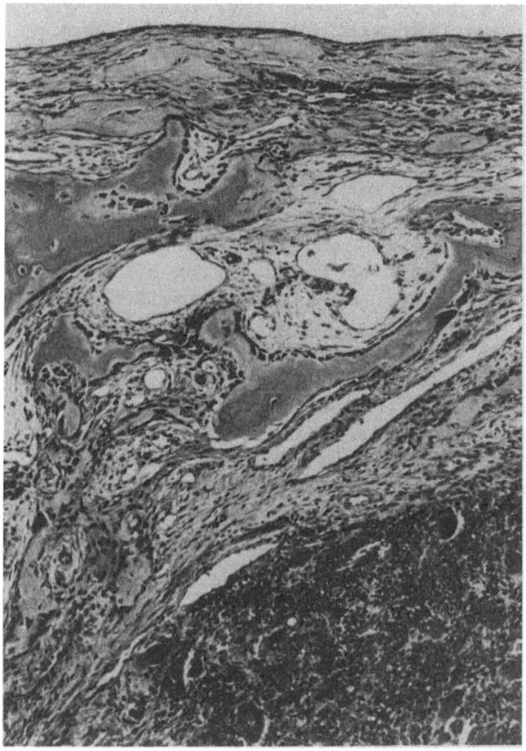

Fig. 5. The tumor is seen at the lower right hand corner of the frame. Trabeculae of newly formed bone with a prominent row of plump osteoblasts identify reactive new bone. The large numbers of distended endothelial lined blood vessels present a striking feature of the reactive periosteal envelope of the giant cell tumor. (H&E x 250)

Lichtenstein were among the first to separate true giant cell tumor of bone from the variety of other conditions that contain giant cells as part of a totally different lesion.

The question of a malignant giant cell tumor is particularly difficult. This malignant tumor cannot be diagnosed with assurance unless one is able to specifically identify ordinary or benign giant cell tumors within a lesion that also show areas of unequivocal cytologic malignancy. However, the malignancy may take the form of osteosarcoma or fibrosarcoma. Biologically, this may not bear any relationship to the giant cell tumor-like areas that one might be able to identify in these tumors. In these cases the benign giant cell tumor-like areas are but an incidental and confusing component. To further complicate matters, giant cell tumors can metastasize even though microscopically one is unable to identify any evidence of malignancy. The metastases, when examined, will also have the totally benign appearance of the primary lesion. Fortunately, such occurrences are rare. The Mayo Clinic experience reported by Dahlin records two cases out of 264 patients with metastasizing giant cell tumors. My own experience with two such cases indicates that these patients follow a long protracted course of recurrences and metastases, and inadequate initial treatment was an important factor in the evolution of these unfortunate circumstances.

While it seems possible that multiple giant cell tumors may occur in one patient, it is far more likely that a patient with multiple tumors that histologically appear to be giant cell tumors may, in fact, be suffering from a metabolic problem such as hyperparathyroidism. This should always be considered in cases of multiple giant cell tumors and appropriate clinical investigation undertaken to exclude this possibility.

The histologic appearance of the tumor, its size, the amount of necrosis, the number of mitotic figures, the degree of cellular pleomorphism in the mononuclear cell component of the tumor, and the abundance of vascularity have all been used as factors in the prognosis of the behavior of giant cell tumors. None of these has individually or as a group provided a reliable indicator of prognosis. An interesting feature of giant cell tumors is the intrusion of giant cells into vascular channels within the tumor and particularly at the margins of the tumor. These have been found to correlate neither with aggressive behavior of the tumor nor with the presence of metastases. Published reports of case studies using electron microscopy, tissue culture, and immunoperoxidase staining for specific enzyme contents of the cells have neither clarified, to date, the nature of the tumor cell nor assisted in understanding the biology of the lesion.

Radiologic Aspects of Giant Cell Tumors

J. A. LIVER

Introduction

Any attempt to review the radiologic aspects of giant cell tumors is plagued by two problems. One is the confusion among pathologists as to what actually is a giant cell tumor, and the other is the fact that giant cell tumors are rare. This results in a variety of statistics relating to bone sites, age and sex incidences, relation to other conditions, and aggressiveness of the tumor.

Classic Description

Textbooks on radiology give a description of a typical giant cell tumor (Fig. 1) which is easy to recognize:

- age incidence between 20-40 years
- equal incidence in men and women
- 75% occurrence in the bones adjacent to the knee or in the distal end of the radius
- geographic bone destruction involving the end of the bone and often extending almost to the articular surface
- well-defined junction with normal bone, usually without sclerotic rim
- eccentric expansion of the bone end
- no tumor calcification
- uneven bone destruction due to the lobulated outline of the tumor mass that may result in internal ridges of bone giving the radiologic appearance of septa
- the overlying bone becoming thinned and the tumor possibly breaking through to form a soft tissue mass; however, subperiosteal new bone formation is rare, even in the presence of a pathologic fracture.

Uncommon Sites

Sacrum

The sacrum is the most common site for giant cell tumor after the classic sites noted above, and 6.5% of all giant cell tumors occur in the sacrum

Current Concepts of Diagnosis and Treatment
of Bone and Soft Tissue Tumors.

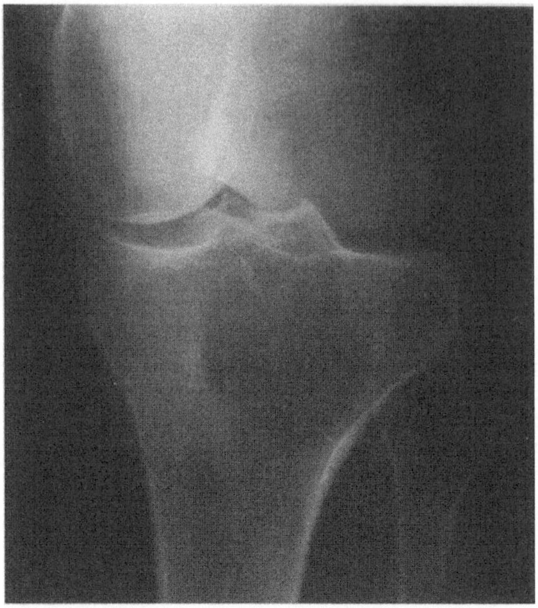

Fig. 1. Giant cell tumor of upper end of tibia.

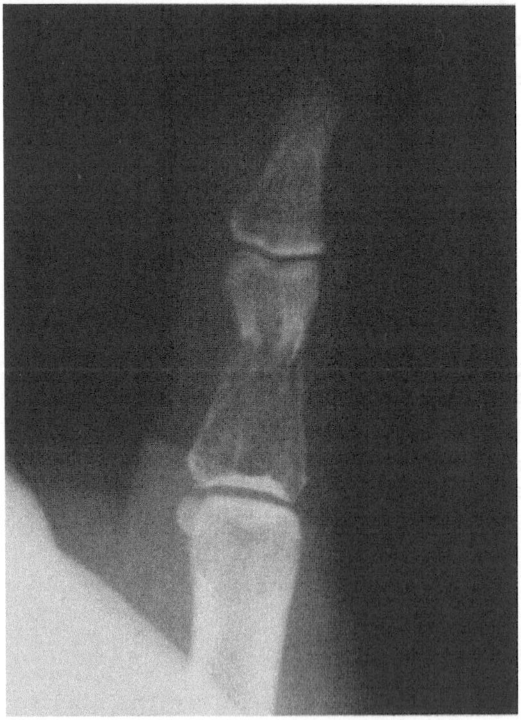

Fig. 2. Giant cell tumor of proximal phalanx of thumb.

(Smith et al, 1979). It is the second most common primary tumor of the sacrum after chordoma. A posterior sacral mass is unusual as the tumor tends to grow anteriorly into the pelvic cavity; it is therefore not recognized early and the smallest tumor in the above series had a diameter of 5 cm. The margins of the tumor are clearly delineated radiologically. The tumor readily extends across the sacroiliac joint into the ilium.

Short tubular bones

Involvement of the short tubular bones (Fig. 2) is rare except for giant cell tumors of multicentric origin. In one series of five patients with multiple giant cell tumors, all had involvement of the short tubular bones (Peimer et al, 1980).

Skull and facial bones

These are also rare sites for giant cell tumor, except as a complication of Paget's disease (Jacobs et al, 1979). Giant cell tumor in Paget's disease may occasionally be seen in the spine but not in the extremities. This complication seems to be limited to the polyostotic form of Paget's, and may in some cases be familial.

Metaphysis

It is unusual for a giant cell tumor to be seen in children before the epiphyses have closed but when it does occur, the tumor is invariably in the metaphysis. It has therefore been concluded that giant cell tumor is a tumor of the metaphysis which rapidly crosses the fused physis in the adult.

Growth and Malignant Change

The usual radiologic criteria of malignancy do not apply in giant cell tumors:

Rate of growth: The natural history of giant cell tumor is uncertain but some evidence suggests that the tumor initially grows rapidly and then growth slows (Lorentzon et al, 1980). The rate of growth as assessed radiologically does not correlate with the histologic grading.

Destruction of cortex: In giant cell tumors, the cortex is often broken through with spread of the tumor into the soft tissues. Yet, the tumor remains essentially benign albeit with a strong tendency to recur if not totally removed -- the so-called primary malignant giant cell tumor. On follow-up films of treated tumors this cannot be distinguished from the secondary malignant giant cell tumor which is probably pathologically a fibrosarcoma induced as a result of radiotherapy (Nascimento et al, 1979).

Pulmonary metastases as seen on chest radiography: These may not be true metastases and may be due to tumor cell embolization after curettage. They do not influence survival and have therefore been given the name benign metastasing giant cell tumors.

Special Radiographic Procedures

Tomography

This procedure is particularly useful when other shadows obscure bone detail, such as in the investigation of tumors of the sacrum.

Computed tomography

The newer machines give an excellent demonstration of bone and soft tissue detail. Knowledge of soft tissue extension of the tumor is particularly important if total surgical removal of the tumor is planned.

Arteriography

In the investigation of bone tumors in general it has been shown that arteriography may help to:

- differentiate benign from malignant tumors
- define the extent of the lesion
- select the best site for biopsy

In the case of a giant cell tumor, arteriography will show tumor neovascularity in most cases, only 10% being avascular as a result of tumor necrosis (Prando et al, 1979). These authors found that 20% of apparent extraosseous extensions on arteriography were false positives, due to either a thin rim of bone over the tumor which was not recognized on the roentgenograph or hyperemic synovial tissue resembling a tumor mass. In the case of tumors of the sacrum, arteriography is particularly helpful before surgery in identifying the blood supply of the tumor. (Smith et al., 1979).

Nuclear medicine scan

A technetium-99m scan is the simplest way to diagnose multicentric giant cell tumors.

Differential Diagnosis

Brown tumor of hyperparathyroidism. This is a particularly important diagnosis because of the entirely different treatment of this condition. The lesion itself may be radiologically identical with giant cell tumor. The radiologist has to look for evidence of subperiosteal bone resorption and, of course, the blood chemistry is entirely different.

Chondromyxoid fibroma. This fibroma has a denser sclerotic rim than giant cell tumor and does not extend as closely to the articular surface of the bone.

Fibrosarcoma. When this neoplasm breaks through the cortex, there is a periosteal response which is unusual in the case of giant cell tumors.

Osteoblastoma. In the spine, this tumor is indistinguishable from a giant cell tumor.

Cystic lesions (aneurysmal bone cyst, unicameral bone cyst). These conditions are characterized by an intact cortex, and occurrence in younger age groups. Involvement of the bone end is rare.

Villonodular synovitis. The radiographic appearance is very similar to giant cell tumor, but there is a much greater tendency for involvement of both sides of the joint than in giant cell tumor.

Enchondroma in a short tubular bone. Enchondroma often contains cartilage calcification and does not extend to the bone end.

References

1. Jacobs, Thomas P.; Michelsen, Jost; Polay, Janet S.; d'Adamo, Ann C.; and Canfield, Robert E.: Giant cell tumor in Paget's disease. Cancer 44, 742, 1979.

2. Lorentzon, Ronny; Lundstrom, Bo; and Larsson, Sven-Erik: Growth rate of giant-cell tumor of bone: Radiographic and clinical considerations. Clin. Orthop. 149, 299, 1980.

3. Nascimento, Antonio G; Huvos, Andrew G; and Marcove, Ralph C.: Primary malignant giant cell tumor of bone. Cancer 44, 1393, 1979.

4. Peimer, Clayton A.; Schiller, Alan L.; Mankin, Henry J.; and Smith, Richard J.: Multicentric giant-cell tumor of bone. J. Bone Joint Surg. 62A, 652, 1980.

5. Prando, Adilson; deSantos, Luis Alonso; Wallace, Sidney; and Murray, John A.: Angiography in giant-cell bone tumors. Radiology 130, 323, 1979.

6. Smith, Julius; Wixon, David; and Watson, Robin Caird: Giant-cell tumor of the sacrum. J. Canad. Assoc. Radiologists 30, 34, 1979.

Unicameral Bone Cysts - Round Table Discussion

I. PATHOLOGY

M. Orizaga

Introduction

Unicameral bone cyst is relatively common and usually discovered during the first two decades of life. It is most frequently located in the proximal diaphysis of the humerus, the femur and tibia, in that order of frequency. The lesion is a benign fluid-filled unicameral cyst of unknown etiology occurring in long bones, predominantly in men. Cysts that arise in unusual sites, such as the flat bones, generally occur in adults, are silent and discovered accidentally.

Clinical signs and symptoms are pain and swelling or stiffness of the nearest joint. Fracture occurs in about 30% of patients with cysts in the long bones (Dahlin, 1978; Mirra, 1980; Spjut et al, 1971).

Radiologic Features

Unicameral bone cyst is a symmetrical lytic lesion in metaphyses that usually abuts the epiphyseal plate. Most cases (70%) involve either the proximal humerus or proximal femur. Usually, fusiform widening of the bone is observed in the involved area due to expansion of the cystic zone which is surrounded by a thin rim of newly-formed periosteal bone. Other cases may show either little or no expansion or trabeculation through the lesion, which generally reaches maximum size before adulthood. Serial roentgenograms have shown that the cyst grows away from the region of the physis, ending finally near the center of the shaft. This has been referred to as the latent stage, usually found in older children. Mirra (1980) has described the so-called "fallen fragment" as pathognomonic of solitary bone cyst. This represents a small portion of the fractured cortex that falls to the bottom of the fluid-filled cyst. Since the solitary bone cyst is the only unilocular fluid-filled cyst of bone, it is considered to be pathognomonic but seen only in less than 10% of patients. Many others show evidence of pathologic fracture with or without displacement.

Lesions that can resemble solitary bone cyst are: lytic osteoblastoma, fibrous dysplasia, osteomyelitis, lytic osteosarcoma, nonossifying fibroma, eosinophilic granuloma, and aneurysmal bone cyst.

Current Concepts of Diagnosis and Treatment
of Bone and Soft Tissue Tumors.

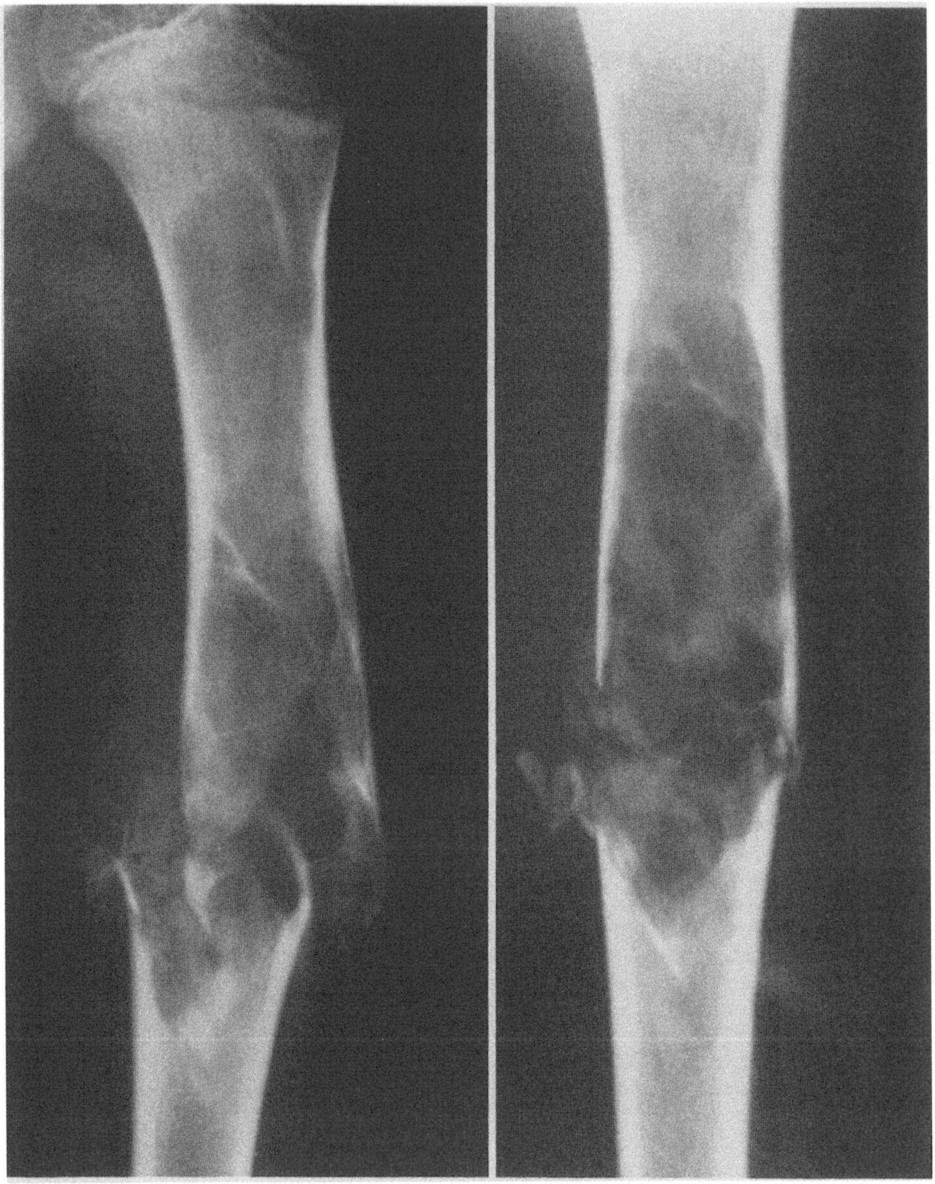

Fig. 1a

Fig. 1. The sequential phases of healing of unicameral bone cyst after MPA injection: a) pathologic fracture in unicameral bone cyst; b) the "frosted glass" opacity of the cyst; c) consolidation with dense sclerotic bone; d) normal morphologic appearance.

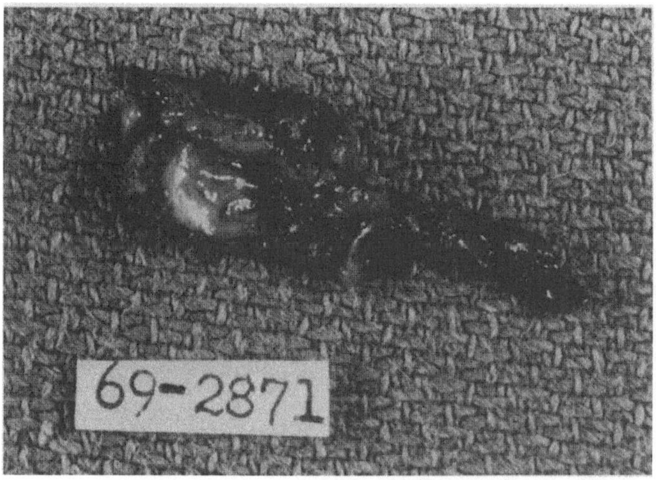

Fig. 1b

Gross Pathology

Descriptions have been gathered mostly from roentgenogram appearances and surgical findings. The overlying periosteum and surrounding soft tissues are intact except when fracture has occurred. The thinned-out cortex bulges in the area of the cyst and, where thinnest, may be fluctuant and bluish due to the presence of fluid filling the cyst. Most cysts are unilocular (unicameral) and the fluid contents are serous (straw-colored) or serosanguinous, and are sometimes under great pressure. The lining is usually thin, fibromembranous, no more than 1 mm thick, and white-gray (Fig. 1). If fracture has occurred, the cyst is filled with blood or with fibrinous or fibrous membranes that may give the cyst a multilocular appearance. In some cases, however, subsequent to the fracture, the cyst cavity may be completely obliterated by fibro-osseous repair-type callus which may be mistaken for fibrous dysplasia or osteoblastoma and, if exuberant, with fibrosarcoma or osteosarcoma; in these cases the lining of the cystic portion of the lesion may be fleshy and up to 1 cm thick. The chemical constituents of the fluid resemble more those of serum than synovial fluid (Cohen, 1960).

Microscopic Pathology

The typical bone cyst is thin-walled (usually less than 1 mm thick). The cyst wall is composed of fibrous tissue containing slivers of osteoid or woven bone, less compact than the cortical bone, with well vascularized intertrabecular tissue (Fig. 2). Mononuclear cells and osteoclast-like giant cells are also present. The lining is composed of a single to double or triple layer of epithelial or mesothelial-like cells, probably of synovial type (Mirra, 1978) (Fig. 3).

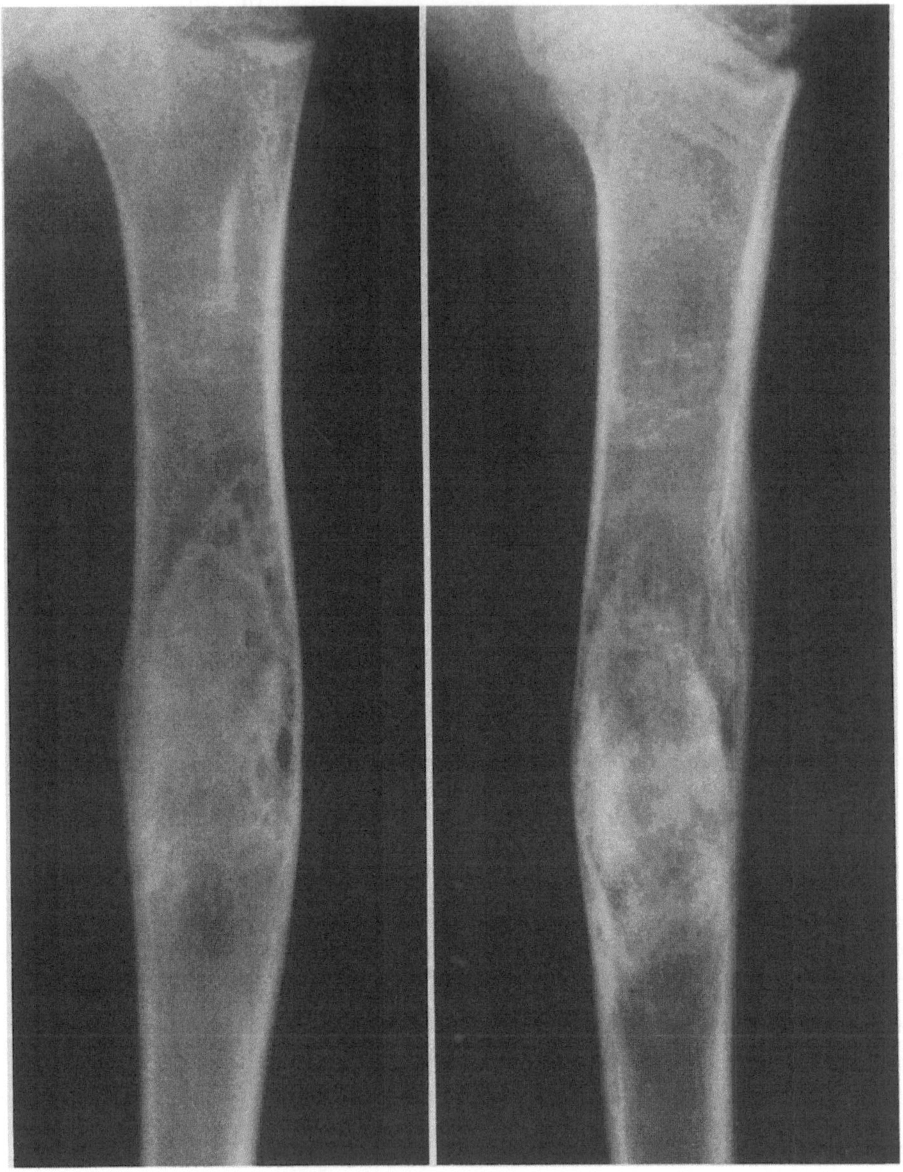

Fig. 1c

If a fracture has occurred, one observes in 15% of patients subperiosteal newly-formed bone, necrosis, inflammation and repair, deposits of hemosiderin and cholesterol clefts. The fibro-osseous repair tissue may be so abundant that the lesion may be confused with fibrous dysplasia, osteoblastoma or ossifying fibroma. Cartilage is rarely seen. Care must be taken to rule out osteosarcoma.

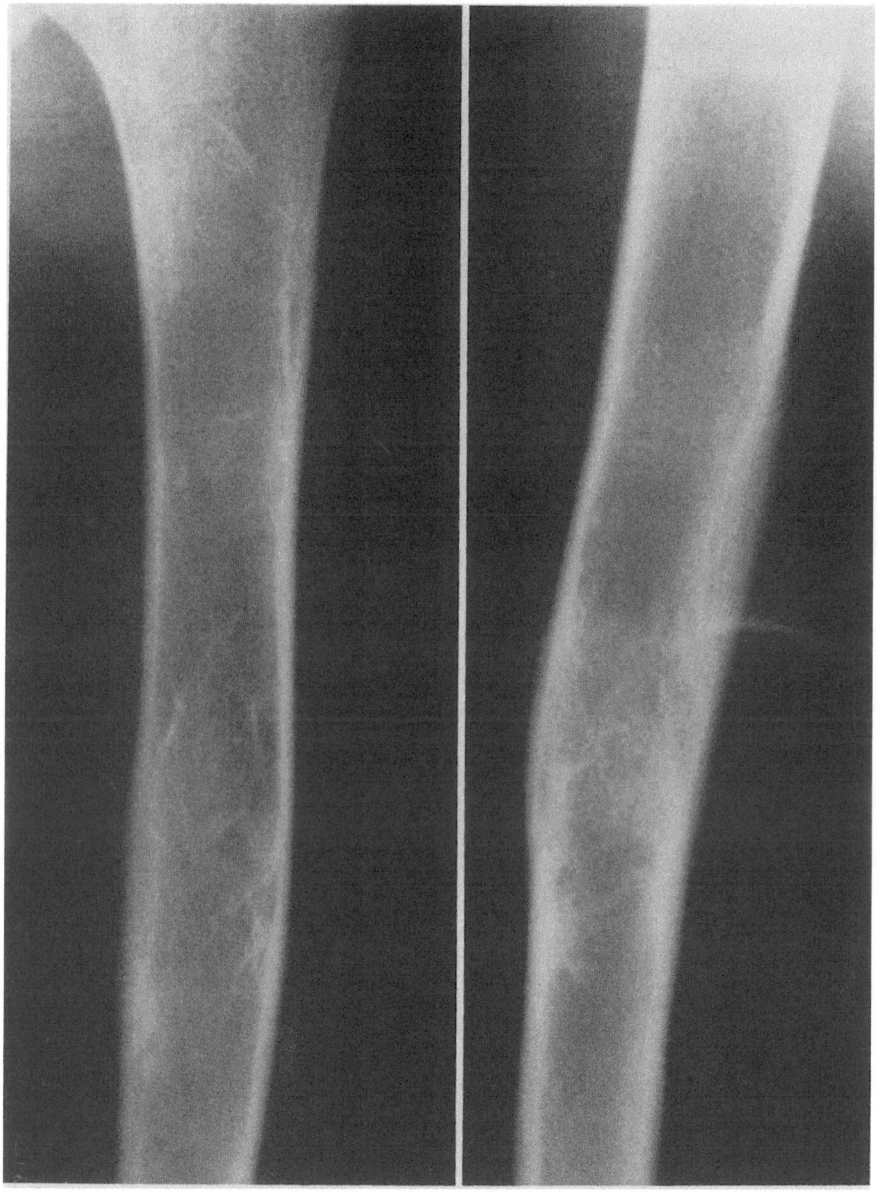

Fig. 1d

Etiology and Pathogenesis

Various pathogenetic mechanisms have been proposed (Coley and Higinbotham, 1934; James et al, 1948; Mirra, 1980), which include:

1. Cystic change in a traumatic hematoma.

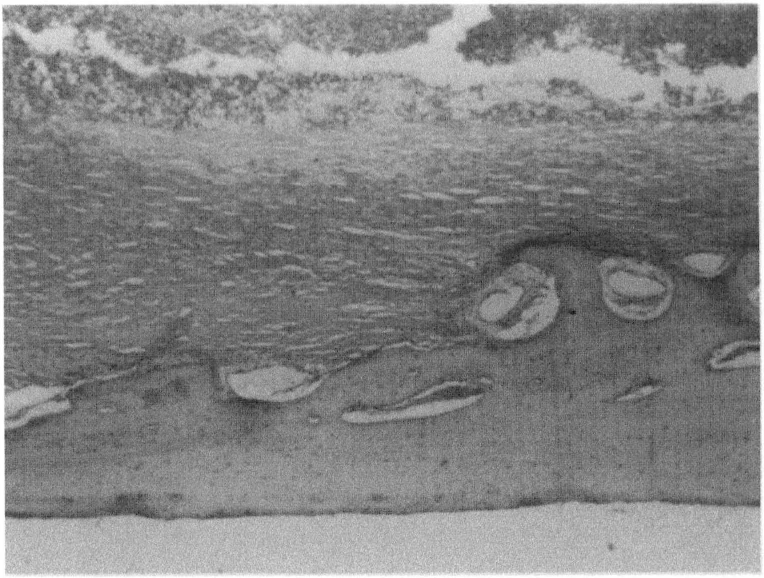

Fig. 2. Healing with residual osteolysis.

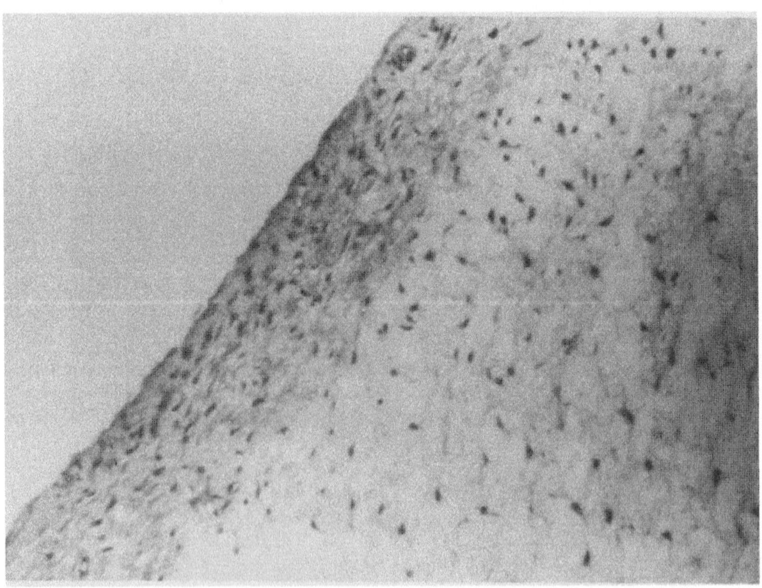

Fig. 3. Recurrence after healing with peripheral rim of osteolysis.

2. End result of a low grade infection or inflammatory process.
3. Faulty calcium metabolism.
4. Local disturbance of bone growth at epiphyseal line, perhaps combined with trauma.
5. Localized abnormal activity of osteoblasts.
6. Cystic degeneration of a preexisting benign tumor such as nonossifying fibroma with blockage of lymphatics (Broder, 1968; Cohen, 1960; 1970).
7. Fibrous proliferation in an area of rapid resorption of bone with blockage of sinusoidal vessels and accumulation of interstitial fluid.
8. Cystic degeneration in a lipoma.
9. Error in embryonic, fetal or infantile development.
10. Congenital rest of synovial tissue (intraosseous synovial cyst or synovial rest) displaced or trapped into the thin cortical metaphyseal end of bones at the synovial capsular reflexion (Mirra, 1980).

In Mirra's view any proposed theory of pathogenesis must explain the following facts:

- why over 70% of the cysts are discovered in childhood
- why 95% of them involve the metaphyses
- why some 70% occur in the proximal humerus and femur
- why its fluid is rich in protein
- the structure of the cyst wall and the type of its lining cells
- its tendency to recur
- its benign nature

In an electron microscopic study carried out on a solitary bone cyst with production of a cementum-like substance, Mirra (1978) noted that the lining cells were arranged in three or four layers, that these cells had the characteristics of mesodermal epithelium, and that the luminal cells possessed microvilli and filopodia as seen in synovial cells. Beneath these cells were an incomplete basal lamina; in addition, the cells were not tightly joined together and contained occasional desmosomes.

Mirra explains the location of the cysts by pointing out that a synovial implant occurs in an intraosseous position at the point where the synovium and bone are in closest approximation to each other, ie, at the reflexion of the capsule with the bone. In long bones, the former inserts in the metaphyseal end, exactly where most simple bone cysts occur. In humans and animals, the capsule and its synovial recess have the greatest surface area of contact with the thin metaphyseal bone in the two bones with the greatest rotatory motion, the areas in which over 70% of the solitary bone cysts occur.

References

1. Broder, H.M.: Possible precursor of unicameral bone cysts. J. Bone Joint Surg. 50(A), 503, 1968.

2. Cohen, J.: Simple bone cysts. Studies of cyst fluid in 6 cases with a theory of pathogenesis. J. Bone Joint Surg. 42(A), 609, 1960.

3. Cohen, J.: Etiology of simple bone cyst. J. Bone Joint Surg. 52(A), 1493, 1970.

4. Coley, B. and Higinbotham, N.L.: Solitary bone cyst. Ann. Surg. 99, 432, 1934.

5. Dahlin, D.C.: Bone Tumors. General Aspects and Data on 6221 Cases. 3rd Ed. p. 385, Charles C. Thomas, Springfield, 1978.

6. James, A., Coley, B., and Higinbotham, N.L.: Solitary unicameral bone cyst. Ann. Surg. 57, 137, 1948.

7. Mirra, J.M.: Bone Tumors; Diagnosis and Treatment. p. 462. J.B. Lippincott, 1980.

8. Mirra, J.M., Bernard, G.W., Bullough, P.G., Johnston, W., and Mink, G.: Cementum-like bone production in solitary bone cysts. (So-called "cementoma" of long bones. Report of three cases. Electron microscopic observations supporting a synovial origin to the simple bone cyst.) Clin. Orthop. 135, 295, 1978.

9. Spjut, H.J., Dorfman, H.D., Fechner, R.E. and Ackerman, L.V.: Tumors of Bone and Cartilage. Second Series: Armed Forces Institute of Pathology, p. 317, 1971.

II. STEROID INJECTION

J. D'Astous

Introduction

The simple or unicameral bone cyst is a benign cystic lesion frequently found in the metaphysis of growing children. Although simple in name, these cysts have been anything but simple to manage, the most common complications being pathologic fracture and recurrence. A review of many series reveals that only 15% of these cysts fill in spontaneously after fracture. Consequently, in the past, surgical treatment has been advocated as the treatment of choice for these lesions. Curettage and bone grafting not only leave: us with a high recurrence rate (18-41%), but the pediatric iliac crest is often not large enough and other sources of bone must be used (Cohen, 1977). Subtotal excision of the cysts results in a lower rate of recurrence but a higher perioperative fracture rate, and therefore is not as useful an alternative in the weight-bearing limbs. In 1979, Scaglietti et al proposed the use of steroid injections for the treatment of unicameral bone cysts, reporting a 96% success rate.

The purpose of this paper is to compare our results of the treatment of unicameral bone cysts by steroid injection with those of Scaglietti's group.

Materials and Methods

Between 1975 and 1981, 28 cysts in 27 patients, whose diagnoses were radiologically or pathologically consistent with unicameral bone cysts, have been seen at the Children's Hospital of Eastern Ontario. Sixteen of these patients, 10 boys and 6 girls, were treated by injection with methylprednisolone. Nine patients were under the age of ten, five were between 11 and 14 years of age, and two were over 14 years of age. Follow-up ranged from nine months to four years. The location of the cysts was as follows: humerus (12), femur (3), tibia (1). Eleven cysts were classified as active and five as latent. Eleven patients presented initially with a pathologic fracture, while five patients had undergone previous surgery for a cyst.

The treatment protocol was a follows: under fluoroscopic control, two marrow aspiration needles were inserted into the cyst. The cyst fluid was allowed to flow out of the needle and was sent for cytologic examination, culture, and antibiotic sensitivity. Following this, 40 to 200 mg of methylprednisolone was injected into the cyst, the dosage depending on the size of the cavity. Large femoral cysts necessitated single hip spica immobilization for six weeks. If, after three months, there was no evidence of calcification of the cyst on follow-up roentgenograms, the injection was repeated. A favorable response consisted of an increasing radio-

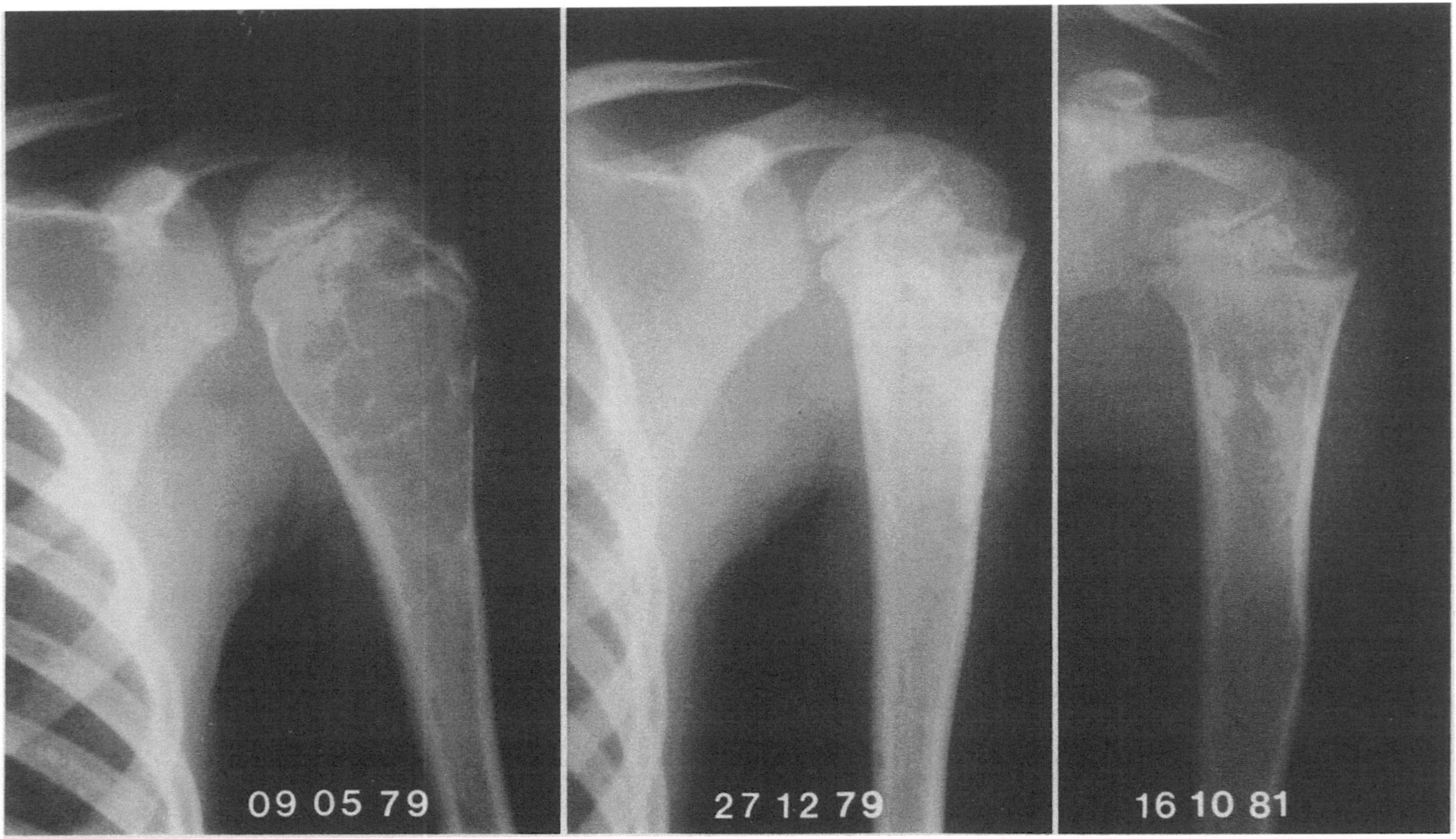

Fig. 1. This 7-year, 4-month old boy suffered a pathologic fracture through a unicameral bone cyst on May 9, 1979. He received a methyl-prednisolone injection into the cyst on June 19, 1979 for persistence of the cyst. Roentgenograms taken ten weeks after the second injection, December 27, 1979, show almost complete healing and remodelling of the proximal humeral shaft. Two years after the second injection, the proximal humerus has completely remodelled, leaving only a small sclerotic area in the distal metaphysis as evidence of the healed unicameral bone cyst.

pacity of the cavity, accompanied by gradual reconstruction of cortical bone at the cyst wall (Fig. 1).

Results

In our series, we had two refractures after injection. Three cysts were considered healed, ten were healing, and we had one failure which was bone grafted after a single injection. Two patients still under active treatment have shown a positive response but will likely require further steroid injections. We believe that our results compare favorably with those of Scaglietti (1979) and other recent series (Campos, 1982; Capanna et al, 1982; Endaloussi et al, 1982; Scaglietti et al, 1982), but there is still a need for long-term follow-up as well as elucidation of the mechanism of action of methylprednisolone on unicameral bone cysts.

In conclusion, we feel that there is a definite place for steroid injections in the treatment of unicameral bone cysts and it is highly possible that steroid injections may in fact replace surgery as the treatment of choice for most unicameral bone cysts.

References

1. Campos, O.: Treatment of bone cysts by intracavity injection of methylprednisolone acetate. Clin. Orthop. 165, 43, 1982.

2. Capanna, R., Dal Monti, A., Gitelis, S., and Campanacci, M.: The natural history of unicameral bone cyst after steroid injection. Clin. Orthop. 165, 204, 1982.

3. Cohen, J.: Unicameral bone cysts, a current synthesis of reported cases. Orthop. Clin. Nth. Am. 8, 715, 1977.

4. Endaloussi, M., Padovoni, J.P., Rigault, P., Finidori, G., Alperovitch, R., and Gardes J.C.: Que peut-on attendre des injections intracavitaires de corticoides dans le traitement des kystes osseux essentiels de l'enfant? Rev. Chir. Orthop. 68 (Suppl. 11), 107, 1982.

5. Scaglietti, O., Marchetti, P.G., and Bartoluzzi, P.: The effects of methylprednisolone acetate in the treatment of bone cyst. J. Bone Joint Surg. 61(B), 200, 1979.

6. Scaglietti, O., Marchetti, P.G., and Bartoluzzi, P.: Final results obtained in the treatment of bone cysts with methylprednisolone acetate (Depo-Medrol) and a discussion of results achieved in other bone lesions. Clin. Orthop. 165, 33, 1982.

III. BONE GRAFTING

H. Zenker and H. Stürz

Introduction

Juvenile bone cysts are benign, tumor-like, expansive growing cystic bone lesions. These lesions appear predominantly in children between seven and 14 years of age, boys being twice as often affected as girls. The most frequent location of the disease is the proximal metaphysis of the upper arm and upper leg. The incidence in the upper arm is the highest, from 38% to 57% (Garceau and Gregory, 1954; Spence et al, 1969), followed by femoral involvement with 35% and tibial disease in 13%. Cysts with immediate contact to the growth plate, which usually is not penetrated, are termed active cysts. They are more frequent during the first decade. In later years, latent cysts predominate, located at the border between metaphysis and diaphysis.

Most cases become evident by a minor trauma leading to a fracture. Rarely are pain or swelling the first symptoms leading to a diagnosis. Age, location and typical radiologic signs are prodromic. The necessity for surgical treatment depends primarily on the loss of stability and function of the involved extremity. Conservative fracture treatment for juvenile bone cysts or therapeutic infraction of the cyst recommended by Hellner (1950) lead to spontaneous healing in only 15% (Garceau and Gregory, 1954). In the remaining 85%, cystic remnants and recurrent cysts can be expected, occurring more frequently in younger patients. According to Garceau and Gregory, between two and three refractures could be observed in these patients up to cessation of growth, and growth disturbances and structural deformities of the extremities were the rule (Neer et al, 1966).

Better results could be achieved with different surgical treatments aimed at removing the cyst as quickly and permanently as possible and reestablishing functional stability of the involved skeletal part without creating disturbing deformities. Opinions are still controversial regarding the optimal surgical treatment. On the one hand, excavation or curettage of the cyst plus subsequent autologous cancellous bone graft is being recommended as treatment of choice (Daum et al, 1975). On the other hand, subtotal resection up to two-thirds or three-fourths of the cystic wall is advised (Boseker et al, 1968; McKay and Nason, 1977). A third opinion recommends complete resection of the cystic bone changes with employment of different bone materials for bridging the gap, especially in the upper arm (Agerholm and Goodfellow, 1965; Harms and Groh, 1978; Mittelmeier, 1965; Witt et al, 1972). The choice of the surgical procedure has to take into consideration the location of the cyst, its extension and the age of the patient.

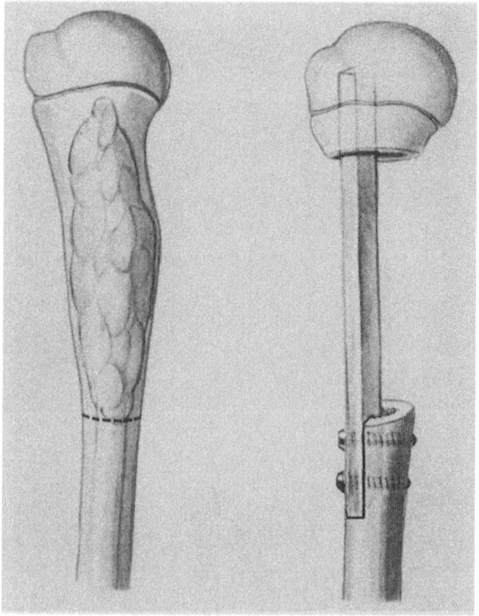

Fig. 1 and 2. En bloc resection of the tumor, impaction of an autologous tibial graft into the humeral head; a recipient site (Lexer fashion) is created in the distal humerus where the graft is fixed with two screws.

Fig. 1

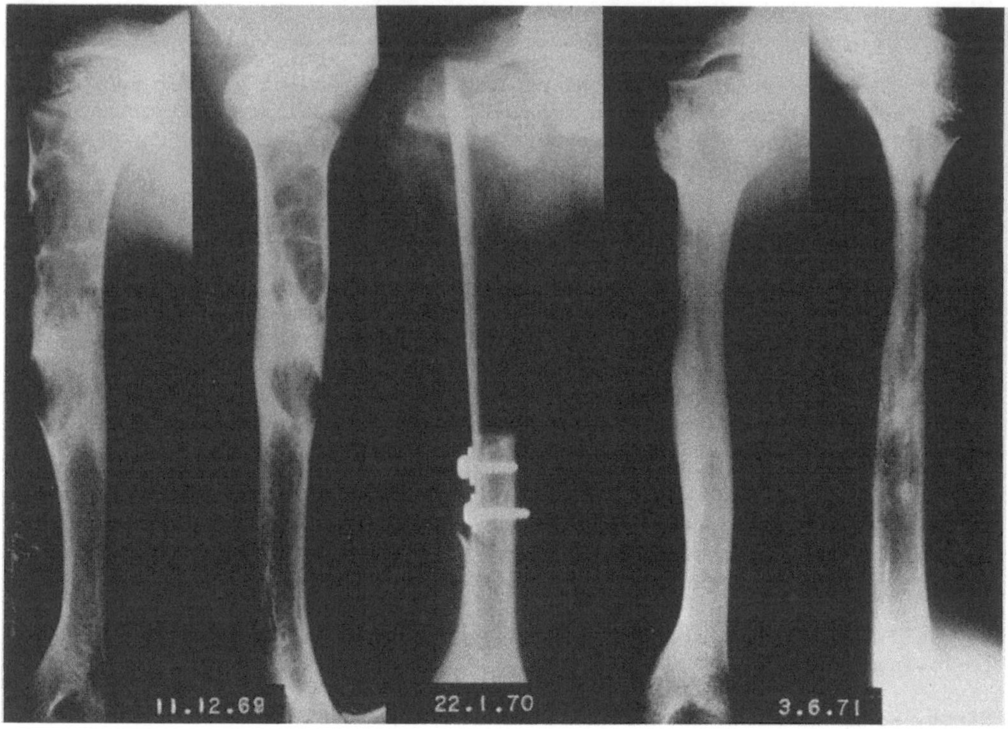

Fig. 2

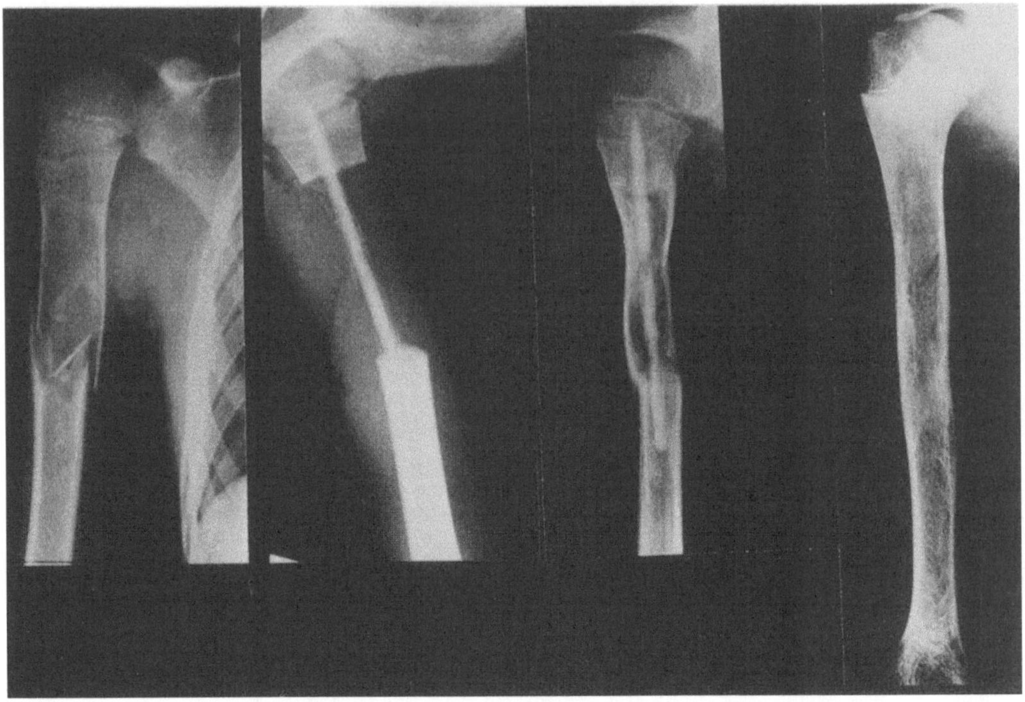

Fig. 3. En bloc resection in the upper arm. Instead of a Lexer recipient site, the tibial graft has been impaled into the medullar canal of the distal humerus without the need for osteosynthetic material.

Curettage and Grafting

On the upper arm, we prefer curettage with a subsequent autologous cancellous bone graft for small juvenile bone cysts. In addition, a cortico-cancellous graft may be used. Yet, this procedure, according to the literature, carries an average risk of 34% recurrence (Stürz et al, 1979). With proper selection of small cysts and careful cleaning and subsequent filling with autologous cancellous bone chips, lower recurrence rates may be achieved, reported to be as low as 12% (Daum et al, 1975). For large cysts and recurrences after previous curettage we do the resection of the involved tissue. This can be done as a subtotal resection with removal of about two-thirds of the cyst's wall or as a complete disruption by removal of the cystic section of bone. With these procedures the median recurrence rate is about 6% (Stürz and Zenker, 1979).

We would like to describe our operative procedure in the upper arm in the following cases:

First, we do an en bloc resection of the cyst, then fashion a notch in the humeral head and impact an autologous tibial graft (Fig. 1). Instead of a tibial graft, a fibular transplant may be used. In the distal humerus

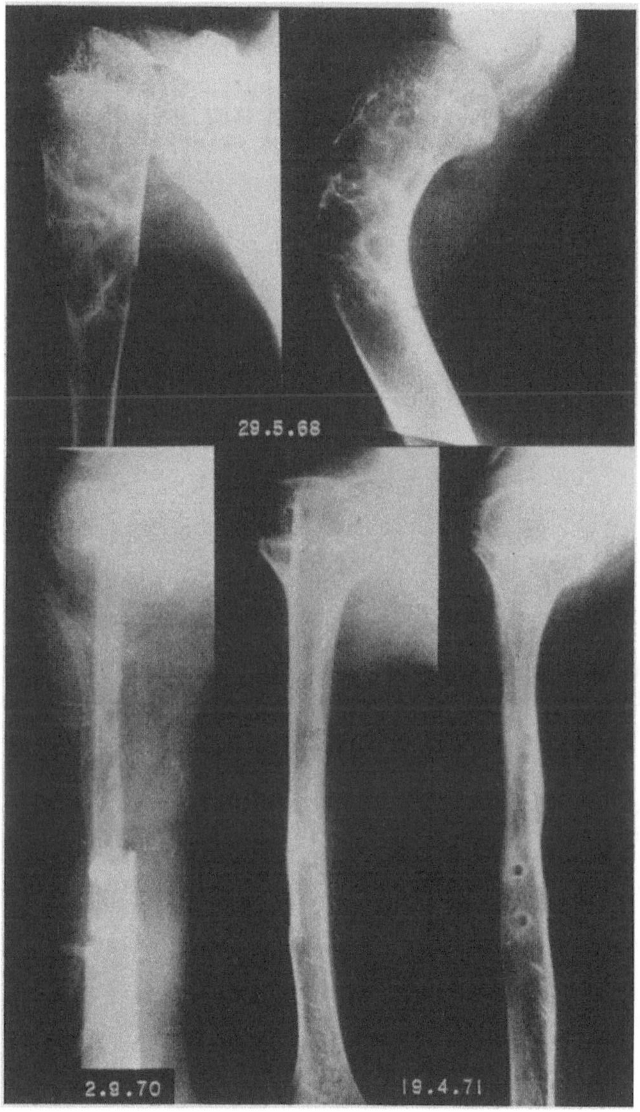

Fig. 4. The en bloc resection of the tumor-bearing bone structure offers the opportunity for correction of an already existing growth disturbance.

we fashion a recipient site, a technique known to us as the Lexer procedure, and fix the graft with two screws (Fig. 2). It is also possible to impale the graft without metallic fixation into the medullary canal. After six to eight weeks, partial recanalization of the medullary canal becomes noticeable (Fig. 3).

The resection of the involved part of the humeral shaft offers the opportunity to correct a pre-existing structural growth disturbance (Fig. 4). After surgery, we place the patient in an abduction cast usually for 13 weeks.

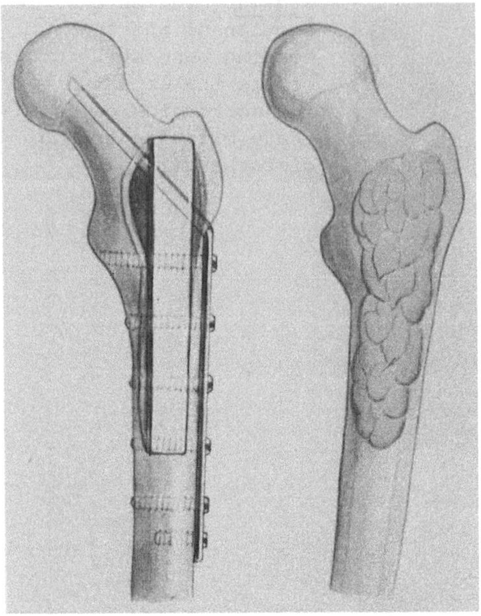

Fig. 5. In the proximal femur a plate osteosynthesis is used for stabilizing a subtotal resection of the inter- or subtrochanteric area and the cavity is filled with autologous bone material.

In 26 patients with surgically treated juvenile bone cysts, we used the above resection technique in 15 patients. Only one case with an active cyst had a recurrence. The functional result was satisfying in all patients.

Shortening of the upper arm frequently is the sequela of a juvenile bone cyst. In our material of 26 patients we encountered shortening of the arm in four patients with five active cysts and two additional arm shortenings in 21 latent cysts. This seems to indicate that it is primarily the contact of the cyst with the growth plate which causes growth disturbances and not the treatment per se. In fact, arm shortenings are reported with conservative treatment in active cysts (McKay and Nason, 1977). We did not observe a significant disturbance of longitudinal growth by implantation of a graft through the growth plate. Our follow-up results emphasize the advantage of resection of juvenile bone cysts in the upper arm as opposed to curettage alone because of a much lower recurrence rate with only slightly prolonged treatment time.

Problems of treatment are more significant when juvenile bone cysts occur in the proximal femur. Within the frame of bone tumors, the prognosis is considered benign in view of life expectancy of the patient. But since it is a disease affecting primarily children and young adults and due to its close proximity to the growth plate, skeletal deformities are frequently observed in this bone disease. In the upper arm, shortening and axial deviation can be easily tolerated without functional loss. Because of its weight-bearing function, the hip joint does not tolerate structural changes without biomechanical functional losses. Remaining prearthritic deformities can become quite significant for the future life of the child.

In the proximal femur a complete resection is usually too complicated for anatomic and biomechanic reasons. For large defects, after subtotal inter- or subtrochanteric resections, a plate osteosynthesis is used for preservation of stability and the cavity is filled with autologous bone from the iliac crest and tibia (Fig. 5). The necessity for a stable internal fixation is greater in this area than in the upper arm after fracture or infraction of the cystic wall has occurred.

Our case material at the orthopedic hospital in Munich includes 24 juvenile bone cysts at the proximal femur. In 15 patients we did a curettage and filling with cancellous bone grafts; in nine patients we did a subtotal resection with osteosynthesis and cancellous bone graft. After curettage we saw three recurrences; after subtotal resection only one. Altogether, the number of recurrences at the femur was considerably lower than at the upper arm. On the other hand, we saw all kinds of structural damage at the proximal femur after healing of the cysts, affecting function of the hip joint. These included differences in length of over 1 cm; changes in the femoral neck angle; deformities of the femoral head; osteonecrosis of the femoral head, and secondary growth disturbances of the acetabulum (Table I).

Table I. Growth disturbances of the hip joint in juvenile bone cysts (n = 55)

Site	No.	Growth disturbances*
Proximal humerus	26	6 (23%)
Proximal femur	24	9 (38%)
Pelvis	2	
Ulna	2	
Tibia	1	

* Leg length differences exceeding 1 cm;
Coxa valga - coxa vara
Femoral head deformity - osteoarthritis
Secondary acetabular dysplasia

To emphasize the problems of juvenile bone cysts near the hip joint we would like to present the following cases:

In a 9-year old boy, a juvenile bone cyst was curetted and filled with cancellous bone. Follow-up examination seven years later shows an overgrowth of the involved leg by 1.5 cm and a coxa valga with a remaining cyst.

After subtotal resection and autologous cancellous bone graft in a 4-year old girl, the follow-up over 10 years shows the development of a severe hip dysplasia after previously normal bilateral hip development (Fig. 6).

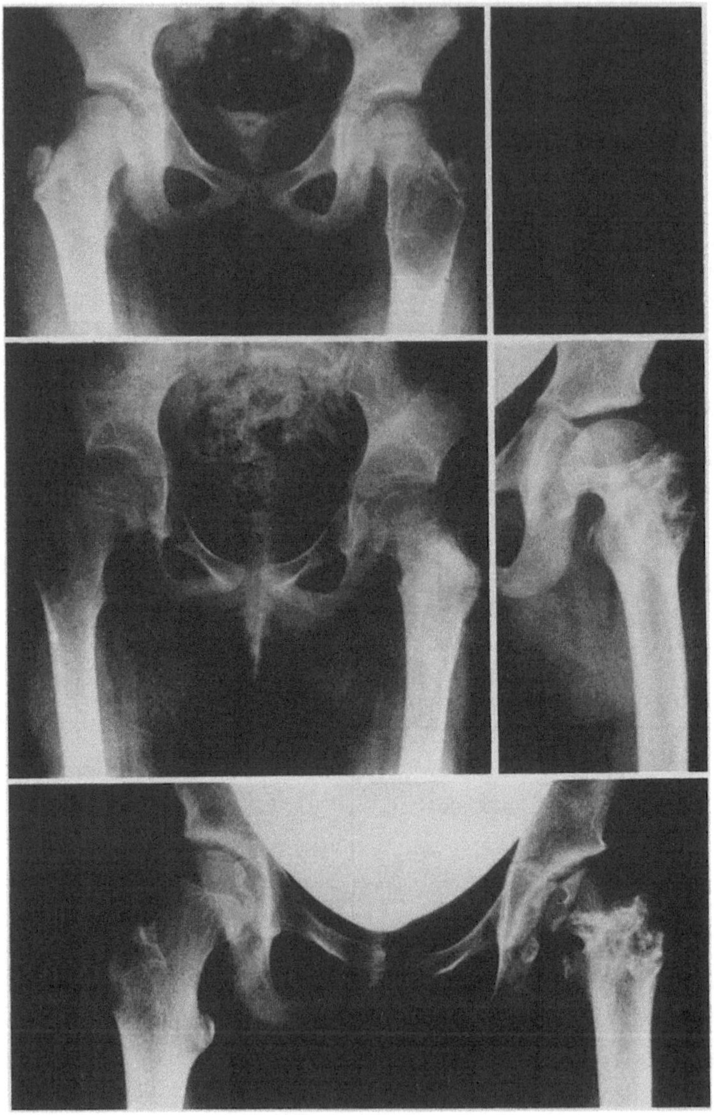

Fig. 6. 4-year old girl. Subtotal resection and autologous cancellous bone graft; over ten years' development of a severe hip dysplasia in a previously normal bilateral hip development.

Another example of the development of severe coxofemoral changes consisting of coxa vara and retroversion of the femoral head and leg shortening of 2.5 cm is shown in Fig. 7. Extension of the cyst far into the femoral neck led to a pathologic fracture.

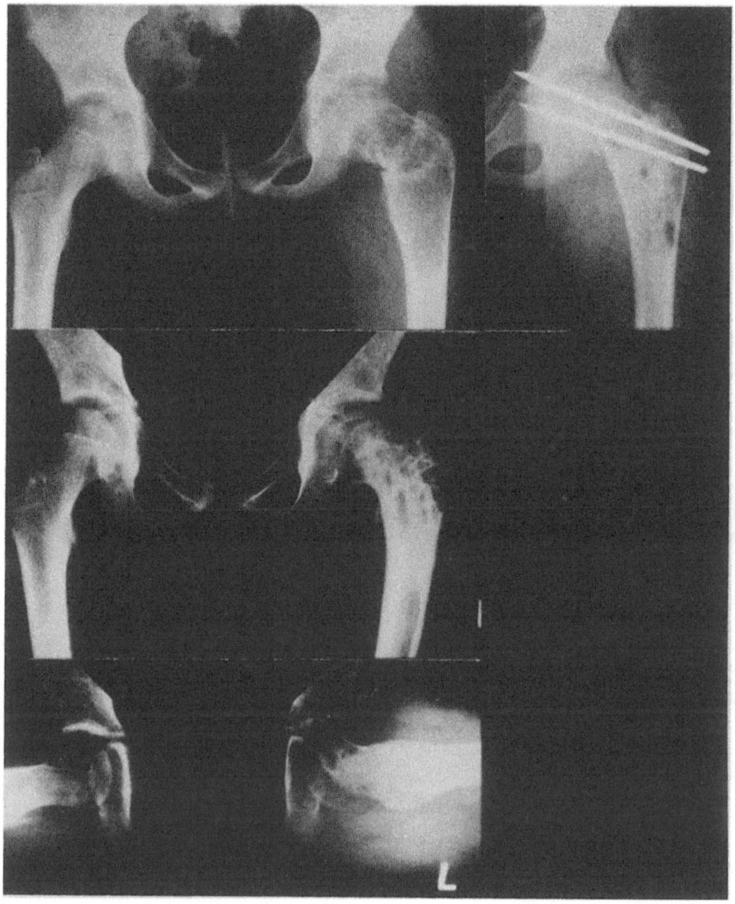

Fig. 7. Example of development of a severe hip dysplasia with coxa vara and retroversion of the femoral head with shortening of the leg by 2.5 cm. The extensive bone cyst led to a pathologic fracture.

Another patient incurred a pathologic fracture because of a juvenile bone cyst in the intertrochanteric region at the age of 13. After adaptation of the fragments and internal fixation followed by curettage of the cyst and cancellous bone grafting, a coxa vara developed with shortening of the leg by 1.5 cm and slight deformation of the femoral head. Now, 28 years later, the patient has an incipient osteoarthrosis with pain on weight bearing.

The review of our 24 patients with juvenile bone cysts in proximal femur shows that the following conditions may lead to an increase in growth disturbances around the hip joint:

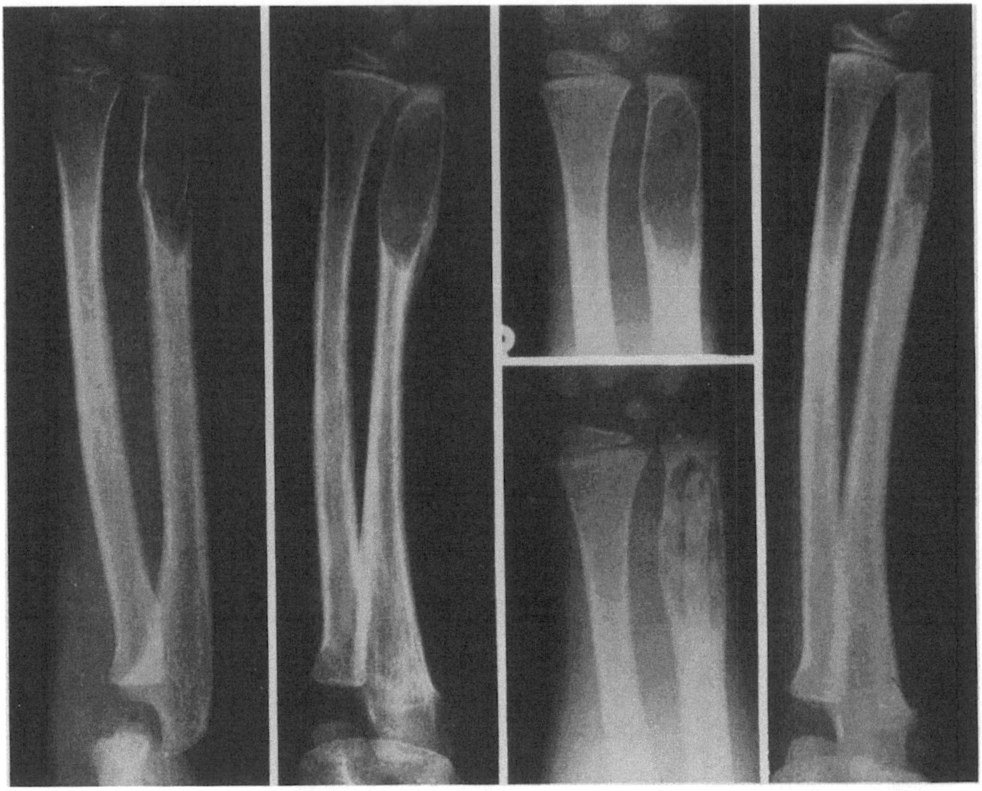

Fig. 8. Fractured juvenile bone cyst of the distal ulna in a 6-year old boy. Conservative treatment failed: therefore, curettage and filling of the defect with autologous corticocancellous material. Healing without growth disturbance.

1. When the patient's age at diagnosis was below 10 years, the likelihood of developing structural and functional disturbances around the hip joint was greater than at an older age (average age in the first group: 10 years; in the latter: 17 years).

2. Of eight cysts which were located within the femoral neck, five developed growth disturbances, while only four out of 12 cysts within the intertrochanteric area led to growth changes; no changes were observed when the cysts were in the subtrochanteric area. Therefore, we conclude that the probability of growth disturbances increases the more proximal the cyst develops within the femoral neck close to the growth plate and the trochanteric apophysis.

3. Out of four patients with pathologic fractures in the femoral neck and intertrochanteric area, three healed with considerable structural damage.

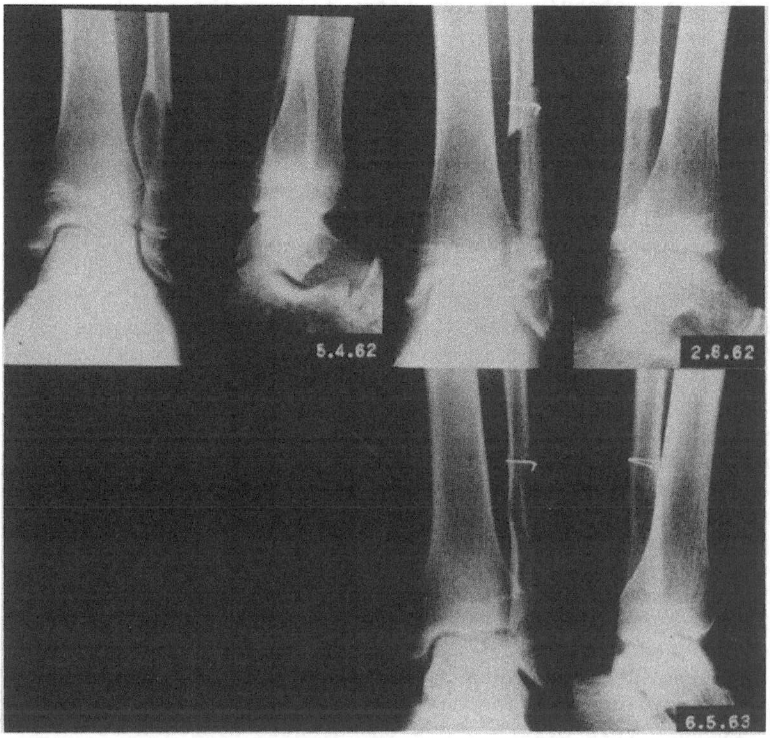

Fig. 9. Juvenile bone cyst in the distal fibula: rare manifestation. En bloc resection and bridging of the defect by an autologous tibial graft. Complete healing.

Discussion

Consequently, for the prognosis of hip function in juvenile bone cysts at the proximal femur we have to consider that

- with patients under the age of ten,
- with extension of the cyst into the femoral neck close to the growth plate,
- and with a pathologic fracture of a juvenile bone cyst at the femoral neck

structural and functional disturbances will result in over 50% of patients. These patients need special care not only until the cyst has healed but until growth cessation. Occasionally, the need for corrective surgery after healing of the cyst may arise.

The following descriptions of rare locations of juvenile bone cysts demand a therapeutic approach according to the individual problem.

A cyst in the *roof of the acetabulum* healed after filling the cavity with cancellous bone.

Juvenile bone cysts in the *ulna* are extremely rare (0.5 to 1.5%). We saw two bone cysts in two children between ages five and six and followed their course of healing over 12 and 16 years respectively, until cessation of growth (Stürz and Witt, 1979). Figure 8 shows the infraction of a cyst in a 6-year old boy. Conservative treatment did not result in healing; therefore, curettage and filling of the defect with cancellous bone was undertaken. Healing followed without growth disturbance.

The *distal fibula* is another rare location. We saw one case which healed after complete resection and bridging of the defect with an autologous tibial graft (Fig. 9).

Our experience so far shows that total or subtotal resection of all bone surrounding the cyst has the lowest recurrence rate. With the description of our approach and results we attempted to show how, by remodelling of autologous material, the morphologic appearance of the original bone structure can be regained in a very short time, especially in the upper arm of growing children and adolescents.

When compared with our surgical results, conservative treatment fared less well. Remarkable in this context seems the paper by Scaglietti and coworkers from Florence and Pisa (1979) whose group could achieve complete healing in 90% of 72 juvenile bone cysts at different sites by local injection of a corticosteroid crystal suspension -- once or several times -- into the cyst. Perhaps this offers a new therapeutic approach to achieve similar results without surgery.

Summary

Juvenile bone cysts are usually regarded as tumor-like bone lesions within the frame of bone tumors. They are benign, but due to their proximity to the growth plate in adolescents, growth disturbance with structural damage may result. Their most frequent location is the proximal humerus, next to the femur. While we curet small cysts and fill them with autologous cancellous bone, we perform complete resections in cases of recurrence or extensive cyst formation.

Our procedure for humeral bone cysts is en bloc resection of the tumor, creation of a notch within the humeral head, impaction of an autologous tibial graft into the head after first creating a recipient site (Lexer fashion) in the distal humerus, and fixation of the graft with two AO screws. Another method is impaction of the graft into the medullary canal of the distal humerus without resorting to metallic osteosynthetic material. Postoperative fixation in an abduction splint is held for 12 weeks on the average. We attribute great importance to the uninvolved, thick juvenile periosteal coverage. After implantation of the graft, this periosteal layer is closed carefully. Follow-up of our results in 26 patients with juvenile bone cysts of the humerus after curettage and grafting with cancellous bone showed a recurrence rate of 55%; after treatment with resection, the recurrence rate was only 7%.

Therefore, complete removal of the bone cyst by resection seems to be the safest way to prevent recurrences of juvenile bone cysts in the upper arm. We occasionally use this same subtotal resection on the proximal femur. In an evaluation of 22 patients, the recurrence rate was lower than in the humerus, depending on the size and the age of the patient. The older the patient, the lower was the recurrence rate. The location of a juvenile bone cyst in the ulna is extremely rare: 0.5 to 1.5%. We were able to observe patients over long periods. Healing was accomplished by subtotal resection and bridging of the gap with cancellous bone from the iliac crest or by curettage and filling of the defect with an autologous graft and compact bone from the tibia. Healing resulted without structural or functional damage.

References

1. Agerholm, J.C. and Goodfellow, J.W.: Simple cysts of the humerus treated by radical excision. J. Bone Joint Surg. 47(B), 714, 1965.

2. Boseker, E.H., Bickel, W.H., and Dahlin, D.C.: A clinicopathologic study of simple unicameral bone cysts. Surg. Gynec. Obstet. 127, 550, 1968.

3. Daum, R.H., Buhr, H., Pflugfelder: Chirurgische Therapie und Ergebnisse von juvenilen Knockenzysten. Klin. Pädiat. 187, 518, 1975.

4. Garceau, G.J., and Gregory, F.: Solitary bone cyst. J. Bone Joint Surg. 36(A), 267, 1954.

5. Harms, J., and Groh, P.: Ergebnisse der Resektion, autologen Spanplastik und Uberbruckungsosteosynthese juveniler Knockenzysten. Arch. Orthop. Traumat. Surg. 92, 285, 1978.

6. McKay, D.W., and Nason, S.S.: Treatment of unicameral bone cysts by subtotal resection without grafts. J. Bone Joint Surg. 59(A), 515, 1977.

7. Mittelmeier, H.: Resektion and freie Spanplastik zur Behandlung rezidivierender Knochenzysten des Humerus. Langenbecks Arch. klin. Chir. 309, 122, 1965.

8. Neer, C.S., Francis, K.C., Marcove, R.C., Terz, J., and Carbonara, P.N.: Treatment of unicameral bone cyst, J. Bone Joint Surg. 48(A), 731, 1966.

9. Scaglietti, O., Marchetti, P.G., and Bartolozzi, P.: The effects of methyl prednisolone acetate in the treatment of bone cysts. J. Bone Joint Surg. 61(B), 200, 1979.

10. Spence, K.F., Sell, K.W., and Brown, R.H.: Solitary bone cyst. Treatment with freeze -dried cancellous bone allografts. J. Bone Joint Surg. 51(A), 87, 1969.

11. Stürz, H., and Witt, A.N.: Juvenile Knochenzysten der Ulnar. Arch. Orthop. Traumat. Surg. 94, 105, 1979.

12. Stürz, H., Zenker, H., and Buckl, H.: Total subperiosteal resection treatment of solitary bone cysts of the humerus. Arch. Orthop. Traumat. Surg. 93, 231, 1979.

13. Stürz, H., and Zenker, H.: Ergebnisse der Resektionsbehandlung juveniler Knochenzysten am Oberarm. Orthop. Praxis 9, 722, 1979.

14. Witt, A.N., Walcher, K., and Zenker, H.: Die Resektionsbehandlung rezidivierender juveniler Knochenzysten. Arch. Orthop. Unfall-Chir. 74, 105, 1972.

IV. TREATMENT COMPARISON: STEROID INJECTION versus SURGERY

M. Campanacci and R. Capanna

Introduction

For the past 20 years, the surgical treatment of simple bone cysts has been undisputed. The need for more radical surgery, however, than curettage and grafting through a small opening of the cystic wall has arisen to reduce the incidence of recurrences. Regardless of the type of surgical procedure performed, the surgeon is faced with potential problems such as infections, growth plate injury, postoperative pathologic fractures, prolonged immobilization, and hospitalization. In 1974, Scaglietti, on the basis of a brilliant clinical experiment, suggested local injections of methylprednisolone acetate (MPA) for the treatment of simple bone cysts. The procedure of MPA injection is technically simple, can be done on an outpatient basis, and requires no immobilization of the involved extremity.

In this article we review the results obtained with nonsurgical and surgical treatments at the Rizzoli Institute, University of Bologna, Italy.

Materials and Methods

A retrospective review was undertaken in 259 patients treated by curettage and grafting through a limited opening in the cortex. Follow-up information including roentgenograms was available one to 35 years after the last surgical procedure in 178 patients. One hundred and forty-one additional patients were treated by MPA injection between 1974 and 1981, and were followed from one to eight years (mean: 4 years).

In 102 of these patients, a definite diagnosis of simple bone cyst was made based on the age of the patient, the site, the radiographic appearances and the analysis of the aspirated fluid; in 39 patients having had a recurrence, a histologic report was available from a previous operation. The technique of MPA injection has been described in various articles.

The two groups (surgical and nonsurgical) were comparable with respect to age, bone involvement and location of the cysts.

Treatment results were classified according to the same criteria proposed by Neer (1966), with a slight modification. Four responses were noted:

1) *healed* - the cyst completely filled in with bone, and the cortical margin thickened;

2) *healed with residual osteolysis* - the cyst was filled with bone and the cortical margin thickened, but there were still small residual areas of osteolysis within the cyst;

3) *recurrence* - the cyst initially consolidated with bone. However, large areas of osteolysis and cortical thinning subsequently returned;

4) *no response* - the cyst showed no evidence of response to MPA treatment.

The last two responses (recurrence or no response) represented a failure of treatment.

Results

Results with surgical treatment were equal to those with MPA injection (Table I). After the conventional curettage and packing of the cavity with bone grafts, the recurrence rate was high (33%) and about one-fourth of the patients required a second operation. The incidence of recurrence apparently increased after a second (46%) or a third operation (55%). Recurrences after surgical treatment were encountered more frequently in cysts that were in contact with the growth plate. The quality of the grafts (auto- or homologous) and the cyst size had no significant bearing on the recurrence rate. When the curettage technique is used, the packing should be complete because unfilled spaces apparently favor recurrences. The packing, however, should be loose because hard packing impedes revascularization.

Table I. Results with MPA* injection compared with surgical treatment

	Totally healed	Healed with residual osteolysis	Recurrence	No response
Surgical treatment				
1st	83(46%)	38(21%)	57(33%)	---
2nd	100(56%)	44(25%)	34(19%)	---
3rd	107(61%)	46(26%)	25(13%)	---
MPA treatment				
1st	60(42%)	36(26%)	32(23%)	13(9%)
2nd	71(50%)	36(26%)	21(15%)	13(9%)

* methylprednisolone

Twenty-four (14%) patients, after surgical treatment, had a limb length discrepancy, possibly due to a surgical lesion of the growth plate. In patients treated with MPA injection, a rather typical sequence of cyst consolidation is noted radiographically. After the first injections of MPA, cystic expansion ceases and the cortical margin thickens (Phase 1). Approximately six months (range 2-14 months) after the onset of treatment (Phase 2), the cyst has a "frosted glass" type of opacity. If this type of consolidation occurs, the MPA injections are terminated. Approximately 12 months (range 4-26 months) after onset of treatment (Phase 3), the cyst becomes consolidated with dense sclerotic bone. Only rarely does the bone segment show complete recovery of its normal structural features. Mainly cysts in a diaphyseal location tend to regain a normal morphologic appearance.

The temporal progression from one phase to another does not necessarily occur simultaneously in all areas of the cyst, especially with respect to multiloculated cysts. The cystic zones that tend to be refractory to treatment are the poles and the area adjacent to the growth plate. Although multiloculated cysts may show delayed consolidation in certain subcysts, this is not a poor prognostic sign because most of the cystic cavities will eventually fill with bone.

When the cyst was totally consolidated with bone after MPA treatment the incidence of recurrence was zero.

When cyst consolidation was incomplete (persistent areas of osteolysis in the cyst; intracystic bony septae with frosted glass areas; irregular bone filling of the cyst; a rim of osteolysis at the periphery of the cyst) the incidence of recurrence was 47%.

Eleven patients who had incomplete consolidation were given additional corticosteroid injections and the cysts totally healed. Eleven patients who had recurrence had further MPA treatment with incomplete consolidation. For this reason we recommend maintaining close radiographic surveillance of those patients whose cysts consolidate and who have any of the abovementioned risk factors. If there is any evidence of cyst recurrence, additional corticosteroid injections are given into the persistent osteolytic areas. Close radiographic surveillance is unnecessary in patients who have incomplete cystic consolidation after MPA treatment.

The incidence of failures (recurrence and no response) after MPA treatment was higher in patients over 15 years of age (Table II), in cysts located near the growth plate (Table III), and in cysts of large size (Table IV). Monoloculated cysts had a better result with MPA treatment than multiloculated cysts (Table V). The larger multiloculated cysts did not respond as well to MPA treatment as the smaller monoloculated cysts. With the larger, more septated cysts the draining procedure and obtaining

Table II. Results of MPA* treatment in different age groups

Age groups	MPA	Totally healed	Healed with residual osteolysis	Recurrence	No response
0-5	20	40%(8)	30%(6)	10%(2)	20%(4)
6-10	61	58%(35)	18%(11)	16%(10)	8%(5)
11-15	51	49%(25)	33%(17)	16%(18)	9%(1)
> 15	9	33%(3)	23%(2)	11%(1)	33%(3)

* methylprednisolone

Table III. Results of MPA* treatment according to location of cyst.

Location	No.	Totally healed	Healed with residual osteolysis	Recurrence	No response
Diaphyseal	31	74% (23)	7% (2)	3% (1)	16% (5)
Metaphyseal diaphysis	40	38% (15)	40% (16)	15% (6)	7% (3)
Contact to growth plate	70	47% (33)	26% (18)	20% (14)	7% (5)

*methylprednisolone

Table IV. Results of MPA* treatment according to size of cyst

Size	No.	Totally healed	Healed with residual osteolysis	Recurrence	No response
$\leq$ 12 cm^2	44	57% (25)	23% (10)	9% (4)	11% (5)
12-21 cm^2	48	56% (27)	29% (14)	11% (5)	4% (2)
> 21 cm^2	49	41% (20)	24% (12)	23% (11)	12% (6)

*methylprednisolone

Table V. Results of MPA* treatment according to loculation.

	No.	Totally healed	Healed with residual osteolysis	Recurrence	No response
Monoloculated	68	66% (45)	21% (14)	7% (5)	6% (4)
Multiloculated	73	36% (26)	30% (22)	22% (16)	12% (9)

*methylprednisolone

an equal distribution of the MPA throughout the entire cyst are more difficult.

No differences were found between cysts located in the humerus and those located in the femur (Table VI). There was no difference between cysts previously treated surgically and those receiving initial treatment with MPA.

Table VI. Results of MPA* treatment according to type of bone.

	No.	Totally healed	Healed with residual osteolysis	Recurrence	No response
Humerus	97	52% (50)	24% (23)	16% (16)	8% (8)
Femur	37	48% (18)	30% (11)	11% (4)	11% (4)
Other	7	43% (3)	29% (2)	14% (1)	14% (1)

*methylprednisolone

Two patients had a pathologic fracture during the course of MPA treatment and 11 patients after a recurrence.

At the conclusion of treatment, 28 patients had a limb-length discrepancy, consisting of shortening of the involved limb in 22 patients and lengthening of the involved limb in six patients. Most of these 28 patients had a history of a pathologic fracture (11), a surgical procedure (7), or both (7). Hence, the limb-length discrepancy was probably secondary to these factors. The other three patients who received MPA treatment had no obvious reason for a limb-length discrepancy. Two of these patients had only 1 cm shortening in the involved limb which may have been an idiopathic limb-length discrepancy. The other, an eight-year old patient who had a multiloculated unicameral bone cyst of the proximal femoral metaphysis, had substantial limb shortening (4.5 cm) secondary to epiphyseal necrosis.

No infection related to the MPA treatment was noted. Only four patients required a surgical procedure after recurrence of the cyst.

Discussion

It is well known that recurrences occur in 30% of patients treated by curettage and grafting through a limited opening in the cortex (Baker et al, 1970; Boseker et al, 1968; Mastragostino and Sanguinetti, 1960; Monesi and Florio, 1964; Neer et al, 1966; Neer et al, 1973; Ottolenghi et al, 1969; Spence et al, 1969).

Surgery by wide subperiosteal excision of the cyst wall for more than half its circumference, with bone grafts and fixation (the latter is necessary especially in the femur) reduces the incidence of recurrences to less than 10% (Agerholm and Goodfellow, 1965; Boseker et al, 1968; Creyssel, 1976; Fahey and O'Brien, 1973; Ravaglia, 1973; Rigault et al, 1975; Sanguinetti, 1975). This surgical procedure is preferable to the first (which should be discarded) because it is more rational and effective. It is, however, relatively extensive and involves surgery requiring lengthy immobilization in a plaster cast and no weight bearing in the lower limbs for a long time.

Conservative treatment by MPA injection is without doubt effective and often produces a complete and final cure (Corrado and Passaretti, 1976; Gualtieri, et al, 1976; Scaglietti, et al, 1976; 1979). Even when the effect is only partial, the treatment in most cases obviates surgery by stopping the growth of the cysts and producing a sufficient increase in mechanical resistance in the affected bone. Even if surgery will be indicated after treatment with MPA -- and we believe this is rare -- the treatment delays the procedure, which is favorable to the success of the operation and avoids possible damage to the growth cartilage (Campanacci et al, 1975).

References

1. Agerholm, J.C., and Goodfellow, J.W.: Simple cysts of the humerus treated by radical excision. J. Bone Joint Surg. 47(B), 714, 1965.

2. Baker, D.M.: Benign unicameral bone cyst: A study of forty-five cases with long-term follow-up. Clin. Orthop. 71, 140, 1970.

3. Boseker, E.H., Bickel, W.H., and Dahlin, D.C.: A clinicopathologic study of simple unicameral bone cysts. Surg. Gynecol. Obstet. 127, 550, 1968.

4. Campanacci, M., De Sessa, L., and Bellando Randone, P.: Cisti ossea (revisione di 275 osservazioni. Risultati della cura chirurgica e primi risultati della cura incruenta con metilprednisolone acetato). Chirurgia Organi di Movimento 62, 471, 1975.

5. Campanacci, M., De Sessa, L., and Trentani, C.: Cura incruenta della cisti ossea con iniezioni locali di metilprednisolone acetato secondo Scaglietti. Giornale Italiano di Ortop. e Traum. 3, 27, 1977.

6. Capanna, R., Dal Monte A., Gitelis, S., and Campanacci, M.: The natural history of unicameral bone cyst after steroid injection. Clin. Orthop. Rel. Res. 166, 204, 1982.

7. Corrado, E.M., and Passaretti, U.: Prime esperienze terapeutiche con cortisonici nelle cisti ossee giovanili. Chirurgia degli Organi di Movimento 63, 239, 1976.

8. Creyssel, J.: Discussion. Lion Chir. 73, 347, 1976.

9. Fahey, J.J., and O'Brien, E.T.: Subtotal resection and grafting in selected cases of solitary unicameral bone cyst. J. Bone Joint Surg. 55(A), 59, 1973.

10. Gualtieri, I., Gualtieri, G., and Montefusco, E.: Risultati ottenuti nel trattamento delle cisti ossee mediante infiltrazione con acetato di metilprednisolone. Osp. Ital. Chir. 29, 155, 1976.

11. Mastragostino, S., and Sanguinetti, C.: I trapianti auto-omoed etero-plastici nel trattamento delle cisti ossee. Arch. Putti Chir. Org. Mov. 13, 96, 1960.

12. Monesi, B., and Florio, L.: Le recidive nelle cisti ossee solitarie operate. Arch. Putti Chir. Org. Mov. 19, 175, 1964.

13. Neer, C.S., II, Francis, K.C., Johnston, A.D., and Kierman, H.A.: Current concepts of solitary unicameral bone cyst. Clin. Orthop. 97, 40, 1973.

14. Neer, C.S. II, Francis, K.C., Marcove, R.C., Terz, and Carbonara, P.N.: Treatment of unicameral bone cysts. A follow-up study of one hundred seventy-five cases. J. Bone Joint Surg. 48(A), 731, 1966.

15. Ottolenghi, C.E., Schajowict, F., and Raffa, J.: Le kyste osseux essentiel unioculaire. Etude de clinique et anatomopathologique de 123 cas. Rev. Chir. Orthop. 55, 287, 1969.

16. Ravaglia, M.: La resezione sottoperiostale nelle cisti ossee solitarie del bambino. Scalpello 2, 51, 1973.

17. Rigault, P., Alain, J.L., Padovani, J.P., and Plumerault, J.: Les kystes osseoux essentiels de l'extremité superieur du femur chez l'enfant. Notre expérience de 12 cas. Rev. Chir. Orthop. 61, 221, 1975.

18. Sanguinetti, C.: Una particolare tecnica operatoria per le cisti ossee solitarie. Giornale Italiano di Ortop. Traum. 1, 353, 1975.

19. Scaglietti, O.: L'azione osteogenetica dell'acetato di metilprednisolone. Bull. Scien. Med., Bologna 146, 159, 1974.

20. Scaglietti, O., Marchetti, P.G., and Bartolozzi, P.: Sull'azione topica del corticosteroid in microcristalli in alcune lesioni dello scheletro. Arch. Putti Chir. Org. Mov. 27, 9, 1976.

21. Scaglietti, O., Marchetti, P.G., and Bartolozzi, P.: The effects of methylprednisolone acetate in the treatment of bone cysts. Results of three years follow-up. J. Bone Joint Surg. 61(B), 200, 1979.

22. Spence, K.F., Sell, K.W., and Brown, R.H.: Solitary bone cyst: treatment with freeze dried cancellous bone allograft. A study of one hundred seventy-seven cases. J. Bone Joint Surg. 51(A), 87, 1969.

Spinal Reconstruction in Tumor Management

C. C. EDWARDS

The Role of Surgical Resection/Reconstruction

There is increasing evidence that complete surgical resection is the treatment of choice for such common primary spinal tumors as chordoma (Francis, 1975; Sundaresan et al, 1979), giant cell tumor (Dahlin, 1977; Dahlin et al, 1970; DiLorenzo et al, 1980; Goldenberg et al, 1970; Jaffe, 1958; McGrath, 1972; Schwimer et al, 1981; Stener, 1977), chondrosarcoma (Francis, 1975; Marcove et al, 1972), and osteoblastoma (Francis, 1975; Huvos, 1979; Jackson, 1978). Chordomas typically grow slowly and metastasize late. Yet, they are associated with a 96% fatality rate when treated with radiation and/or local resection (Huvos, 1979). Death is usually caused by local compression or invasion since only 10% of these tumors eventually metastasize (Francis 1975). Hence, chordomas should be ideal for early and complete surgical removal.

Like the chordoma, giant cell tumors also grow slowly and metastasize late (Dahlin et al, 1970; Goldenberg et al, 1970; Johnson and Dahlin, 1959; Murphy and Ackerman, 1956). When treated with radiation and/or curettage, the recurrence rate is up to 20% in the upper spine (Dahlin, 1977; DiLorenzo et al, 1980; Schwimer et al, 1981) and a 50% recurrence rate is reported with occasional fatalities for tumors about the sacrum (Dahlin et al, 1970; Goldenberg et al, 1970; Mnaymneh et al, 1964; Smith et al, 1979). Hence, early and complete resection is probably the treatment of choice for giant cell tumors of the spine (Schwimer et al, 1981; Stener, 1977). Chondrosarcoma and osteoblastoma of the spine are also suitable for complete resection. Chondrosarcoma is not very responsive to radiation and is notorious for local recurrence and seeding following less than complete resection. Osteoblastomas are benign, but over 20% of them recur if excision is not complete (Francis, 1975; Huvos, 1979; Jackson, 1978).

Chordomas, chondrosarcomas, and some giant cell tumors tend to invade adjacent tissues to varying degrees. Hence, complete resection must include variable margins of normal tissue. To achieve adequate margins, significant portions or even entire regions of the spine must sometimes be removed and subsequently reconstructed.

On the other hand, radical resection of metastatic lesions in the spine is usually not warranted. For cases without neurologic deficit, radiation therapy, occasionally supplemented with stabilization procedures, usually

Current Concepts of Diagnosis and Treatment of Bone and Soft Tissue Tumors.

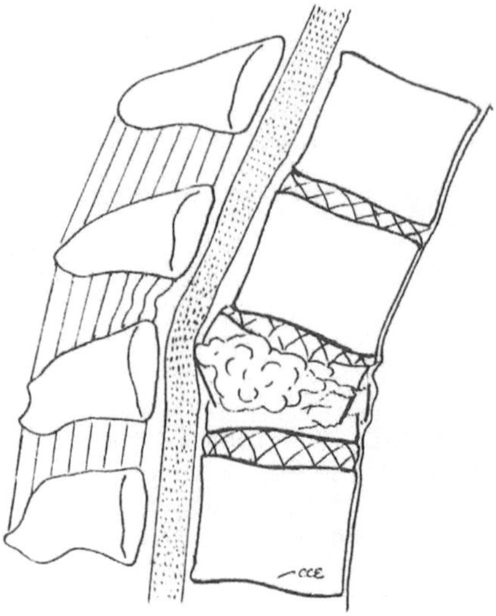

a

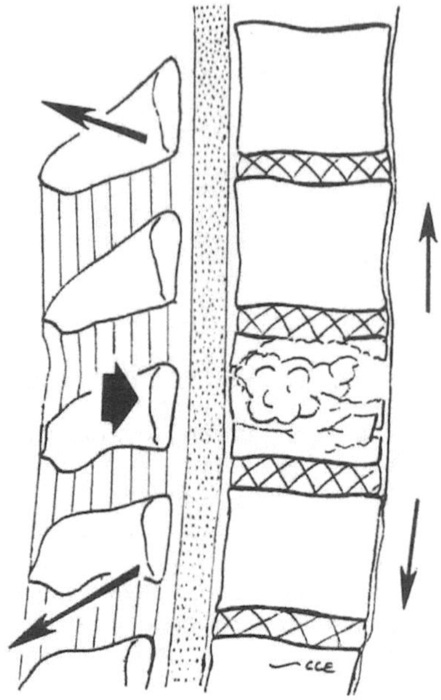

b

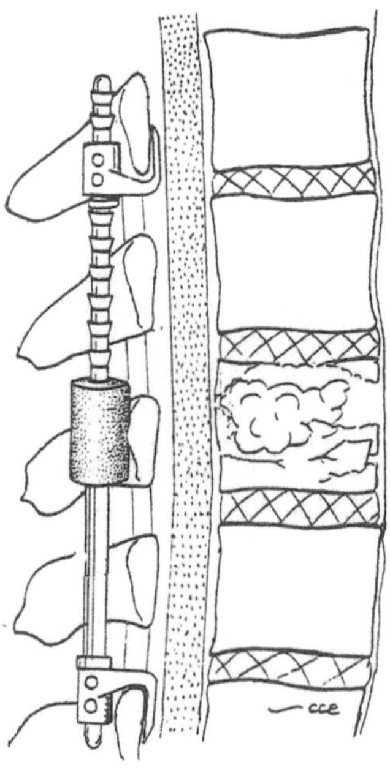

c

Fig. 1. Indirect cord decompression and posterior rod-sleeve stabilization.
a) Neurologic symptoms and pain are usually due to anterior pressure from pathologic fractures and resulting kyphosis.
b) Hyperlordosis and distraction will often relieve anterior pressure and provide relative cord decompression.
c) The distraction rod-sleeve method provides hyperextension and distraction as well as sufficient stability in flexion and rotation to support the spine until fusion, despite complete destruction of the vertebral body.

provides satisfactory palliation. When present, neurologic (especially motor) deficit is usually due to pressure against the anterior cord by the tumor. Kyphosis, secondary to vertebral collapse, tethers the cord over the anterior lesion (Fig. 1a). Elimination of kyphotic angulation and subsequent stabilization of the diseased portion of the spine in an anatomic, or preferably slightly hyperlordotic posture, will relieve anterior pressure from the cord and will usually diminish, and sometimes eliminate, the neurologic deficit (Fig. 1b). Accordingly, for metastatic lesions with subtle or reasonably stable incomplete neurologic deficits, we first restore spinal alignment, obtain rigid posterior fixation, and then irradiate the metastasis (Fig. 1c). If sufficient resolution of the neurologic deficit does not occur, we proceed with vertebral body resection as a second stage. We find less morbidity and fewer fixation failures with this staged approach than with direct primary anterior corporectomy and only anterior fixation.

For cases with a rapidly progressing neurologic deficit from direct tumor pressure, however, we agree with surgeons advocating primary vertebral body resection, anterior fixation, and then radiation therapy (Bucy, 1963; Harrington, 1981). This will accomplish more effective decompression than indirect posterior methods if significant tumor mass is present within the spinal canal. Resection before radiation also reduces tumor bulk which is said to make subsequent radiation therapy more effective (Bucy, 1963; Harrington, 1981).

The Feasibility of Complete Spinal Tumor Excision

Although complete resection of many spinal tumors is considered optimum treatment, until recent years this procedure has been thought "impossible or unfeasible" (Dahlin, 1977; DiLorenzo et al, 1980; Francis, 1975; Huvos, 1979; Kambin, 1966; Smith et al, 1979). The reason is largely because spinal tumors tend to occur in such inaccessible areas as the anterior sacrum or vertebral bodies and are often quite large when first recognized. For example, the two most common primary spinal tumors, chordomas and giant cell tumors (Edwards, 1983), tend to form large presacral soft tissue masses before symptoms become pronounced and often escape detection on initial plain radiographs (Federle et al, 1980; Nakagawa et al, 1977; Smith et al, 1979).

The former barriers to complete resection of large primary spinal tumors -- inadequate exposure, control of blood loss, and subsequent spine reconstruction -- have all been gradually overcome by the efforts of many surgeons. In the 1950's, MacCarty and Hays both achieved unprecedented resection of large tumors involving the distal four sacral vertebrae (Hays, 1953; MacCarty et al, 1952). Complete removal of one or more vertebral bodies was introduced for the cervical spine by Scoville (Scoville et al, 1967) and extended to the lumbar spine by Lievre in 1968 (Lievre et al, 1968). Another important step was the use of the staged anterior then posterior approaches to permit better anterior tumor dissection and vascular control before removal. Using this procedure, successful resection of tumor masses up to 10 cm affecting the lower sacral vertebrae was accomplished by Loçalio (Localio et al, 1967). In

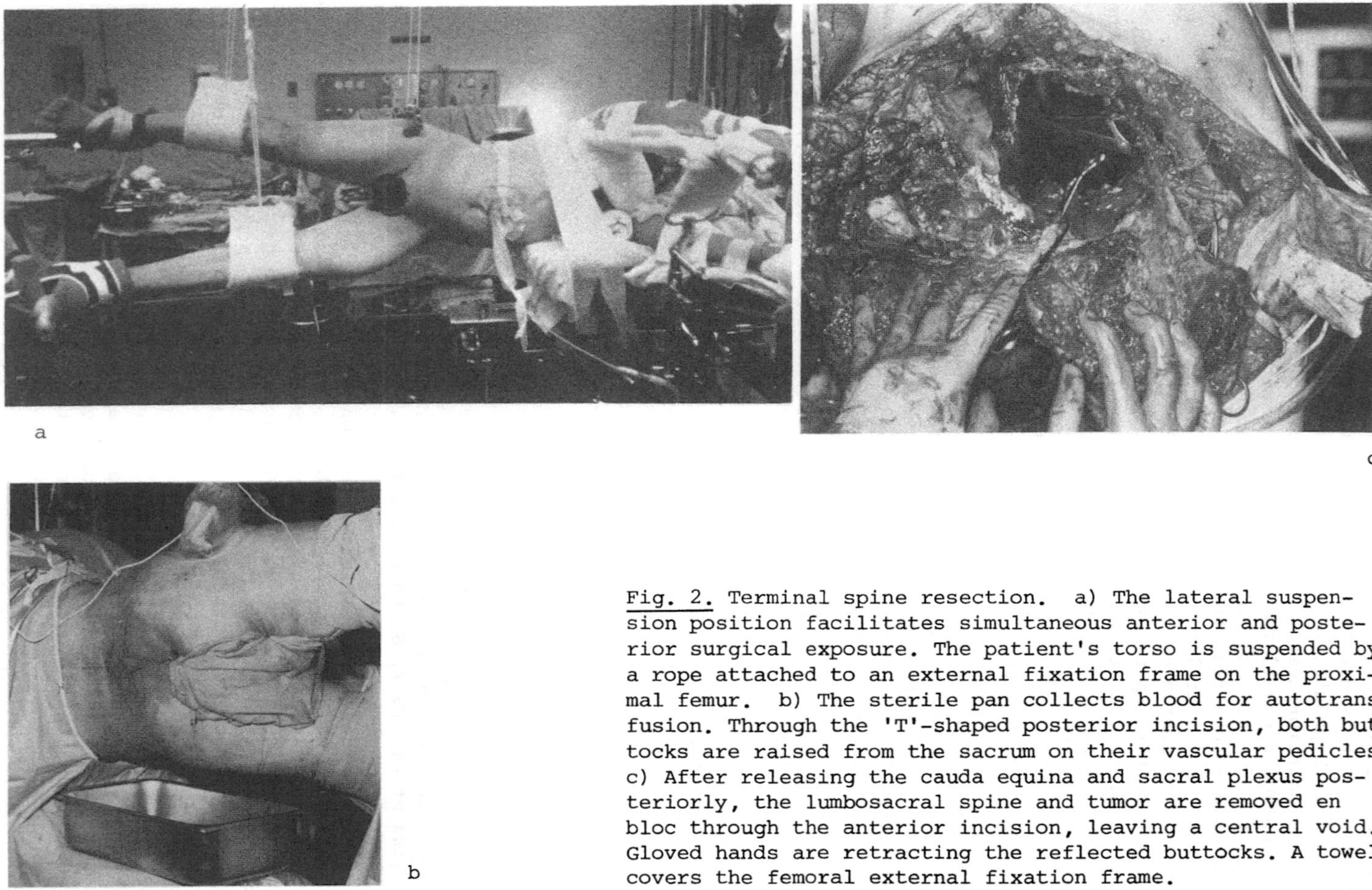

a

b

c

Fig. 2. Terminal spine resection. a) The lateral suspension position facilitates simultaneous anterior and posterior surgical exposure. The patient's torso is suspended by a rope attached to an external fixation frame on the proximal femur. b) The sterile pan collects blood for autotransfusion. Through the 'T'-shaped posterior incision, both buttocks are raised from the sacrum on their vascular pedicles. c) After releasing the cauda equina and sacral plexus posteriorly, the lumbosacral spine and tumor are removed en bloc through the anterior incision, leaving a central void. Gloved hands are retracting the reflected buttocks. A towel covers the femoral external fixation frame.

1971, Stener reported the use of anterior and posterior staged procedures to successfully remove a giant cell tumor, together with the T11, T12, and L1 vertebrae (Stener and Johnson, 1971). In 1979, we performed a three and one-half level (T12-L3) complete resection of the spine to remove a 14 cm malignant giant cell tumor. This was accomplished in a one-stage procedure through combined anterolateral and posterior incisions (Edwards and DeSilva, 1981; Elliott, 1979).

In 1980, we removed the entire sacrum and lower one or two lumbar vertebrae in several tumor cases, an operation we call "terminal spine resection" (Edwards and DeSilva, 1981). To obtain simultaneous anterior and posterior surgical exposure for lumbosacral tumors, we use the lateral suspension position (Fig. 2a). We first pass AP transfixion pins through the proximal femur and construct a lateral external fixation frame. The patient's pelvis is suspended by attaching the frame to either a ceiling hook or the overhead beam of a Chick-Langren table with a rope. The patient's chest is supported by the upper portion of the table and the feet are placed in traction stirrups. The areas between the rib cage and midthighs can then be prepped, permitting circumferential surgical access and collection of blood in a sterile pan under the patient for autotransfusion (Fig. 2b).

Resection of the entire sacrum is accomplished by anterior and posterior surgical teams working simultaneously. Through an abdominal incision, the tumor mass is dissected anteriorly and feeding vessels are ligated. A second surgical team raises the buttocks and its overlying skin from the sacrum as two flaps based on the superior gluteal artery. They next dissect the cauda equina and roots posteriorly through a tumor-free interval, if present. Thereafter, the tumor and resected spine and/or sacrum are rotated anteriorly away from the preserved cord or roots (Fig. 2c). With this procedure, we have resected primary spinal tumors as large as 33 cm in diameter with preservation of the cauda equina and retention of useful distal neurologic function (Edwards and DeSilva, 1983). At present, few large primary spinal tumors remain truly inoperable. Nevertheless, the risks and complications of massive spine resections are considerable and their successful reconstruction can be challenging.

Spine Stabilization

Stabilization implants and techniques, and our understanding of spine biomechanics, are all evolving at a rapid pace. Although experience and data are limited, techniques are now available which enable the orthopedic surgeon to stabilize, replace, or bypass almost any portion of the pathologic or resected spine. Hence, surgery for malignant or borderline-malignant spinal tumors should now be aimed at complete en bloc resection with adequate margins whenever possible. Using this approach, large malignant giant cell tumors, chordomas, chondrosarcomas, and a solitary myeloma have already been successfully removed from the spine without recurrence (Drobni and Kudasz, 1975; Edwards and DeSilva, 1983; Hamdi, 1969; Simeone, 1975; Stener, 1977; Stener and Johnson, 1971).

Stabilization of the pathologic or resected spine may require posterior fixation, anterior fixation, or both. In general, if the vertebral body has not been surgically removed and if the facets are intact, posterior stabilization alone will suffice. The affected spinal segments are fixed in relative lordosis to transmit axial loads through the facets and unload the pathologic vertebral body. On the other hand, if the vertebral body has been resected and no posterior laminectomy has been performed, anterior reconstruction alone will suffice. However, where both anterior and posterior bone and/or ligamentous structures have been removed or previously injured, combined anterior and posterior reconstruction is necessary.

Reconstruction after Partial Vertebral Destruction or Resection

Cervical spine

Following vertebral body collapse or destruction in the cervical spine, two posterior methods provide secure fixation and restoration of lordosis. The first method combines: a) tight interspinous wiring with b) guy-wires from every other facet laterally to the most caudal spinous process to be fused to prevent rotation (Robinson and Southwick, 1960) and a tension-band from the most cranial lamina to the most caudal spinous process to counteract the flexion moment resulting from the weight of the head (Edwards et al, 1982b). If a laminectomy has been performed, a Southwick lateral rib strut can be substituted for the interspinous wire portion of the fixation (Robinson and Southwick, 1960)(Fig. 3a,b,c). The second method is to combine the flexion tension-band with a curved metal rod (Murphy and Southwick, 1983) wired to each facet to maintain the desired lordosis and provide rotatory fixation.

Both of the above methods depend on intact facets and hyperlordosis to restore vertebral height and reduce anterior tumor pressure against the cord. If the tumor involves the facets or lateral masses, the fixation technique must include distraction. Until more sophisticated implants become available, distraction can be accomplished by wiring the facets to conformable threaded rods (as Harrington compression rods). The wires, and hence vertebrae, can be raised cephalad by advancing the position of nuts supporting each wire on the rod (Fig. 3d,e).

For patients with a life expectancy of more than two years, iliac bone grafting is added to any of the constructions described above to achieve lasting stability. For patients with more limited life expectancy, methylmethacrylate bone cement can be used for supplemental fixation to decrease the need for postoperative external immobilization. However, if methacrylate is used in place of a bone graft in patients with a short life expectancy, it is important to make numerous drill holes through the cortex of each segment and/or to insert facet wires (Fig. 3c) or multiple K-wires so the cement will achieve reliable fixation at each level. It has been clearly demonstrated that simply pressing cement over lightly decorticated lamina and a single interspinous wire achieve no better fixation after several months than use of the wire alone (Dunn, 1977; Panjabi et al, 1977; Whitehill et al, 1983).

Following resection of a cervical vertebral body, either bone or methycrylate anterior reconstructions have proven satisfactory if close attention is paid to technical detail (Cantu, 1974; Dunn, 1977; Scoville, et al, 1967). Bone should be used following removal of primary tumors and

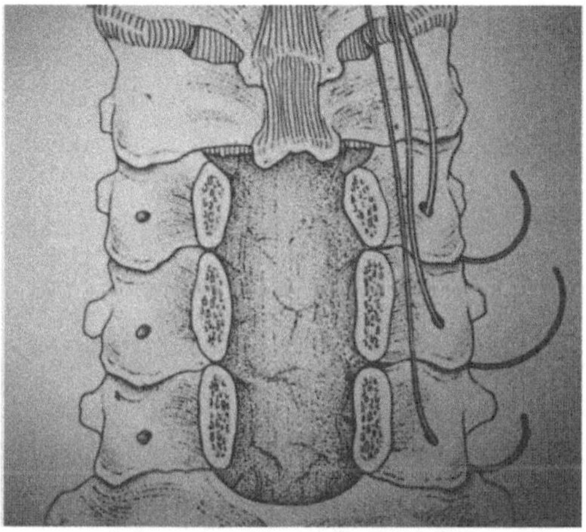

a

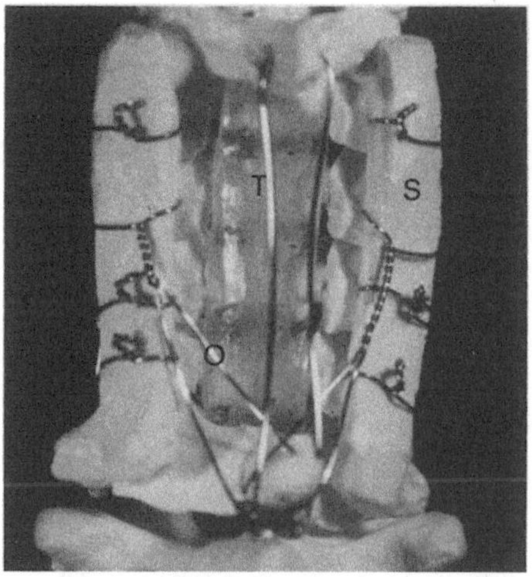

b

Fig. 3. Posterior cervical reconstruction with the multiwire technique. a&b) Effective fixation can be accomplished following laminectomy and tumor destruction of multiple vertebrae by combining oblique wires (O), a bone or metal strut wired to the facets (S), and a posterior tension band (T). Wires pass through holes in the inferior facets and out the facet joints. The facet wires afix the lateral strut and extend caudally, together with the posterior tension band to the first intact major spinous process.

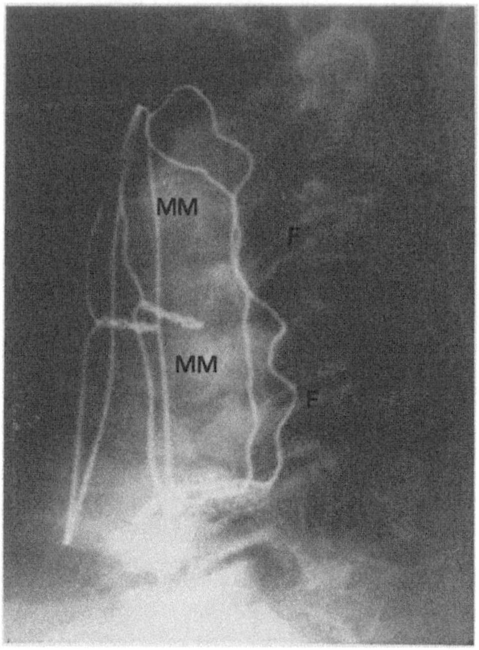

c

Fig. 3
c) The multiwire technique uses the facets as a fulcrum (F) to unload the collapsed vertebral bodies in this patient with vertebral body destruction from metastatic breast carcinoma. Supplemental methylmethacrylate cement (MM) and a collar provide adequate protection for patients with limited life expectancy.

d

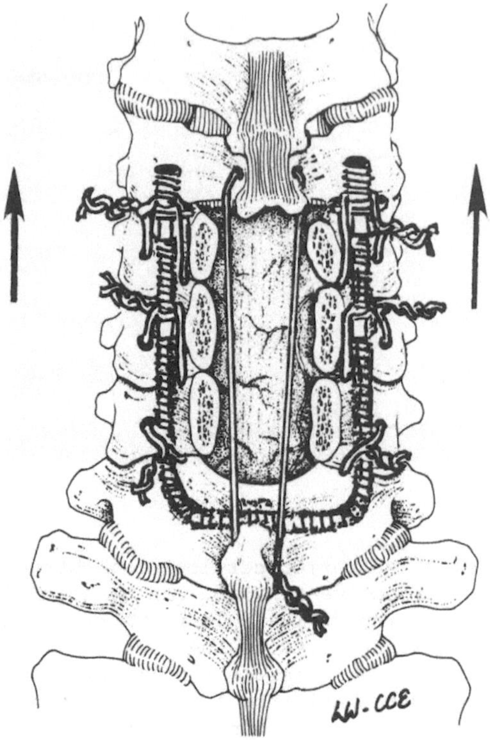

e

Fig 3 (cont'd). d&e) When there is destruction of the facets (arrow) as well as the vertebral bodies, distraction is required in addition to multiwire fixation. A threaded rod is contoured for the desired lordosis and wired to the facets. To distract the spine, the facet wires are elevated by advancing nuts and washers on the threaded rod.

where life expectancy of three years or more is anticipated. A bicortical iliac graft is cut to shape and longitudinally transfixed with one or two Steinman pins. These pins fit into holes predrilled in the normal body above and slots in the vertebral body below; the anterior slots are plugged with cement to fix the pin position (Barrasso and Keggi, 1983). Alternatively, a small plate on the anterolateral spine can be used to fix the graft and the position of adjacent vertebrae (Tscherne, 1982).

A similar technique, using methacrylate to replace the bone, was first recommended by Scoville for patients with short life expectancy to minimize the need for postoperative immobilization and eliminate the added surgery to obtain iliac graft (Scoville et al, 1967). Anterior cement loosening and dislodgement can be reduced by using wires or other fixation devices embedded within the cement to connect the normal bodies on either side of the defect (Dunn, 1977; Harrington, 1981; White and Panjabi, 1978). After removal of multiple cervical vertebral bodies, prosthetic replacements have been used successfully in China (First Medical College of

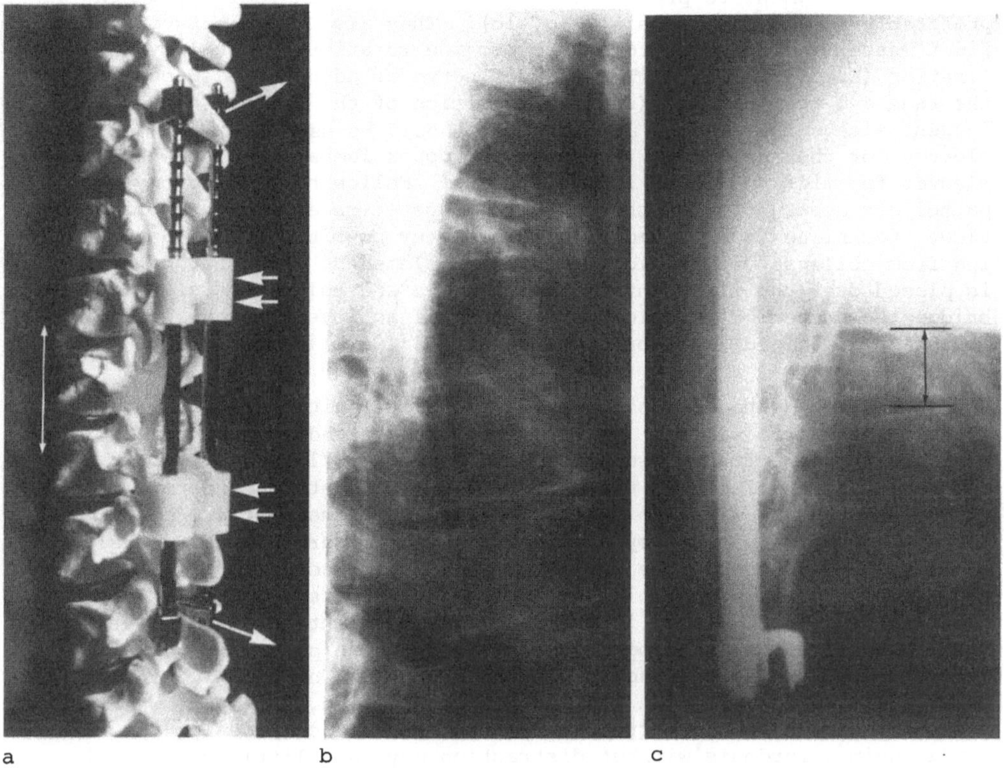

Fig. 4. Rod-sleeve stabilization. a) The bridging sleeve technique is useful after laminectomy. One pair of sleeves is placed over the first intact lamina above and below the posterior defect. b&c) In this case, vertebral height and anatomic alignment were restored one month following metastatic vertebral body collapse. There was substantial improvement in the patient's level of pain and neurologic symptoms prior to subsequent radiation treatment.

Shanghai, 1974) and Japan (Ono and Tada, 1975). Japanese workers recently reported that a polyethylene prosthesis cemented in place of multiple cervical bodies is less prone to loosening than when methacrylate alone is used (Ono et al, 1983).

Thoracic and lumbar spine

The forces traversing the thoracic and lumbar spine are much greater than in the cervical spine. Accordingly, either posterior wire fixation or anterior body grafting alone are usually inadequate following kyphotic angulation. However, reduction can be reliably maintained with posterior rod-sleeve stabilization (Edwards et al, 1982a). Double distraction rods affixed to the spine with hooks restore vertebral height. Spinal rod-sleeves placed over the rods push anteriorly through the pedicles of the most apical segment to provide 3-point loading and anatomic alignment or preferably some hyperlordosis (Fig. 1c). They also wedge between the facet masses and spinous process to provide rotational stability and rigid fixation (Edwards, 1982). The sleeves serve as adjustable spacers between the rods and the anatomic (reduced) position of the lamina; hence, the largest sleeve that can be accommodated should be used, ie, the medium sleeves for thoracolumbar, the large for upper lumbar, and the eliptical sleeves for mid- and lower lumbar lesions. Following laminectomy or pathologic destruction of the pedicle or posterior elements, a bridging sleeve technique is employed to eliminate any possibility of canal narrowing from collapse of pathologic posterior elements. One pair of sleeves is placed over the first intact lamina above and below the defect to bridge the posterior pathology and yet still achieve the same complete reduction and rigid fixation (Edwards et al, 1982b)(Fig. 4a,b,c).

An alternative method for stabilizing the pathologic thoracic or lumbar spine is segmental wiring. In 1971, Stener reported the use of AO plates wired to thoracic transverse processes (Stener, 1971). A more satisfactory method is now available in which Luque rods are bent to conform to the desired lordosis and wired to the lamina of each segment after approximate spinal length is restored by a distraction outrigger. Spinal rod-sleeves can also be placed over the Luque rods to enhance rotatory stability and provide more precise control over localized lordosis. The principle limitation of segmental wiring is the inability to maintain significant distraction across the pathologic segments. When the posterior cortex of the vertebral body remains intact or in the absence of neurologic deficit, lordosis alone will suffice in lieu of net distraction. However, when the posterior cortex of the vertebrae is destroyed and compressing the cord of cauda equina, lordosis without distraction may have little beneficial effect on indirectly decompressing the cord.

Both distraction rod-sleeve and Luque rod-sleeve fixation methods provide sufficient fixation to permit mobilization without a cast in cases where vertebral destruction or removal is incomplete. We have found that the former problem of hook dislodgement following distraction rod treatment can be almost eliminated by incorporating four technical considerations:

1) using new spinal hooks with an anatomic shoe configuration (Fig. 1c, 4);
2) using rod-sleeves to increase the rigidity of fixation;
3) selecting rods that extend at least 1 cm beyond the upper hooks; and
4) adding a small compression rod check rein for cases where the hooks must go below L4 or above T4 (Edwards et al, 1984).

The midline compression rod check rein is fixed with hooks under the most cranial lamina and around the most caudal spinous process to be fused (Murphy et al, 1980). This counteracts any flexion which might result in spine elongation and hook dislodgement. Regardless of the method employed, iliac bone graft is used for long-range security of fixation.

When the goal is resection of a tumor involving only the vertebral body in an otherwise intact spine, the surgeon may proceed anteriorly without posterior fixation. However, after resection of a vertebral body in the lower thoracic or lumbar spine, it is important to provide some means for long-term maintenance of vertebral space height. Without supplemental internal fixation, late collapse will likely occur with the partial graft resorption that accompanies the normal creeping substitution phase of bone graft healing. Various fixation alternatives are available to supplement the vertebral body bone graft. These include anterior plates, the Dunn distraction device (Dunn, 1982), distraction rods encased within cement (Harrington, 1981), custom prosthetic devices with fixation by means of axially-directed screws (Hamdi, 1969), or intramedullary rods (Edwards et al, 1982b). The most secure alternative is to supplement conventional anterior interbody fusion with one-level posterior compression rod fixation. Certainly, posterior fixation is essential in cases of previous laminectomy or ligamentous injury at the same level as the vertebral resection. When anterior bone grafting alone is used in such cases, we often find eventual loss of anterior fixation with late pain and deformity (Harrington, 1981).

Reconstruction after Complete Vertebrectomy

The reconstruction plan after one-level vertebrectomy varies with the level and extent of spinal resection. In the upper and midthoracic spine there is protection from the rib cage and a lack of major forces acting upon the spine. Hence, conventional anterior grafting (fibula, rib, or iliac bone struts screwed or slotted into place) combined with either secure anterior or posterior internal fixation or long-term casting will suffice.

For reconstruction following complete vertebrectomy in the thoracolumbar or upper lumbar spine, both secure anterior and posterior internal fixation are needed for predictable results. For one-level vertebrectomies, meticulously constructed interconnected bone grafts have succeeded for the anterior portion of the construction (Stener, 1977). However, we recommend using supplemental anterior fixation to maintain the distance between the vertebral bodies during the period of creeping substitution. Fixing the position of the vertebral bodies anteriorly also greatly reinforces the rotational stability contributed by the rod-sleeves posteriorly. Any of the anterior implants described under thoracolumbar vertebral body replacement are satisfactory for these purposes.

For reconstruction following multiple-level complete vertebrectomy, both anterior and posterior fixation are essential. Without a versatile and rigid posterior fixation system, accurate spinal alignment is difficult to achieve and maintain (Edwards, 1983; Hamdi, 1969; Stener and Johnson, 1971). Without a supplemental anterior fixation implant, late graft collapse is likely (Stener and Johnson, 1971). Accordingly, to reconstruct multiple spinal segments, we have used a custom prosthesis anteriorly combined with double rod-sleeve fixation posteriorly.

The anterior prosthesis for the thoracic spine can be constructed from standard materials available in the operating room. After tumor and vertebral removal, we span or bridge the distance between the vertebral bodies above and below the defect with one or two threaded rods. Each rod is fitted with a pair of washers and nuts. The rod is inserted into the adjacent vertebral bodies after making a hole through the subchondral plate of the body above and a slot in the body below. The washer platforms are then advanced proximally and distally by the nuts to correct any kyphosis and to maintain the distance between the vertebrae. Next,

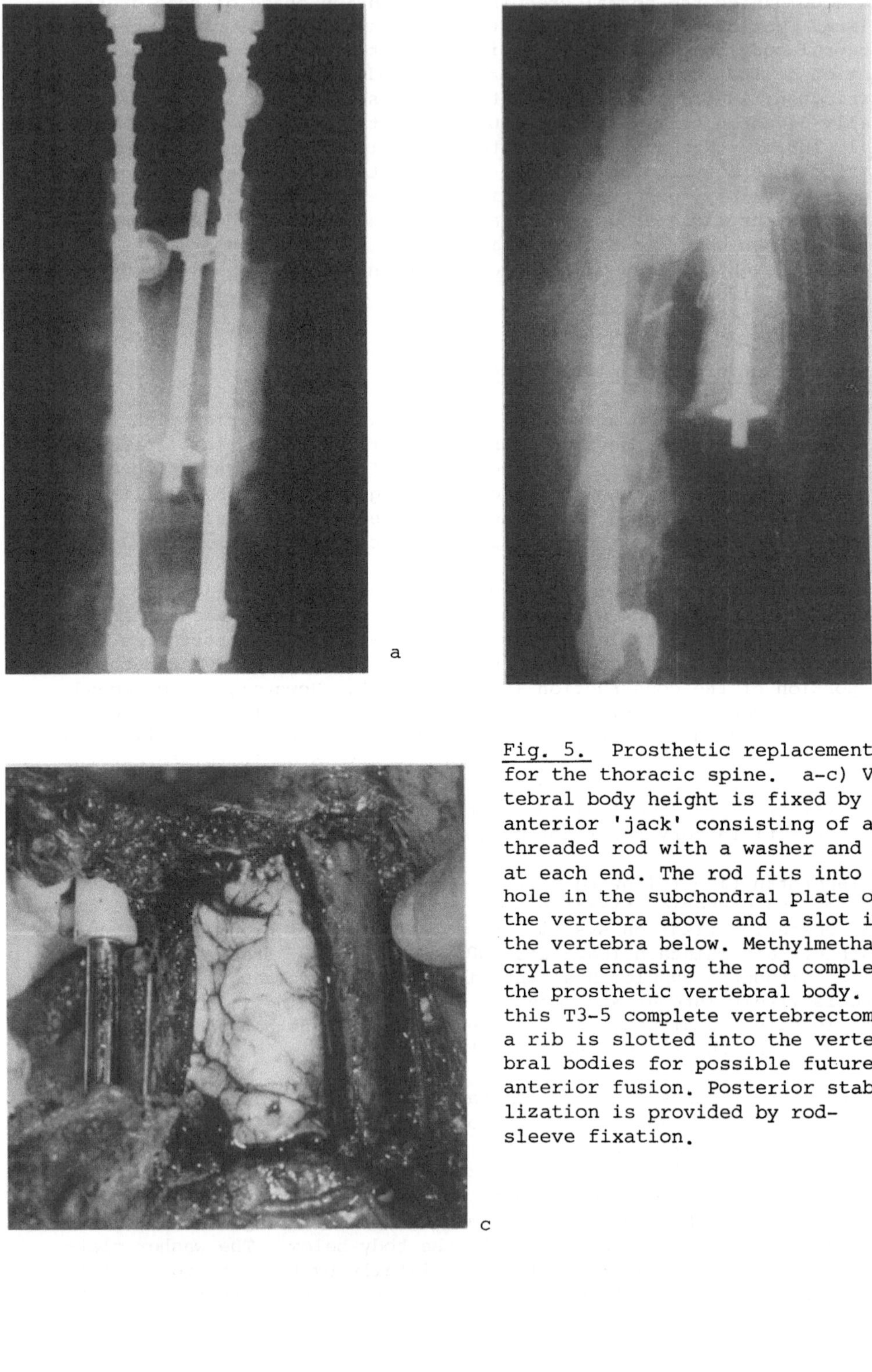

Fig. 5. Prosthetic replacement for the thoracic spine. a-c) Vertebral body height is fixed by an anterior 'jack' consisting of a threaded rod with a washer and nut at each end. The rod fits into a hole in the subchondral plate of the vertebra above and a slot in the vertebra below. Methylmethacrylate encasing the rod completes the prosthetic vertebral body. In this T3-5 complete vertebrectomy, a rib is slotted into the vertebral bodies for possible future anterior fusion. Posterior stabilization is provided by rod-sleeve fixation.

a foil mold is held in place anterior to the spinal cord. Methylmethacrylate is molded to form the vertebral body prosthesis -- embedding the fixation rods. We have used this method successfully in reconstructing up to three consecutive levels (T3, 4, 5) following complete vertebrectomy (Fig. 5a,b,c).

Without the stress-shielding effect of the rib cage and with greater moments to contend with, methylmetacrylate may not have sufficient strength for multiple-segment replacements in the thoracolumbar region. For this reason, we have used a custom prosthesis featuring retractable intramedullary rods which can be jacked into methacrylate-lined holes within the vertebrae above and below the defect (Fig. 6a,b). The prosthesis should be constructed to permit addition of fibular or rib struts and/or cancellous bone graft. This custom anterior prosthesis combined with posterior stabilization has been successful in replacing three and one-half vertebrae (T12 - L3), although length of follow-up is short (Edwards and DeSilva, 1983; Elliot, 1979)(Fig. 6c,d).

For posterior reconstructions of vertebrectomy defects, we have had excellent results using rod-sleeve fixation. Although either compression or distraction rods can be used, compression rods are more effective in blocking unwanted spine flexion. Bridging spinal rod-sleeves are placed above and below the resection defect. They provide both rigid 3-point AP fixation to shield the anterior prosthesis from flexion moments and also medial-lateral fixation to counteract rotational moments (Edwards et al, 1982b) (Fig. 6b).

In three situations after complete vertebral resection, spine reconstruction can be greatly simplified or even bypassed. Firstly, shortening should be considered after one-level lumbar vertebrectomies provided the roots of the resected segment are already sacrificed or afunctional. The inferior facets of the body above the resection will articulate with those below. Reconstruction then merely requires conventional anterior and posterior grafting combined with posterior single-level compression rods. Secondly, Gunterberg has shown that sufficient mechanical strength of the lumbosacral connection permits ambulation following resections of all sacral vertebrae below the superior half of S1, if the proximal half of the sacral ala are not disturbed (Gunterberg et al, 1976).

Thirdly, we have found it is not necessary to reconstruct the spine following terminal spine resection (complete removal of the sacrum plus one or more lumbar vertebrae) (Edwards and DeSilva, 1983). However, we have learned that a diverting uretero-ileostomy and colostomy should be performed before the resection. Otherwise, tumor margins are comprised, the chance of inadvertent bowel or ureteral injury during tumor resection is high, and kinking or scar constriction of the ureters is likely where they traverse the massive dead space following vertebrectomy and/or sacrectomy. We recommend carrying out the diversions at least two weeks in advance of the tumor resection. This will permit healing of the ureteral and bowel anastomoses before the passive catabolic insult of spine resection; it will also reduce time and blood loss during the tumor resection operation.

After terminal spine resection we place the patient on a rotating bed (Rotobed) for three months (Fig. 7a,b). This allows gradual shortening,

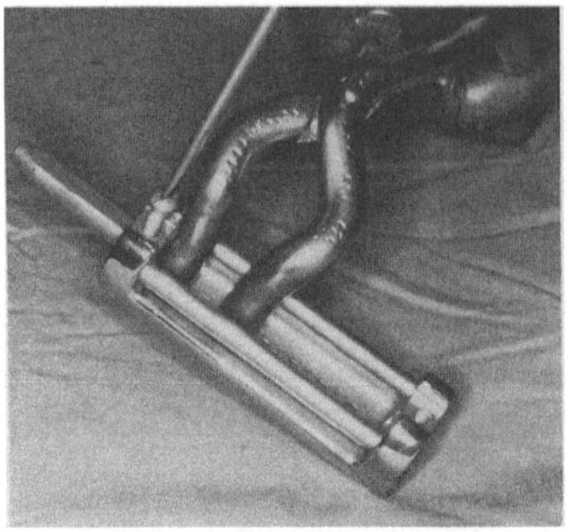

a

Fig. 6. Thoracolumbar spine replacement using a custom metallic prosthesis. a) Retractible rods are inserted into adjacent vertebral body holes with a special jack-type instrument to fix the anterior prosthesis. b) Posterior compression rod-sleeve fixation counteracts flexion moments. The sleeves also wedge between the spinous processes (S) and the facets (F) to provide rotational stability. c&d) Following T12-L3 resection, this patient was treated on a Stryker frame until prosthetic spine replacement.

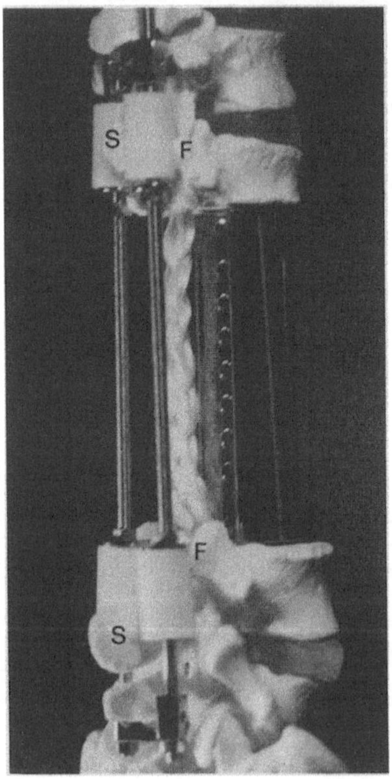

b

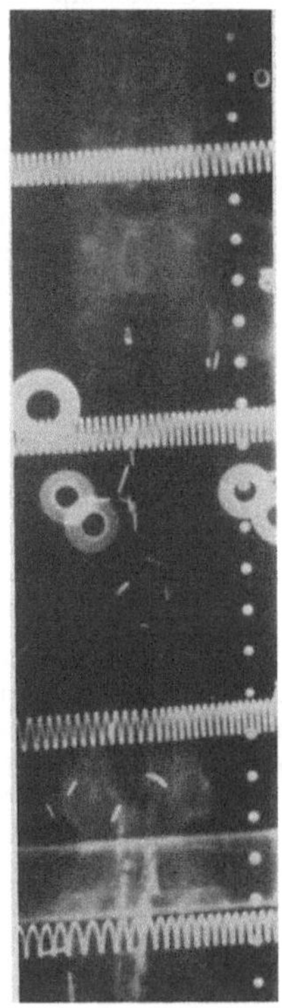

c

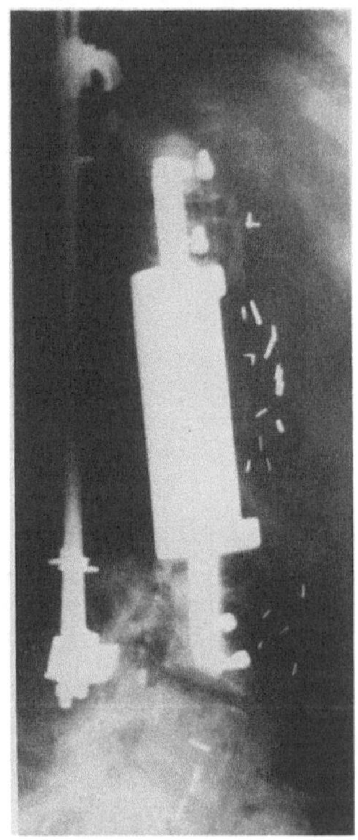

d

prevents excess motion across the unprotected cauda equina and, yet, provides the position changes needed to support respiratory physiology. Axial stability is finally achieved via the natural development of a biologic sling (Edwards and DeSilva, 1981); this sling is formed by muscles attaching the pelvic ring to the distal lumbar vertebrae and from

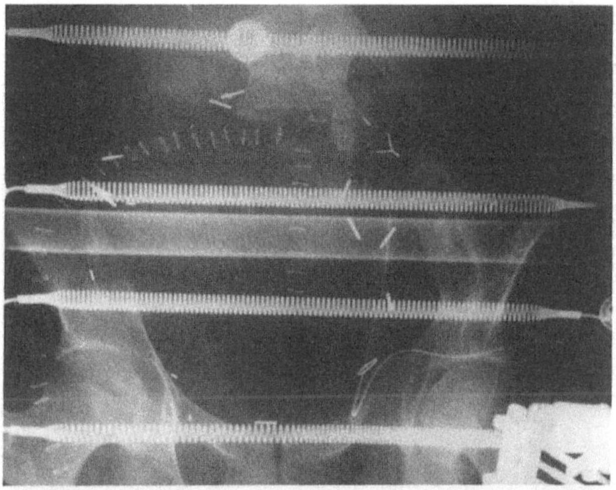

a

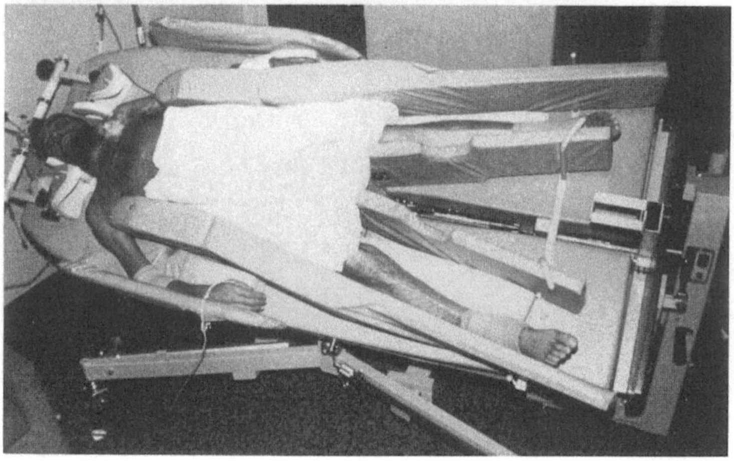

b

Fig. 7. Adaptation following terminal spine resection. a) Radiograph on a Stryker frame immediately after L4-S5 resection and preservation of the right sacral plexus. b) The patient (GB) was subsequently treated on a rotating bed to protect neurovascular structures and allow gradual settling of the lumbar spine into the pelvis.

associated scar (Fig. 7c,d). The patient can then be mobilized upright with a removable plastic orthosis to provide angular support. This concept has been used in two of our patients with surprisingly good results. One patient (R.G.) regained only protective sensation distally. However, the other patient (G.B.) had full motor and sensory recovery of his leg despite the 9-inch circumferential dissection of his cauda equina and the absence of a lumbosacral spine (L4 - S5). At two-year follow-up he was able to walk short distances using a cane and corset (Fig. 7e).

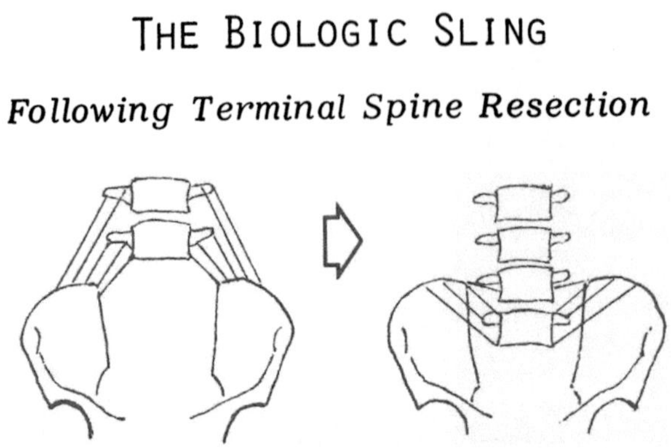

c

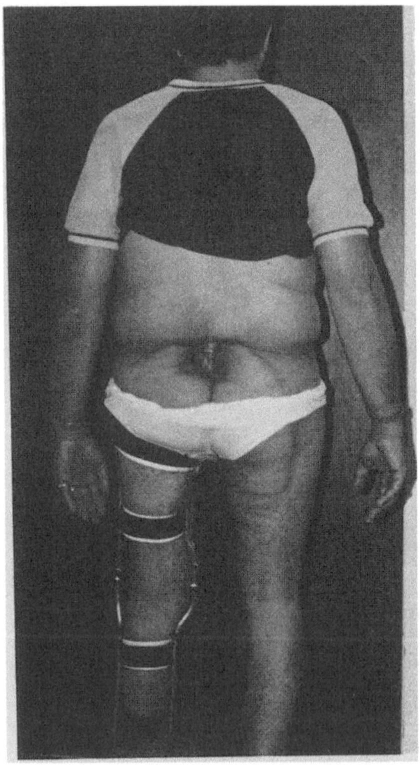

e

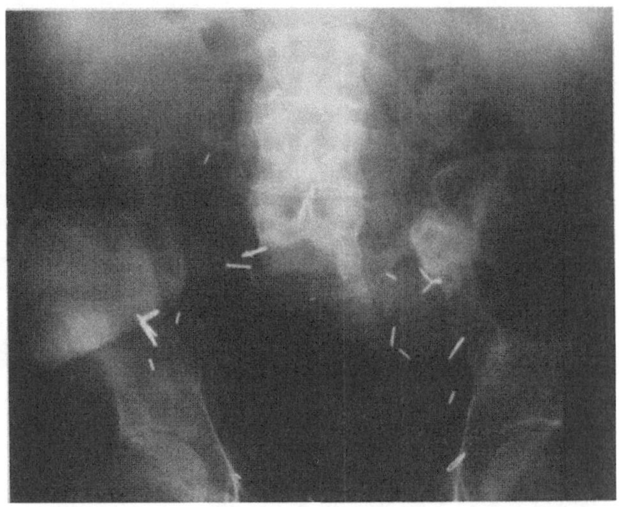

d

Fig. 7 (cont'd) c) and d) After shortening, the muscles and scar between the pelvis and spine form a 'biologic' sling to support the torse during ambulation. e) At 2-year follow-up the wound is well healed and the patient can stand without support and walk with a cane and a corset.

It should be noted that the largest of the spine resection/reconstruction procedures discussed here are associated with significant risk and postoperative complications. Due to the magnitude of surgery, some major complications are unavoidable, but others can be prevented with good preoperative planning. For example, computerized tomography with selective use of metrizamide myelography, arteriography, and contrast pyelograms will enable the surgeon to gain a three-dimensional concept of the tumor and nearby vital structures before surgery (Edwards, 1983). As discussed earlier, bowel and ureteral diversion procedures done well in advance of terminal spine resection can prevent major complications. It is well established that preoperative radiation complicates dissection and aggravates bone graft and wound healing (Harrington, 1981). Hence, radiation should be deferred until after surgery whenever possible. Postoperative chemotherapy markedly depresses incorporation of bone grafts. Accordingly, stabilization constructs which depend primarily on a bone graft fixation should be avoided if chemotherapy is anticipated (Watts, 1980).

The spine stabilization and reconstructive techniques described above have served us well to date. Although our experience with some of these methods in patients with tumors is limited to a few cases and our follow-up is short, none of the constructs has failed to date, and we have not yet experienced late loss of spinal alignment. For metastatic vertebral body lesions, posterior cervical wire techniques and distraction rod-sleeve procedures more distally have restored spinal alignment, permitted resumption of patient mobility, and have often provided indirect cord decompression and neurologic improvement. Following vertebral body removal, anterior implant-methacrylate or implant-graft replacement is satisfactory if posterior structures are intact. After complete vertebrectomy, combining the appropriate anterior prosthesis with posterior compression rod-sleeve fixation appears to provide at least adequate short-term reconstruction of even multiple-level resections. We hope that these various methods of spine reconstruction, combined with improved resection techniques, will facilitate more complete surgical removal of selected primary spine tumors and reduce their unduly high rate of recurrence and fatality.

References

1. Barrasso, J.A., and Keggi, K.S.: Vertebral body excision in the treatment of cervical disc disease, spondylosis and spinal stenosis. J. Bone Joint Surg. Orthop. Trans. 7, 114, 1983.

2. Bucy, P.C.: The treatment of malignant tumors of the spine. A review. Neurol. 13, 938, 1963.

3. Cantu, R.C.: Anterior spinal fusion using methylmethacrylate. Int. Surg. 59, 110, 1974.

4. Dahlin, D.C.: Giant-cell tumor of vertebrae above the sacrum. Cancer 39, 1350, 1977.

5. Dahlin, D.C., Cupps, R.E., and Johnson, E.W. Jr.: Giant cell tumor: A study of 195 cases. Cancer 25, 1061, 1970.

6. DiLorenzo, N., Spallone, A., Nolletti, A., et al: Giant cell tumors of the spine: A clinical study of six cases, with emphasis on the radiological features, treatment, and follow-up. Neurosurg. 6, 29, 1980.

7. Drobni, S., and Kudasz, J.: Abdominoperineal resection for enormous presacral cysts and tumors. Am. J. Proctol. 26, 33, 1975.

8. Dunn, H.K.: The role of methyl methacrylate in the stabilization and replacement of tumors of the cervical spine: A project of the Cervical Spine Research Society. Spine 2, 15, 1977.

9. Dunn, H.K.: The operative correction of congenital kyphoscoliosis. J. Bone Joint Surg. Orthop. Trans. 6, 13, 1982.

10. Edwards, C.C.: The spinal rod sleeve: Its rationale and use in thoracic and lumbar injuries. J. Bone Surg. Orthop. Trans. 6, 11, 1982.

11. Edwards, C.C.: The diagnosis and surgical management of spinal tumors using computed tomography. In, Computed Tomography of the Spine. (Ed.) Post, J.D. Williams and Wilkins, Baltimore, 1983.

12. Edwards, C.C., and DeSilva, J.B.: Total resection of multiple lumbar and sacral segments in tumor surgery. Proc. 16th Ann. Scoliosis Res. Society. p. 140, 1981.

13. Edwards, C.C., and DeSilva, J.B.: Resection of the sacrum and multiple lumbar vertebrae in tumor surgery. J. Bone Joint Surg. Orthop. Trans. 7, 1983.

14. Edwards, C.C., DeSilva, J.B., and Levine, A.M.: Early results using spinal rod sleeves in thoracolumbar injury. J. Bone Joint Surg. Orthop. Trans. 6, 345, 1982a.

15. Edwards, C.C., Levine, A.M., Murphy, J., and DeSilva, J.B.: New Techniques in Spine Stabilization (technical monograph), Zimmer USA Publications, Warsaw, Indiana, 1982b.

16. Edwards, C.C., York, J.J., and Levine, A.M.: Prevention of spinal hook dislodgement. J. Bone Joint Surg. Orthop. Trans. 7, 1984 (to be published).

17. Elliott, J.: Orthopedist is pioneer in prosthetic design. (Medical News) JAMA 242, 1831, 1979.

18. Federle, M.P., Moss, A.A., and Margolin, F.R.: Role of computed tomography in patients with "sciatica". J. Computer Asst. Tomography 4, 335, 1980.

19. First Medical College of Shanghai, Dept. of Orthopedics: Treatment of tumors of the cervical spine by excision and prosthetic replacement: Report of 5 cases. Chinese Med. J. 1, 5, 1974.

20. Francis, K.C.: Tumors of the spine. In, The Spine (Vol. II). Eds: Rothman, R.H., and Simeone, F.A. W.B. Saunders Co., Philadelphia, 1975.

21. Goldenberg, R.R., Campbell, C.J., and Bonfiglio, M.: Giant-cell tumor of bone: An analysis of 218 cases. J. Bone Joint Surg. 52(A), 619, 1970.

22. Gunterberg, B., Romanus, B., and Stener, B.: Pelvic strength after major amputation of the sacrum. Acta Orthop. Scand. 47, 635, 1976.

23. Hamdi, F.A.: Prosthesis for an excised lumbar vertebra: A preliminary report. Can. Med. Assoc. J. 100, 576, 1969.

24. Harrington, K.D.: The use of methyl methacrylate for vertebral-body replacement and anterior stabilization of pathological fracture-dislocations of the spine due to metastatic malignant disease. J. Bone Joint Surg. 63(A), 36, 1981.

25. Hays, R.P.: Resection of the sacrum for benign giant cell tumor: A case report. Ann. Surg. 138, 115, 1953.

26. Huvos, A.G.: Bone Tumors: Diagnosis, Treatment, and Prognosis. W.B. Saunders, Philadelphia, 1979.

27. Jackson, R.P.: Recurrent osteoblastoma: A review. Clin. Orthop. 131, 229, 1978.

28. Jaffe, H.L.: Tumors and Tumorous Conditions of the Bones and Joints. Lea & Febiger, Philadelphia, Philadelphia, 1958.

29. Johnson, E.W., and Dahlin, D.C.: Treatment of giant-cell tumor of bone. J. Bone Joint Surg. 41(A), 895, 1959.

30. Kambin, P.: Giant-cell tumor of the thoracic spine with pathological fracture and paraparesis: A method of stabilization: A case report. J. Bone Joint Surg. 48(A), 779, 1966.

31. Lièvre, J.A., Darcy, M., Pradat, P., et al: Tumeur à cellues géantes du rachis lombaire spondylectomie totale en deux temps. Revue du Rhumatism et des Maladies Osteo-articulaires 35, 125, 1968.

32. Localio, S.A., Francis, K.C., and Rossano, P.G.: Abdominosacral resection of sacrococcygeal chordoma. Ann. Surg. 166, 394, 1967.

33. MacCarty, C.S., Waugh, J.M., Mayo, C.W. et al: The surgical treatment of presacral tumors: A combined problem. Proc. Staff Meet. Mayo Clin. 27, 73, 1952.

34. Marcove, R.C., Mike, V., Hutter, R.V.P., et al: Chondrosarcoma of the pelvis and upper end of the femur: An analysis of factors influencing survival time in 113 cases. J. Bone Joint Surg. 54(A), 561, 1972.

35. McGrath, P.J.: Giant-cell tumor of bone: An analysis of 52 cases. J. Bone Joint Surg. 54(B), 216, 1972.

36. Mnaymneh, W.A., Dudley, H.R., and Mnaymneh, L.G.: Giant-cell tumor of bone: An analysis and follow-up study of 41 cases observed at Massachusetts General Hospital between 1925 and 1960. J. Bone Joint Surg. 46(A), 63, 1964.

37. Murphy, W.R., and Ackerman, L.V.: Benign and malignant giant-cell tumors of bone: A clinical-pathological evaluation of 31 cases. Cancer 9, 317, 1956.

38. Murphy, M.J., and Southwick, W.O.: Spinal instrumentation for stabilization and fusion of the cervical spine. J. Bone Joint Surg. Orthop. Trans. 7, 119, 1983.

39. Murphy, M.J., Panjabi, M.M., and Southwick, W.O.: Treatment of unstable thoracolumbar fracture dislocations with combined Harrington compression and distraction rods. Trans. Orthop. Res. Soc. 5, 1980.

40. Nakagawa, H., Huang, Y.P., Malis, L.I., et al: Computed tomography of intraspinal and paraspinal neoplasms. J. Computer Asst. Tomography 1, 377, 1977.

41. Ono, K., Fuji, T., and Okada, K.: Pathologic basis of surgical salvaging of the spine with metastases or invasion of cancer. J. Bone Joint Surg. Orthop. Trans. 7, 118, 1983.

42. Ono, K., and Tada, K.: Metal prosthesis of the cervical vertebrae. J. Neurosurg. 42, 256, 1975.

43. Panjabi, M.M., Hopper, W., White, A.A., et al: Posterior spine stabilization with methylmethacrylate: Biomechanical testing of a surgical specimen. Spine 2, 241, 1977.

44. Robinson, R.A., and Southwick, W.O.: Indications and technics for early stabilization of the neck. South. Med. J. 53, 565, 1960.

45. Schwimer, S.R., Bassett, L.W., Mancuso, A.A., et al: Giant cell tumor of the cervico-thoracic spine. Am. J. Roentgen. 136, 63, 1981.

46. Scoville, W.B., Palmer, A.H., Samra, K., et al: The use of acrylic plastic for vertebral replacement or fixation in metastatic disease of the spine. A technical note. J. Neurosurg. 27, 274, 1967.

47. Simeone, F.A.: Intraspinal neoplasms. In, The Spine (Vol. II), Eds: Rothman, R.H. and Simeone, F.A., W.B. Saunders, Co., Philadelphia, 1975.

48. Smith, J., Wixon, D., and Watson, R.C.: Giant cell tumor of the sacrum: Clinical and radiologic features in 13 patients. J. l'Assoc. Canad. Radiologistes 30, 34, 1979.

49. Stener, B.: Total spondylectomy in chondrosarcoma arising from the seventh thoracic vertebra. J. Bone Joint Surg. 53(B), 288, 1971.

50. Stener, B.: Total spondylectomy for removal of a giant-cell tumor in the eleventh thoracic vertebra. Spine 2, 197, 1977.

51. Stener, B. and Jonsen, O.E.: Complete removal of three vertebrae for giant-cell tumour. J. Bone Joint Surg. 53(B), 278, 1971.

52. Sundaresan, N., Galicich, J.H., Chu, F.C.H., et al: Spinal chordomas: A clinical review. J. Neurosurg. 50, 312, 1979.

53. Tscherne, H.: Operative treatment of cervical spine injuries. J. Bone Joint Surg. Orthop. Trans. 6, 389, 1982.

54. Watts, H.G.: Introduction to resection of musculoskeletal sarcomas. Clin. Orthop. 153, 31, 1980.

55. White, A.A., and Panjabi, M.M.: Clinical Biomechanics of the Spine. J.B. Lippincott, Co., Philadelphia, 1978.

56. Whitehill, R., Reger, S.I., Barry, J.S., et al: A biomechanical analysis of the use of methylmethacrylate as instantaneous posterior fusion mass: A canine in vivo experimental model. J. Bone Joint Surg. Orthop. Trans. 7, 119, 1983.

The Role of Arteriography in the Diagnosis of Bone and Soft Tissue Tumors

Z. MATEJOVSKY, H. ZIDKOVA, and J. KOLAR

Introduction

The role of arteriography in the diagnosis of musculoskeletal lesions, which was extensively used 10 to 15 years ago, has been diminished to a certain extent by the development of computed tomography. In some centers, the latter method is currently preferred and arteriography is used only for exceptional cases. Since for our diagnostic procedures both methods are performed in most patients with bone and soft tissue tumors, we have developed our own concept for the indication and evaluation of arteriography.

Materials and Methods

Between 1965 and 1982, we have performed arteriography in 874 patients who were admitted to the clinics that are cooperating in the Prague investigative study of musculoskeletal tumors. These patients had been diagnosed or suspected of having tumorous lesions.

The contrast material was applied routinely by percutaneous arterial injection, in some patients by direct puncture (mainly in the upper extremity), but in the majority by catheterization through the iliac artery. Serialographs were taken at various intervals depending on the site and suspected type of lesion, with an average of 12 pictures per investigation. For evaluation of the arteriographic findings, we consider it essential to obtain pictures of the full arterial, capillar and venous phase. In certain instances, each of these phases can provide valuable information. In extremity lesions, we prefer to perform simultaneous biplane arteriography which is especially helpful for the planning of surgical intervention. Pharmacoarteriography was performed only in several exceptional cases. We do not consider the use of vasoconstrictors and vasodilators essential for the evaluation of musculoskeletal lesions. Injection of a local anesthetic before injecting the contrast material may be useful to prevent arterial spasm.

Arteriography in contrast to computed tomography is an invasive method, reported to cause a variety of adverse effects. In our series of 874 patients we observed no complications, neither fatalities nor any permanent sequelae. We encountered eight (0.9%) allergic reactions and 12 (1.3%) local reactions such as prolonged bleeding or severe hematoma after cathe-

Current Concepts of Diagnosis and Treatment
of Bone and Soft Tissue Tumors.

terization, or extravasation of the contrast material after direct puncture. In one case the tip of the guide wire broke but it was located radiographically and surgically removed. All investigations in adults were performed under local anesthesia. Children under 12 to 14 years of age were administered general anesthesia.

Results

By viewing the vascular network of a musculoskeletal lesion, arteriography may permit a precise diagnosis and helps in planning the surgical procedure, providing the following information:

- Tumorous or other cause of the lesion.
- Malignancy of the lesion may be confirmed but not excluded.
- In some cases, it may help to classify the tumor.
- By showing displacement of vessels, the extent of the lesion in soft tissue and its arterial and venous blood supply, arteriography may be a valuable aid in planning the surgical procedure.
- In tumors, an uneven distribution of pathologic vascularization may point to sections with highest proliferative activity and, thus, aid the surgeon in selecting the most appropriate site for biopsy.

Arteriography is most informative if evaluated together with other clinical and radiologic findings. Although certain patterns for arteriographic diagnosis may be established with increasing experience, exceptions may be encountered in some cases. Arteriography, therefore, cannot replace histologic diagnosis.

The value of arteriography can be further demonstrated in three diagnostic situations:

1. Differential diagnosis of Ewing's sarcoma

Between 1965 and 1982, we treated 68 patients with histologically confirmed Ewing's sarcoma of which 58 had been confirmed by arteriography. Of these 58 patients, 47 (81%) had positive findings of malignant vascularization (Fig. 1 and 2), and only five were negative. Necrosis of the tumorous tissue and tumors without extraosseous spread were responsible for the negative findings. In the same period, 87 patients with other osseous lesions were admitted for suspected Ewing's sarcoma. Of these, 63 patients had hematogenous osteomyelitis and 14 patients had eosinophilic granuloma. No malignant type of vascularization was observed by arteriograph in these patients. Especially in hematogenous osteomyelitis and large eosinophilic granulomas, which posed the most difficult differential diagnostic problems, no malignancy was seen. The reactive type of inflammatory vascularization is specific and will not be mistaken by an experienced radiologist for signs of malignancy. According to our experience, arteriography will show signs of malignancy in most cases of Ewing's sarcoma, especially in the extremities and in pelvic lesions. In tumors localized in the vertebrae, in ribs and other parts of the trunk, it is sometimes difficult to adequately view the tumor vessels without selective catheterization of the

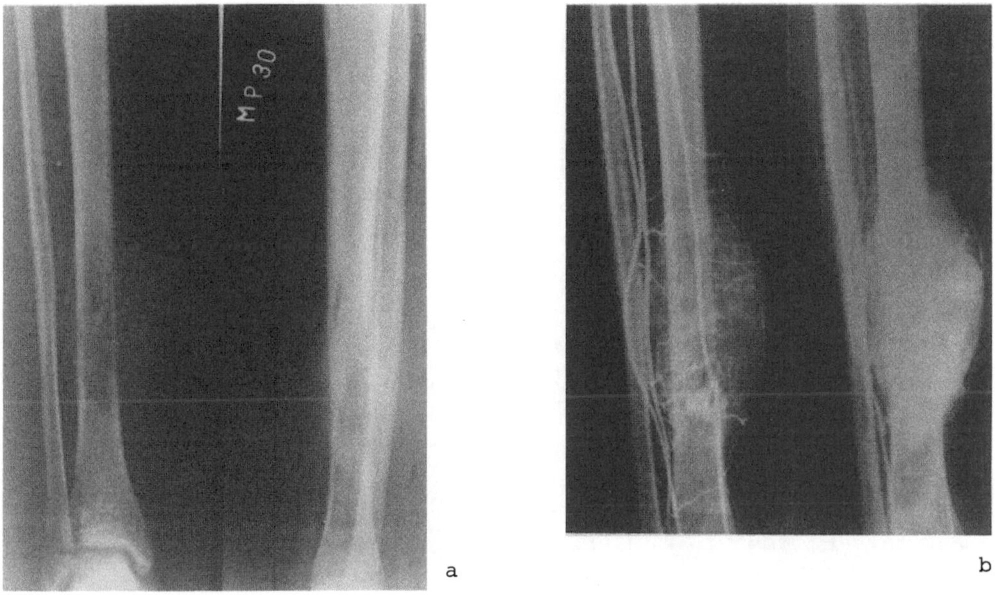

Fig. 1. Ewing's sarcoma. A 13-year old girl was admitted because of suspected hematogenic osteomyelitis. a) The radiograph shows irregular osteolytic changes in the lower part of the tibial shaft with no periosteal involvement. b) Arteriography of the arterial phase demonstrated a highly vascularized soft tissue extension of a tumor with new vessel formation. In the capillar phase, a vivid staining of the tumorous tissue is visible.

specific arteries. In these diagnostic situations, computed tomography can usually provide information about the extent of bone destruction and surrounding soft tissue involvement, but aids less in assessing the lesion.

2. Giant cell tumor of bone

Benign bone tumors, except benign osteoblastoma and osteoid osteoma, usually have poor vascularization. Displacement or occlusion of surrounding vessels may reveal the extent of the lesion. Giant cell tumor of bone is a distinct exception to this rule. Of 46 such tumors treated during the last 18 years, 34 were studied arteriographically. Twenty-nine (85%) of these tumors had a high degree of vascularization; in 12 cases invasion of adjacent soft tissue was shown (Fig. 3). Even giant cell tumors with a favorable clinical course, treated by curetting and grafting with good long-term results, may show an alarming arteriographic picture of high-grade confluent vascularization, which may frequently lead to the wrong diagnosis of malignancy and thus to unnecessary radical surgical treatment (Fig. 4). On the other hand, we have occasionally seen giant cell tumors, untreated for a longer period, and observed regressive changes. This results in a lower degree of vascularization and the differential diagnosis

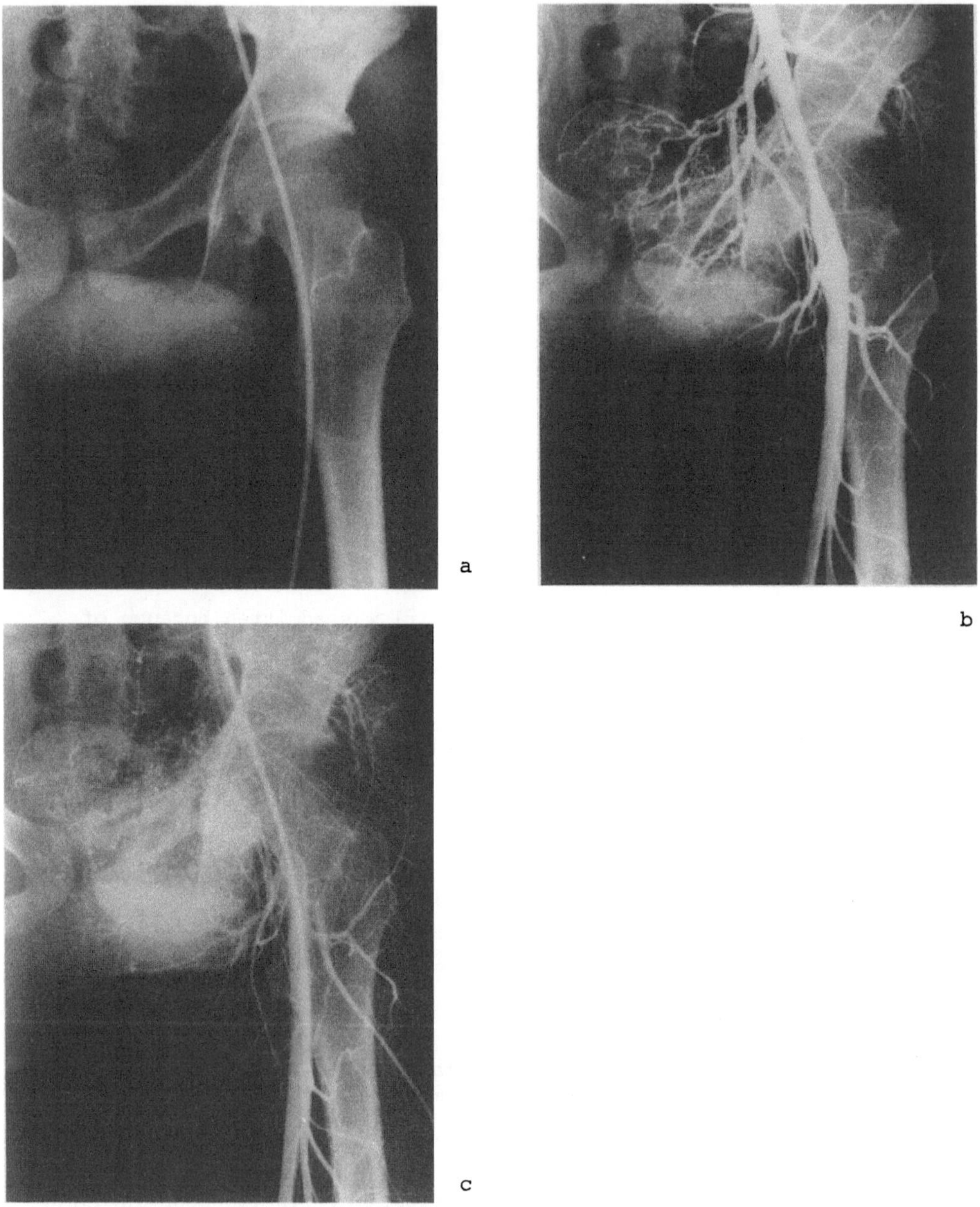

Fig. 2. Ewing's sarcoma. A 17-year old boy was admitted because of pain in the left groin. a) Radiograph shows only a slight periosteal effect on the os pubis. b) Intensive new vessel formation and increased blood supply of the tumor is seen in the arterial phase during arteriography. c) The late arterial and capillar phase with blood pools and staining of the tumorous tissue are clear signs of malignancy. The extension of the tumor into the pelvis and its sharply delineated margins are visible.

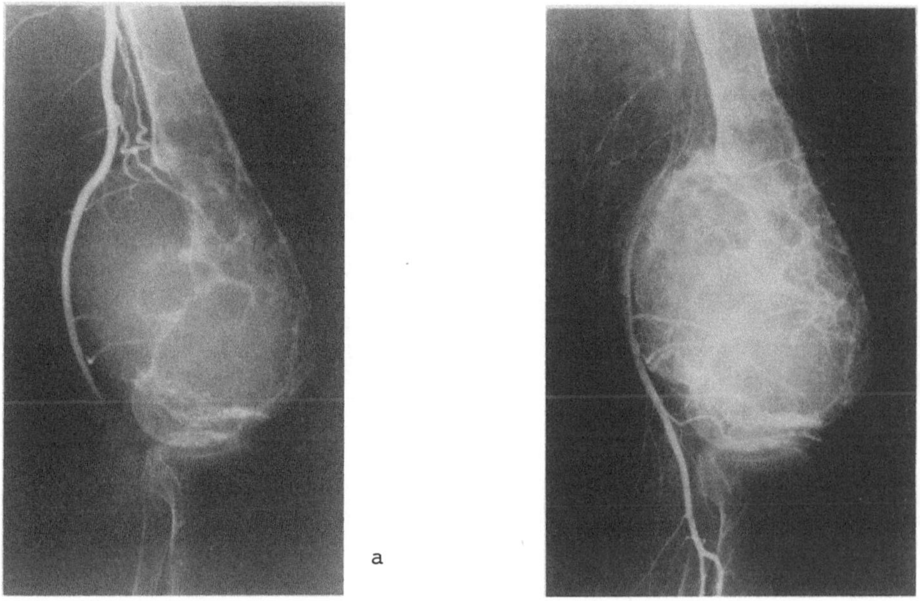

Fig. 3. Giant cell tumor. A 36-year old woman was admitted because of several years' duration of pain in the left knee. a) The beginning of the arterial phase shows a multilobulated osteolytic lesion of the distal femur with extension of the tumor into the popliteal region. b) The late arterial and capillar phase show extensive new vessel formation and staining of the tumorous tissue. The irregularity of staining was caused by cystic degeneration in parts of the tumor.

of an aneurysmal bone cyst may become difficult. An almost impossible task is the early arteriographic diagnosis of malignant transformation of a giant cell tumor.

3. Hypernephroma metastasis in bone

After 40 years of age, nine out of ten malignant tumorous bone lesions in our series were metastatic tumors. Bone metastases of hypernephroma, according to our experience, show a specific arteriographic pattern. These lesions sometimes occur as solitary osteolytic bone tumors, without clinical signs of a small primary tumor, which complicates the diagnosis. Arteriography is usually most helpful since the picture of hypernephroma metastasis in bone is specific for establishing the proper diagnosis. We have documented 74 treated cases of hypernephroma metastases to bone; of these, 54 were confirmed by arteriography, and of these 40 (74%) showed a typical malignant confluent vascularization of the lesion with extremely fast arteriovenous flow of the contrast material and distinct, prolonged staining of the tumorous tissue.

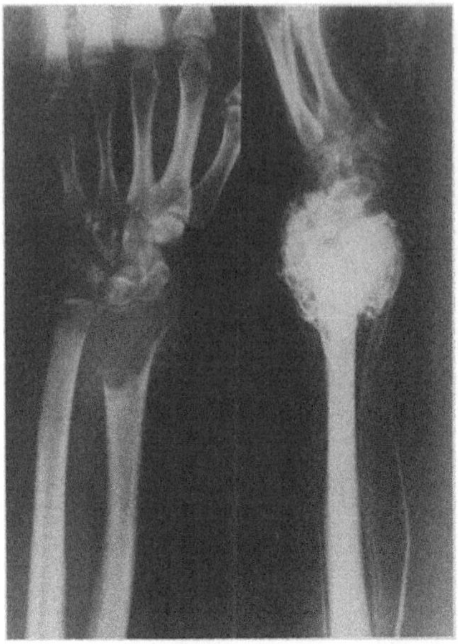

Fig. 4. Giant cell tumor. The radiograph shows a lytic lesion with destruction of cortical bone in a 29-year old man. Arteriography shows all three phases. While arterial filling is still continuing, the tumor extending into surroundig soft tissues shows a vivid confluent staining. At the same time, the filling of veins demonstrates an extremely accelerated arteriovenous flow in the tumorous tissue. The patient had undergone two intralesional surgical procedures. Five years postoperatively, the patient has not had any recurrence.

Discussion

In recent years, computed tomography has proved to be extremely helpful in the diagnosis of musculoskeletal lesions, whereas the assessment of the value of conventional radiographic methods differs among investigators. De Santos et al (1978) did not find computed tomography superior to conventional radiography whereas Levine et al (1979) found that computed tomography revealed more data than angiography and other investigative methods. In our experience, both methods have advantages and both should be applied whenever possible in the preoperative evaluation of musculoskeletal tumors. The specific diagnostic value of arteriography and its importance as a source of information in planning biopsy sites and surgery has been demonstrated by many authors (Hudson et al, 1981; Lechner, 1978; Voegeli et al, 1976). Furthermore, the great variety of angiographic pictures in musculoskeletal tumors has been extensively described by Yaghmai (1979). Most of the published observations of arteriography in bone tumors are based on a variety of tumors, but the three types of tumors discussed in our communication have been given special attention because of interesting differential diagnostic problems (Bowers et al, 1982; Kittredge, 1970; Lechner et al, 1978; Lundström et al, 1977; Prando et al, 1979). The conclusions of these authors mostly concur with ours.

Conclusions

According to our experience with 874 arteriographic investigations performed in musculoskeletal lesions to confirm or exclude the diagnosis of a tumor, this radiodiagnostic method proved to be of great help in establishing the diagnosis of benignancy or malignancy.

In some cases, arteriography may contribute to the identification of the tumor. In addition, it is of great value for the surgeon in planning the biopsy site and surgical procedure. Specialists can reduce the risk of complications. For an early and accurate diagnosis of bone and soft tissue tumors, arteriography should be used together with other diagnostic methods. It cannot be fully replaced by computed tomography which shows precisely the extent of the tumor but gives little information about the blood supply and proliferative activity of the lesion.

References

1. Bowers, T.A., Murray, J.A., Charnsangavej, C., Soo, C.S., Chuang, V.P. and Wallace, S.: Bone metastases from renal carcinoma. J. Bone Joint Surg. 64(A) 749, 1982.

2. Hudson, T.M., Enneking, W.F., and Hawkins, I.F., Jr.: The value of angiography in planning surgical treatment of bone tumors. Radiology 138, 283, 1981.

3. Kittredge, R.D.: Arteriography in Ewing's tumor. Radiology 97, 609, 1970.

4. Lechner, G.: Angiographie und Knochentumor. Z. Orthop. 116, 480, 1978.

5. Lechner, G., Kotz, R., Riedl, P. and Salzer-Kuntschik, M.: Zur Problematik der angiographischen Dignitätsbeurteilung beim Riesenzelltumor. Radiologe 18, 31, 1978.

6. Levine, E., Lee, K.R., Neff, J.R., Maklad, N.F., Robinson, R.G., and Preston, D.F.: Comparison of computed tomography and other imaging modalities in the evaluation of musculoskeletal tumors. Radiology 131, 431, 1979.

7. Lundström, B., Lorentzon, R., Larsson, S.E., and Boquist, L.: Angiography in giant cell tumors of bone. Acta Radiol. Diagn. 18, 541, 1977.

8. Prando, A., de Santos, L.A., Wallace, S., and Murray, J.A.: Angiography in giant cell bone tumors. Radiology 130, 323, 1979.

9. de Santos, L.A., Goldstein, H.M., Murray, J.A., and Wallace, S.: Computed tomography in the evaluation of musculoskeletal neoplasms. Radiology 128, 89, 1978.

10. Voegeli, E. and Uehlinger, E.: Arteriography in bone tumors. Skel. Radiol. 1, 3, 1976.

11. Yaghmai, I.: Angiography of Bone and Soft Tissue Lesions. Springer-Verlag, Berlin-Heidelberg-New York, 1979.

Pelvic Malignancies - Resections of the Pelvic Bones

M. CAMPANACCI and R. CAPANNA

Malignant tumors of pelvic bones, the same as those of the scapula, can often be removed by limited resection obtaining about the same rates of success which would be obtained with a quarter amputation (Eilber et al, 1979; Enneking and Dunham, 1978; Johnson, 1978; Steel, 1978). Pelvic resections, however, are usually more difficult than scapulectomy because of the nonmobility of the bone, the depth and therefore difficulty of exposing some pelvic structures, and the connections with important nerves, vessels and viscera.

This surgical procedure is not only difficult but risky. It requires experience, repeated dissections and mock operations on cadavers, a meticulous preoperative study of the patient, and team work uniting the general surgeon, vascular surgeon, urologist and orthopedic surgeon.

The principal presurgical tests are CAT scan, sometimes isotope scan, angiography, and cystography. It is important to apply a urethral catheter preoperatively so that during the operation the surgeon can palpate the urethra and avoid injuries to it. The common iliac artery can be temporarily clamped to reduce bleeding, but this is rarely indicated and should be combined with heparin infusion and carried out only in young patients.

Schematically, we can distinguish three types of pelvic resections (Table I):

1. Resection of the ileum: either a wedge of the wing; or a total section through the sacroiliac joint or the sacral wing (Fig. 1).

2. Resections of the entire acetabulum, with a variable segment of the ileum and the anterior arch. This resection can be intra-articular (where the joint is opened and the femur dislocated), or extra-articular (where the joint is not opened and the acetabulum is removed en bloc with the femoral head and neck) (Fig. 2).

3. Resections of the anterior pelvic arch, monolaterally or bilaterally, sometimes with the anterior and inferior portion of one acetabulum (but preserving enough of it to maintain a stable joint) (Fig. 3).

Current Concepts of Diagnosis and Treatment
of Bone and Soft Tissue Tumors.

Table I. Types of pelvic resections*

Site	No. of patients	Procedure (number)
Ileum	17	Wedge of iliac (11) Entire segment, through sacrum or sacroiliac joint (6)
Acetabulum	13	Intra-articular (10) Extra-articular (3)
Anterior arch	10	Monolateral (3) Monolateral plus anterior or inferior acetabulum (5) Bilateral (2)

* 40 patients (1971-1981)

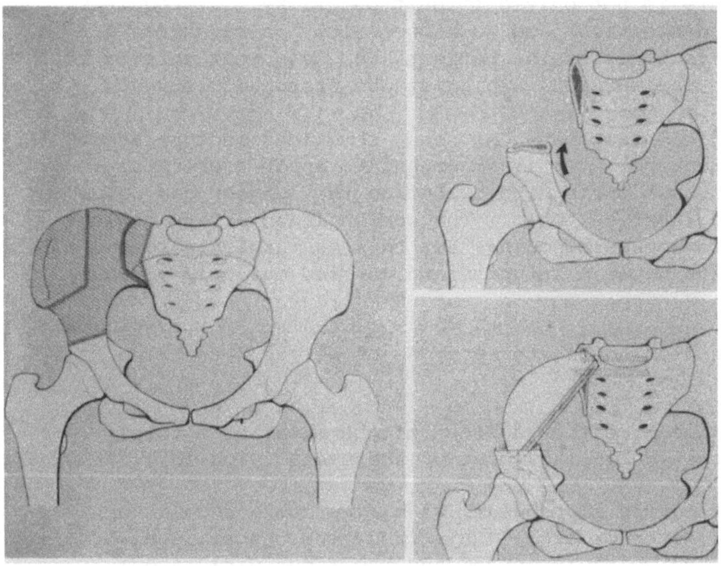

Fig. 1

The more frequent indications are malignant primary low-grade tumors of bone, namely chondrosarcomas. Less commonly, the pelvic resection is used in high-grade bone tumors (some of which are in fact subjected to hindquarter amputation while others are inoperable). The indication in high-grade tumors is primarily Ewing's sarcoma. Most of the recent therapeutic protocols for this tumor recommend the resection of the pelvic

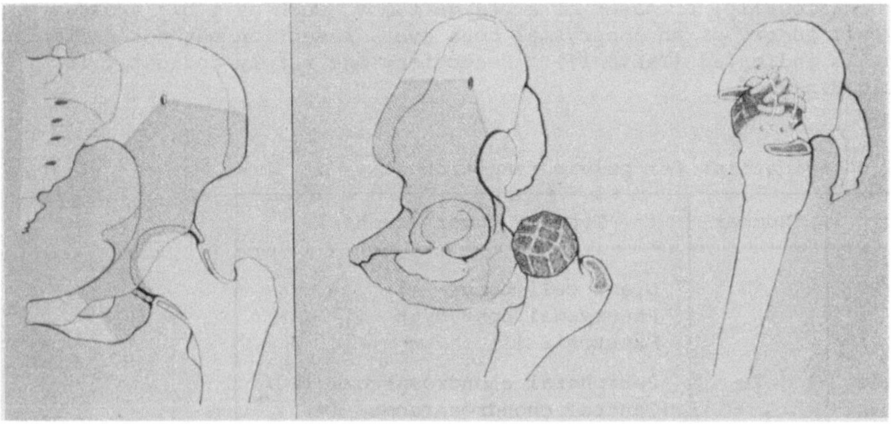

a

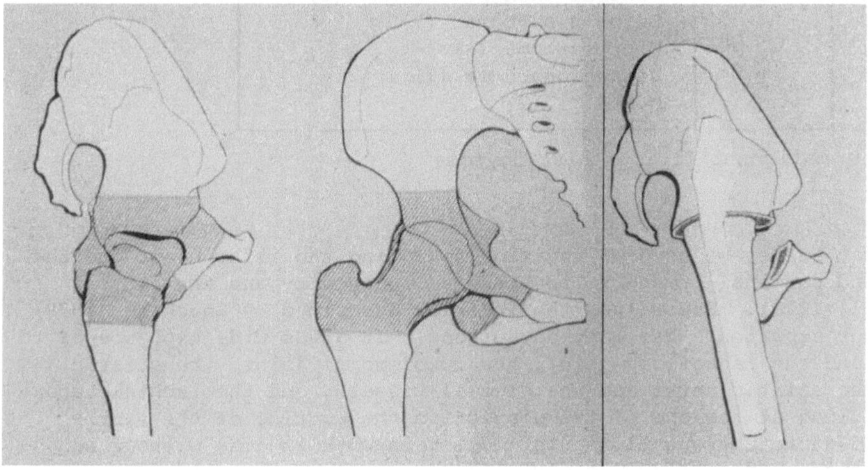

b

Fig. 2

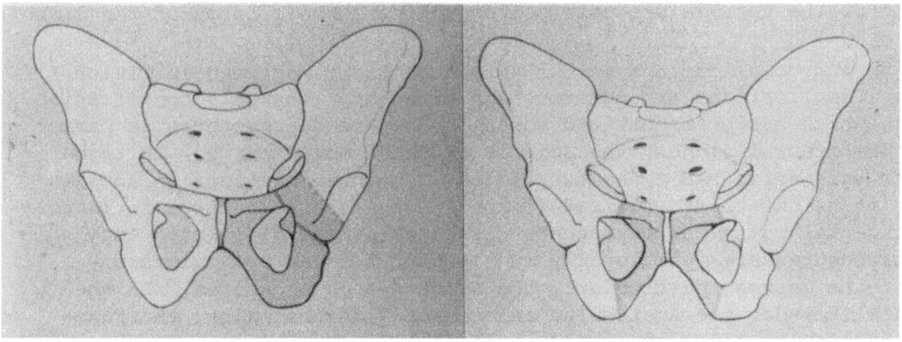

Fig. 3

locus. Occasionally, in cases of a benign tumor, such as a big exostosis, a giant cell tumor, or an aneurysmal bone cyst, resection but not radiation treatment is indicated (Table II). Resections are rarely indicated in metastatic disease.

Table II. Indications for pelvic resection

Tumor	Number	Type of tumor (number)
Benign	3	Giant cell tumor (1) Aneurysmal bone cyst (1) Exostosis (1)
Low-grade	20	Peripheral chondrosarcoma (16) Central chondrosarcoma (4)
High-grade	17	Central chondrosarcoma (6) Ewing's sarcoma (7) Fibrosarcoma (1) Hemangiosarcoma (1) M.F.H.* (2)

* M.F.H. = malignant fibrous histiocytoma

In our first cases, we used an incision following the iliac crest and the inguinal and perianal crease to the ischial tuberosity and the gluteal fold (Steel, 1978). Now we use the incision described by Enneking (1978), which in our experience has many advantages. It gives wide exposure of the ileum and the sacroiliac joint, the coxofemoral joint, the sciatic notches, the sciatic nerve and the gluteal vessels, and the ischial tuberosity, allowing at the end of the dissection the cutting of the sacrospinous and sacrotuberous ligaments. All this work is done without exposing or disturbing the iliac vessels. The exposure of the anterior arch needs a second anterior incision over the pubic rami.

The ileofemoral vascular bundle is not isolated but simply identified, protected and mediolaterally moved. In low-grade tumors this is possible because there is no need for excising the iliac lymphatic nodes. This handling of the vessels prevents postoperative chronic edema of the limb.

Follow-up and complications in our 40 patients are reported in Tables III and IV. Complications are numerous and frequent. The only complication, however, which has permanent and sometimes severe consequences is damage to the lumbosacral plexus, the sciatic or the femoral nerve. In three patients some nerve had to be sacrificed. In three patients an accidental but transient palsy occurred (femoral nerve), and in three patients the accidental palsy was permanent. As Enneking (1978) pointed out, these procedures are functionally much better than an hemipelvectomy provided the nerves, particularly the sciatic, can be spared. If the resection includes the acetabulum and causes a severe palsy, the functional result can be worse than that of an hemipelvectomy.

Table III. Follow-up of patients (average: 4-1/2 years)

Length of follow-up	No. of patients
1-2 years	8
2-3 years	4
3-5 years	5
5-10 years	12
Died at 4 to 21 months	11

Table IV. Complications encountered

Complication	No. of patients	Result
Slough of the wound	3	Healed with musculocutaneous flap
Deep infection	4 (10%)	Healed; tow cases requiring plate removal
Femoral, sciatic or peroneal palsy	6 (15%)	3 transient (femoral)
Nerve purposely sectioned	3	
Opening of iliac vein	1	Sutured without adverse effects
Thrombosis of iliac artery	1	Removed without adverse effects
Chronic edematous limb	6	Regressed in 1 year
Opening of bladder	2	Sutured without adverse effects
Fractured residual pelvis	2	Satisfactory
Medial femoral dislocation	1	Satisfactory
Knee stiffness	1	

Table V shows the results we have achieved. In Ewing's sarcomas, after three courses of chemotherapy, the resection is intended to remove the major bulk of the tumor, not extensive since radiation treatment follows surgery. This table also shows how difficult it is to achieve a high rate of success in high-grade tumors (including Ewing's). This confirms that pelvic resections have a limited indication and little likelihood of success in high-grade tumors.

Table V. Surgical margins

Margin	Type of tumor		
	Benign low-grade	High-grade	Ewing's sarcoma
Wide	17	5	-
Marginal	6	3	2
Intralesional	-	2	5

The oncologic results (Table VI) confirm our previous statements. Good results were obtained in benign and low-grade malignancies, poor results in high-grade and especially in Ewing's sarcoma. These results may be less favorable after a longer follow-up period.

Table VI. Oncologic results

Result	Type of tumor		
	Benign low-grade	High-grade	Ewing's sarcoma
No evidence of disease	22*	6	1
Local recurrence with death	-	-	1
Local recurrence with metastases and death	-	2	3
Metastases and death	1	2	2

* 2 cases after second operation

The functional results have been evaluated in 31 patients (Table VII). As anticipated, they are best in resections of an iliac wedge and of the anterior arch. They are worse when the sacroiliac joint or a coxofemoral joint has been included in the resection. The table also confirms that some of the fair and poor results were due to either femoral or sciatic nerve palsy.

Table VII. Pelvic resections: functional results (31 patients)

Iowa score	Ileum wedge	Entire ileum	Acetabulum	Anterior arch	Anterior arch and part of acetabulum
Excellent (90-100)	5(62.5%)	-	-	4(80%)	4(80%)
Good (80-90)	2(25%)	1(20%)	1(12.5%)	1(20%)	-
Fair (60-80)	-	3(60%)*	6(75%)	-	1(20%)
Poor (<60)	1(12.5%)**	1(20%)	1(12.5%)**	-	-
Total no. of patients (31)	8	5	8	5	5

* 2 patients with femoral nerve palsy
** 2 patients with sciatic nerve palsy

Conclusions

Most pelvic resections are complex procedures. The main indications are low-grade malignancies of the pelvic bones. Pelvic resection can be adopted in high-grade tumors as well. In Ewing's sarcoma it should be combined with radiotherapy and chemotherapy, although the probability of success is much lower. These operations require extensive preoperative study, specific surgical experience, team work between orthopedic and general surgeons, and intensive care facilities. The major causes of complications are damage to the sciatic or femoral nerve. This procedure is not indicated in metastatic tumors, when patients' local or general conditions are poor, and in very obese patients.

References

1. Eilber, F.R., Grant, T.T., Sakai, D. and Morton, D.L.: Internal hemipelvectomy-excision of the hemipelvis with limb preservation. Cancer 43, 806, 1979.

2. Enneking, W.F. and Dunham, W.K.: Resection and reconstruction for primary neoplasms involving the innominate bone. J. Bone Joint Surg. 60(A), 731, 1978.

3. Johnson, J.T.H.: Reconstruction of the pelvic ring following tumor resection. J. Bone Joint Surg. 60(A), 747, 1978.

4. Steel, H.H.: Partial or complete resection of the hemipelvis. J. Bone Joint Surg. 60(A), 719, 1978.

Limb Salvage in Pelvic Tumors

F. H. Sim and W. E. Bowman, Jr.

Introduction

Bone tumors of the pelvis continue to present a difficult therapeutic problem. The prognosis for patients with malignant lesions of the pelvis is worse than that for patients with distal tumors (Dahlin, 1978). Moreover, hemipelvectomy, which has been the standard surgical treatment, is a debilitating and disfiguring procedure with a high incidence of postoperative morbidity. In recent years, interest has revived in limb-saving resection for malignant tumors of the pelvis (Enneking and Dunham, 1978). This has been fostered both by the recent enthusiasm for adjuvant chemotherapy and by improved techniques of oncologic reconstruction. Moreover, the improved techniques for clinical staging, particularly computed tomography, have enhanced the ability to determine the local extent of tumor involvement and have helped in the decision whether amputation is necessary or whether a limb-saving resection can be performed.

Resection Technique

The proper selection of patients is mandatory and the surgical approach demands careful planning. The aggressiveness of the surgical treatment depends on the nature of the lesion. For benign lesions, less radical surgery may be performed and excision-curettage is usually satisfactory in controlling the lesion, with bone grafting for preserving bone stock. For metastatic lesions, the treatment is usually medical unless the acetabulum is compromised. In such cases surgical treatment may be necessary, and the selections range from a Girdlestone procedure to a palliative amputation or acetabular reconstruction. In primary bone tumors, however, the first goal is to eradicate the tumor by adequate surgical resection. An important consideration is the placement of the biopsy wound where it can be completely circumvented at the time of definitive resection. The en bloc resection should be done through normal tissues, the lesion being excised with an envelope of normal surrounding muscle.

Resection of the lesions of the pelvis is performed with the patient positioned for a hemipelvectomy. An extensive exposure provides access to both intrapelvic and extrapelvic structures. The incision begins at the postero-inferior iliac spine, follows the iliac crest and inguinal ligament, then turns distally toward the anterolateral aspect of the thigh. If more exposure is needed, the incision may curve laterally and posteriorly at the junction of the proximal and middle thirds of the thigh. Furthermore, if

Current Concepts of Diagnosis and Treatment
of Bone and Soft Tissue Tumors.

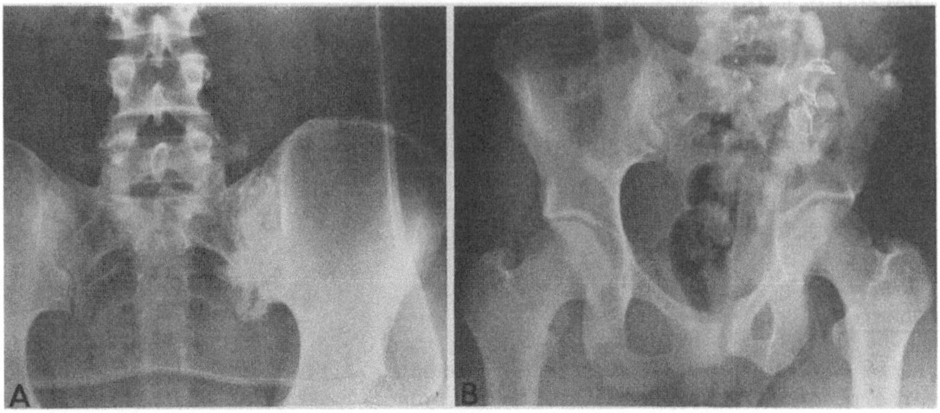

Fig. 1. A, Anteroposterior view of left hemipelvis showing chondrosarcoma involving sacroiliac region. B, After en bloc resection and reconstruction with iliosacral arthrodesis.

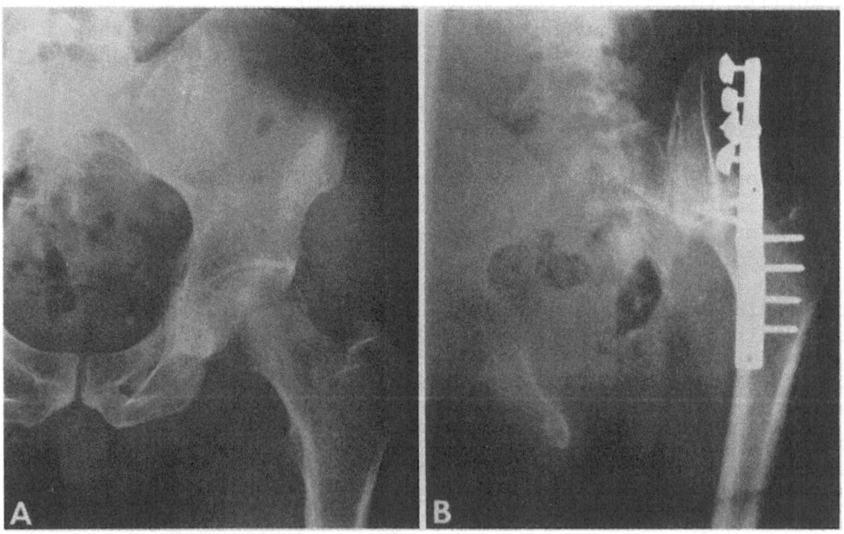

Fig. 2. A, Anteroposterior view of left hemipelvis showing chondrosarcoma involving innominate bone and acetabulum. B, After en bloc resection and reconstruction with femoral iliac arthrodesis. (From Sim, F.H., Bowman, W.E., Jr, and Chao, E.Y.S.: Limb-salvage and reconstructive techniques. In, Diagnosis and Treatment of Bone Tumors. Eds, Sim, F.H., et al. Thorofare, New Jersey, Charles B. Slack Inc. In press. By permission.)

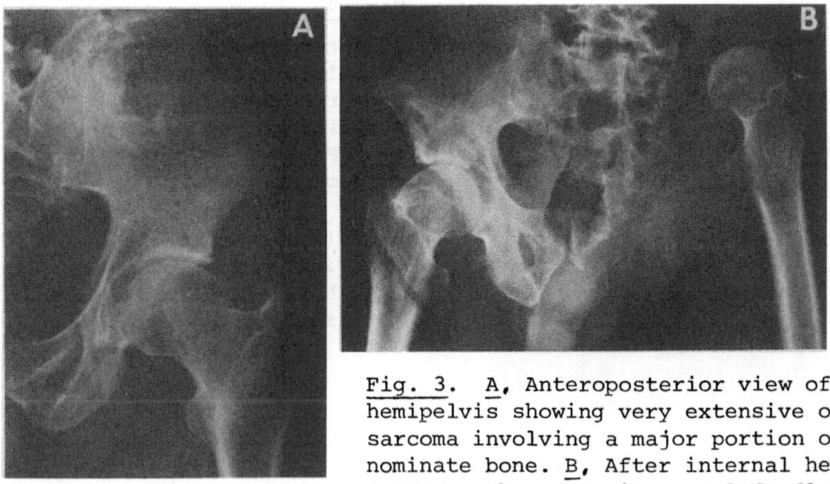

Fig. 3. A, Anteroposterior view of left hemipelvis showing very extensive osteosarcoma involving a major portion of innominate bone. B, After internal hemipelvectomy, the extremity was left flail.

the ischiopubic region has to be exposed, a supplemental incision may be used along the remaining portion of the inguinal ligament. Large flaps are then developed, and the sciatic nerve and femoral neurovascular structures are identified and protected. The location, size, and nature of the lesion dictate the extent of the resection. If the acetabulum is involved, the preoperative assessment should determine whether an extra-articular or intra-articular resection has to be performed. Occasionally, in the presence of a large lesion in the sacroiliac region, hemilaminectomy must be done to allow for adequate resection.

After resection of a pelvic lesion, the method of reconstruction varies according to the anatomic location of the lesion. Reconstruction is more difficult when the lesion involves the acetabulum or the sacroiliac joint. If a partial resection of the iliosacral region is performed and pelvic continuity is satisfactory, no reconstruction is necessary. If a complete iliosacral resection is done, the remaining part of the pelvis may be hinged on the pubic symphysis and then secured to the remaining part of the sacrum or ilium with an arthrodesis (Fig. 1). Care must be taken in these circumstances to allow adequate space for the sciatic nerve to exit the pelvis.

A major problem in reconstruction develops when a lesion involves the acetabular region. This increases the technical difficulty. If only a portion of the acetabulum is removed, the acetabulum may be reconstructed by means of autogenous bone grafts; however, when a lesion involves the entire acetabulum and total acetabular resection is necessary, we prefer to fuse the remaining part of the femur to the ilium or ischium (Fig. 2). For a lesion that requires removal of the major portion of the innominate bone, the femur is usually left flail (Fig. 3).

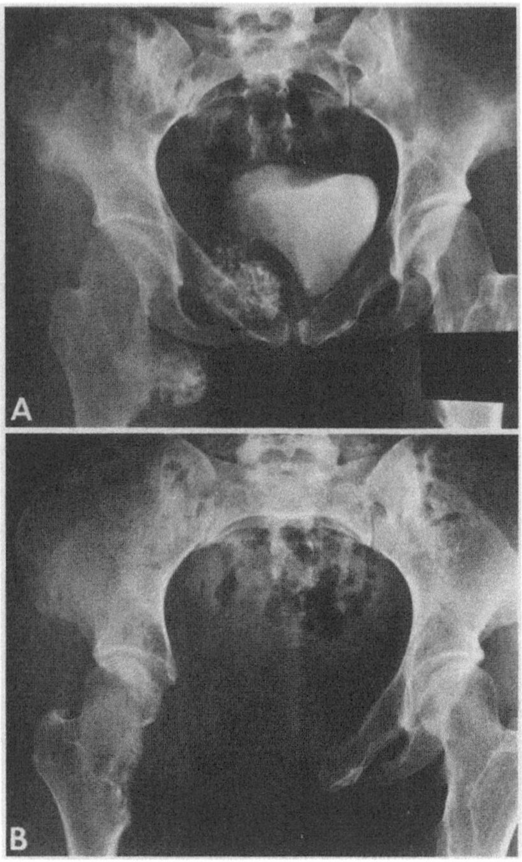

Fig. 4

A, Anteroposterior view of pelvis showing grade 1 chondrosarcoma of right anterior portion of pelvis. Patient had multiple osteochondromatosis.

B, After resection of anterior portion of pelvis. No acetabular reconstruction was necessary.

After resection of lesions of the anterior pelvis, reconstruction is generally not necessary unless a portion of the acetabulum has been resected (Fig. 4).

Clinical Series

Between 1970 and 1981, 160 patients underwent limb-salvage procedures at the Mayo Clinic (Dahlin, 1978). Thirty of these patients had primary malignant tumors involving the pelvis: 22 of 50 patients with chondrosarcomas of the pelvis; five of 37 patients with osteosarcomas, and three of 17 patients with fibrosarcomas. The ages of the 30 patients (17 women and 13 men) ranged from 19 to 62 years, with an average of 34.8. Follow-up ranged from six to 129 months, with an average of 39.6. These figures reflect a highly selected group of patients. There were 12 lesions in the iliosacral region, eight in the acetabular region, and 10 in the ischiopublic region.

We used the system advocated by Enneking et al (1980) in staging these tumors to allow for valid comparisons with other series. With this method, we observed four stage IA lesions, 15 stage IB, and 11 stage IIB. Each stage was subdivided on the basis of pathologic diagnosis. Most of the chondrosarcomas were low grade (stage IA or IB). There were five osteosarcomas, four of which were stage IIB, and three fibrosarcomas, all of which were stabe IIB. We also analyzed the extent of the surgical procedure, as recommended by Enneking et al (1980). In that system, a grade 1 procedure is an intralesional excision involving excision-curettage. A grade 2 surgical procedure is a marginal excision that is very near the pseudocapsule of the tumor. A grade 3 excision is a wide resection that involves more than 2 cm of normal tissue between the margin of the resection and the pseudocapsule of the tumor. In Enneking's system of classification, a grade 4 procedure involves removal of the entire bone and the surrounding musculature. On close scrutiny, we believe that none of our patients underwent a grade 4 procedure. Of the 30 patients, 23 underwent a wide resection, four patients an intralesional resection, and three patients a marginal resection.

One might anticipate a higher incidence of complications in limb-salvage procedures of such magnitude. In our series, four infections occurred, all of which resolved with treatment. Two patients had wound hematomas that required drainage; one patient had a decubitus ulcer; one patient had partial sciatic palsy that resolved; and one patient had a pulmonary embolism.

We rated the final results as good, satisfactory, or poor. To be rated as a good result, the patient had to walk with no support, have no or only occasional pain, and have no limitation of normal daily activity. To be rated as a satisfactory result, the patient had to walk with a cane or single crutch, have moderate pain requiring prescription analgesics, and have moderate limitations of daily activities. Patients rated as a poor result used a walker, complained of severe pain requiring marcotics, had severe limitations of activity, or required an amputation because of complications. Of the patients with adequate follow-up, 21 had good results, five patients had satisfactory results, and three had poor results. The three poor results occurred in the patients whose femurs were left flail after complete acetabular resection.

Comment

Is limb salvage worthwhile and is it safe? To be worthwhile, there must be adequate functional restoration with an acceptably low level of morbidity. The functional result must be analyzed with respect to the location of the tumor and compared with the results of treatment alternatives. In this series of 30 patients, when limb salvage is compared with the alternative of hemipelvectomy, there is no question that the immediate functional results are far superior. The morbidity appears to be acceptable for major reconstructions of this nature. To achieve a safe procedure, local control of the tumor must be obtained without adversely affecting survival. In our series of 30 patients, four patients experienced local recurrences (13.3%). Another four patients had pulmonary metastasis, but only one patient had died at the time of review. We analyzed various prognostic factors to determine their effects on local recurrence and survival. Age and sex were not significant. We continue to see patients in whom limb salvage may be precluded by a poorly positioned biopsy. In this carefully selected group of patients, neither the time nor the site of the biopsy appeared to have any effect on the outcome. In our overall experience with limb salvage, we noted that the most important factor is the surgical stage of the lesion and the adequacy of the surgical resection. We also noted that when a local recurrence developed, survival was affected adversely.

Summary

Evaluation of limb salvage in 30 patients with primary malignant bone tumors of the pelvis indicates that while these procedures are still experimental and more time is needed for follow-up, the early results promise usefulness in functional restoration. The Enneking Staging System correlates well with the prognosis and indicates that the surgical resection must be tailored to the surgical stage of the lesion. The best functional results occur when reconstruction allows a stable lower extremity. The local recurrence rate of 13.3% is still high and suggests the need for effective adjuvant treatment.

References

1. Dahlin, D.C.: Bone Tumors: General Aspects and Data on 6,221 Cases. Third Ed. Charles C. Thomas., Springfield, Ill. 1978.

2. Enneking, W.F., and Dunham, W.K.: Resection and reconstruction for primary neoplasms involving the innominate bone. J. Bone Jt. Surg. (Am.) 60, 731, 1978.

3. Enneking, W.F., Spanier, S.S., and Goodman, M.A.: A system for the surgical staging of musculoskeletal sarcoma. Clin. Orthop. 153, 106, 1980.

Pelvic Malignancies - Resections of the Sacrum

M. CAMPANACCI and R. CAPANNA

This procedure is usually indicated for chordoma. Much rarer indications are benign tumors (giant cell tumor, osteoblastoma) and other low-grade malignant tumors (chondrosarcoma) (Gunterberg and Stener, 1976; Localio et al, 1967; MacCarty et al, 1952; Stener and Gunterberg, 1978).

In the chordomas, diagnosis and adequate treatment are often delayed, sometimes so long that treatment becomes difficult, ineffective, or impossible. Some reasons for this delay are the slow indolent course of the tumor, the difficulty of recognizing radiographically early osteolysis of the sacrum, and the wrong initial therapeutic approaches, such as intralesional or marginal excision or radiation.

For a proper clinical evaluation, it is important to explore the rectum, and to recognize whether the tumor bulges anteriorly, and whether the mucosa is uninvolved and freely movable. If it adheres to the tumor, the rectum must be included in the sacral resection.

A preoperative study with the CAT scan is essential. Usually, the chordoma extends more anteriorly than posteriorly. In the more advanced cases, it may bulge from the lateral sacral margins, narrowing the great sciatic notches. Its peripheral nodes may grow into the muscles, aponeurosis, subcutaneous tissue and even reach the skin. It may grow into the sacral canal, upward beyond the upper radiographic margin of the osteolysis, and downward extruding posteriorly from the opening of the coccygeal foramen. It may reach the lumbosacral disc and sacroiliac joints, and in these cases adequate surgery is either very difficult or impossible.

When performing a biopsy of this tumor, it should be remembered that it may be mucoid and under tension so that, when incised, it leaks and spills liquid neoplasm into the surrounding tissues. A biopsy through the rectal mucosa is certainly wrong because it would subsequently necessitate including the rectum in the resection. A needle biopsy posteriorly is appropriate as it is an incisional biopsy, provided that both biopsy tracks can be subsequently excised en bloc with the tumor. We prefer the intraoperative biopsy by frozen sections.

The resection of the sacrum should be as wide as possible. Sometimes the problems start with the skin, either because of the scars from previous surgery or because some tumoral nodes reach the subcutaneous layer and

Current Concepts of Diagnosis and Treatment
of Bone and Soft Tissue Tumors.

Table I. Complications of sacral resection in patients with chordoma (n=12)*

Complication	Number	End result
Wound slough and infection	5	repaired
Limited wound necrosis	1	repaired
Cerebrospinal fluid fistula	1	healed
Hemorrhage requiring reopening of wound	1	no complications
Opening of the rectal wall	1	repaired

* Follow-up: 12 months - 6-1/2 years (mean: 45 months)

the derma. These cases may need a plastic stage (cutaneous flap). The muscles (gluteus maximus piriformis) should be sectioned at a distance from the tumor, the piriformis possibly near its tendon. The section of the sacrum should be 3-5 cm above the tumor, shaped in the center where the tumor may bulge cranially along the sacral canal.

When the tumor is below the third sacral foramen and the sacroiliac joint, the procedure can be safely done through a single posterior approach. After exposure of the inferior sacroiliac joint and the medial part of the great sciatic notch on both sides, the fingers separate the space between the sacrum and the rectum, and a gauze is passed transversally in this space from one sciatic notch to the other. With the anterior protection of this gauze and of the finger, the sacrum is transected and mobilized. The last step is to dissect the anterior surface of the tumor, protected by the presacral fascia, from the rectum. This dissection is often easy and clean. Sometimes it is difficult and may cause an opening of the rectal wall, which is difficult to repair and may lead to a fistula. But when the rectum adheres to the tumor it is advisable to resect the rectum en bloc with the tumor itself.

When the tumor is more extensive so that it is necessary to resect the sacrum above the third foramen, including part of the sacroiliac joint, a combined anterior and posterior approach is necessary. The purpose of the anterior stage is mainly to ligate and divide both internal iliac arteries and veins. This decreases not only the subsequent bleeding but especially avoids that, during the removal of the sacrum posteriorly, the traction might cause a laceration of an iliac vein. The anterior approach is also used to isolate the anterior surface of the sacrum and start the osteotomy at the proper level above the anterior bulging of the tumor. This anterior stage can be done either intraperitoneally (longitudinal median incision) or extraperitoneally (transverse suprapubic incision).

The posterior incision can be either transversal with the concavity downward, or median and longitudinal. The first procedure has the advantage

Table II. Results of sacral resection

Surgical margins	No.	Deaths	Survival with local recurrence	NED*	Local recurrence	Metastasis	Salvage procedure
Wide	6	-	2 (33%)	4 (66%)	3 (50%)	-	1/3
Marginal	4	2 (50%)	1 (25%)	1 (25%)	3 (75%)	2 (50%)	-
Intra-lesional	2	2 (100%)	-	-	2 (100%)	-	-

* NED - no evidence of disease

of being distant from the anus and therefore less easily contaminated. The second has the advantage of permitting the excision of the skin overlying the coccyx where the tumor could protrude from the opening of the coccygeal foramen.

Our experience includes 12 patients with chordoma. Results of follow-up and complications are recorded in Table I. Wound slough and infection occur frequently but eventually heal either spontaneously or with the use of a cutaneous flap.

Table II is fairly significant despite the small number of patients. Marginal and intralesional resections were almost always followed by a local recurrence. Resections considered to be wide resulted in local recurrence in one-half of the patients.

The neurologic effects due to resection of the sacrum were:

a) Only the first sacral root preserved (1 case)

- Unilateral loss of sacral nerves:
  S_1 anesthesia
  loss of Achilles tendon reflex
- Bilateral loss of sacral nerves:
  perianal, perigenital anesthesia
  bladder/bowel insensibility
  loss of voluntary bladder/bowel evacuation

b) First and second sacral roots preserved (3 cases)

- Unilateral loss of sacral nerves:
  no functional loss
- Bilateral loss of sacral nerves:
  urinary stress incontinence
  liquid stool incontinence
  hypotonic bladder/bowel

c) First, second and third sacral roots preserved (8 cases)
no functional loss.

Conclusions

Chordoma is a treacherous and dangerous tumor. Resection should be extensive. To achieve such a resection, aggressive surgery is justified and the resection of the rectum or a loss of the sphincter's power is the price to pay for the control of the tumor.

Delayed diagnosis, lack of presurgical staging, wrong biopsies, previous inadequate surgery or radiation therapy contribute to the high failure rate.

References

1. Gunterberg, B. and Stener, B.: Pelvic strength after major amputation of the sacrum. Acta Orthop. Scand. 47, 635, 1976.

2. Localio, S.A., Francis, K.C., and Rossano, P.G.: Abdomino-sacral resection of sacro-coccygeal chordoma. Ann. Surg. 166, 394, 1967.

3. MacCarty, C.S., Waugh, G.M., and Mayo, C.W.: The surgical treatment of presacral tumors: a combined problem. Proc. Staff Meet. Mayo Clin. 27, 73, 1952.

4. Stener, B., and Gunterberg, B.H.: High amputation of the sacrum for extirpation of tumors. Principles and techniques. Spine 3, 351, 1978.

Osteosarcoma: Experience at the University of California at Los Angeles

F. R. EILBER, J. ECKARDT, J. MIRRA, E. CAULKINS, and T. WEISENBURGER

Introduction

One hundred and eleven patients with osteosarcoma of the extremity were seen and treated by the Divisions of Surgical Oncology and Orthopedics, University of California at Los Angeles (UCLA) School of Medicine over a 10-year period. We describe our experience with various methods of treating the primary tumor, specifically limb salvage, and the natural history of the disease, with and without adjuvant chemotherapies. We chose this 10-year experience since at least half of the patients have been followed for more than five years. Furthermore, this period represents a relatively modern era in terms of the availability of adjuvant chemotherapeutic agents.

Materials and Methods

From March 1973 to March 1983, 111 patients with osteosarcoma and 45 patients with other malignant tumors of bone - 20 chondrosarcomas, three Ewing's sarcomas, six giant cell tumors, six malignant fibrous histiocytomas, and ten others (two leiomyosarcomas, five undifferentiated tumors, and three metastatic tumors) - were treated. Osteosarcoma represented 70% of the malignant bone tumors. The series consisted of 69 men and 42 women, ranging in age from 3 to 75 years (mean age 16). The primary tumor was located in the distal femur in 66 patients, proximal femur in four, humerus in 16, tibia in 17, ilium in five, scapula in two, and fibula in one. Confirmed histologically, 98 patients had high-grade, three had low-grade intraosseus tumors, three parosteal, and two osteosarcomas developing in Paget's disease. Five patients had metastatic pulmonary disease at the time of their first visit. All patients had routine physical examinations, chest roentgenograms, whole lung tomograms, and radionuclide bone scans performed on initial evaluation. In all patients, histologic examination was confirmed by open biopsy and was reviewed by one of the authors, Dr. Joseph Mirra. Patients with osteosarcoma of the ribs, vertebrae, or mandible were not included in this review.

After surgical or chemotherapeutic procedures, patients were followed at monthly intervals for the first year, at two-month intervals during the second year, and four-month intervals thereafter for the duration of the study. The follow-up information was available for all patients.

Current Concepts of Diagnosis and Treatment of Bone and Soft Tissue Tumors.

From 1973 to 1983, adjuvant chemotherapy was administered with two different protocols. In the first protocol, from 1973 to 1981, patients received doxorubicin* and high-dose methotrexate (Eilber et al, 1980). Doxorubicin* at a dosage of 45 mg/m^2, divided over two consecutive days, was administered postoperatively and followed two weeks later by vincristine, 1 mg/m^2, and high-dose methotrexate at a dosage of 200 mg/kg body weight, followed by citrovorum rescue. Fifty-five of the 98 patients with high-grade intraosseous osteosarcomas received this adjuvant drug combination. From 1981 to 1983, 19 patients with high-grade intraosseous osteosarcoma received the T-7-type regimen described by Rosen, consisting of weekly dosages of 200 mg/kg body weight of methotrexate followed two weeks later by BCD -- bleomycin, 12 mg/m^2, on each of two consecutive days; and actinomycin-D, 450 µg/m^2, on each of two consecutive days (Rosen et al, 1982). Doxorubicin at a dosage of 45 mg/m^2, divided over each of two consecutive days, was administered two weeks later. The duration of both adjuvant chemotherapeutic regimens was 10 months. A third group of 24 patients refused any adjuvant chemotherapy.

Limb salvage

Beginning in 1974, we began a preoperative adjuvant study in which patients received intra-arterially doxorubicin at a dosage of 30 mg by an indwelling arterial catheter over a 24-hour period for each of three consecutive days. One to five days later, the involved bone was irradiated at 350 rad fractions per day. Total radiation dosage was 3500 rad for 40 patients, and 1750 rad for 30 patients.

Surgical removal at the involved bone was performed from two to four weeks after preoperative intra-arterial chemotherapy and radiation therapy. This procedure involved en bloc excision of the diseased bone with proximal marrow margins of 8 cm. Soft tissues were dissected to leave a cuff of normal tissue attached to any extraosseous tumor. Frozen sections were submitted to the pathologist throughout the surgical procedure for confirmation of disease-free margins of soft tissue and marrow. Bony replacement was achieved with cadaver allografts in 19 patients -- eight humeri, one proximal femur, and 11 distal femora. In two patients, a rod and an autogenous bone graft replaced diseased bone (one tibia and one distal femur), and no bony replacements were required for nine patients (eight ilia and one fibula). Forty patients received metallic endoprostheses -- four had proximal femur or total hip replacements, two had total femur replacements, and 30 had distal femoral replacements, 18 with a kinematic rotating hinged knee prosthesis and 12 with a spherocentric knee. Three patients had total replacement of the scapula and proximal humerus for primary tumors of the scapula, and five patients had replacement of the proximal humerus. In patients with distal femoral lesions, extra-articular resection was required for six patients but intra-articular resection was done in the remaining 60 patients.

----

* Adriamycin

During this same time, 36 patients had primary amputation: 11 before the advent of limb-salvage procedures and 27 for other reasons -- tumor site in the tibia (13), age below 11 years (6), and tumor size (9).

Tumor specimens from 78 out of 89 patients who received preoperative doxorubicin and radiation therapy were examined to assess tumor cell necrosis. At least ten sections and five high-power fields per section on all resected tumors were required for the evaluation by Dr. Mirra.

Results of limb salvage

Of the 98 patients with high-grade osteosarcomas, 62 were entered into the limb salvage protocol. Of the 62, three developed local recurrences -- one patient in the tibia after resection of the distal femur; one patient in the ilium after an internal hemipelvectomy; and one patient in the humerus after replacement of the proximal humerus. Recurrence in the distal femur occurred at 18 months, in the ilium at 14 months, and in the proximal humerus at 16 months. The local recurrence in the tibia required amputation, but the other two recurrences were treated by surgical resection of the affected bone.

Complications from bony replacement with the cadaver allografts required reoperation in 57% of the patients and, in fact, all the surviving patients who had cadaver allografts had complications. The majority included fracture of the allograft, loosening of the Sampson rod, and resorption of the graft. None of the cadaver grafts was incorporated by the body. One of the two patients who had rod and autogenous bone grafts required amputation because of rod fracture and nonunion. The complication rate for patients with the metallic endoprostheses that required reoperation was 15%. These complications included stem fractures requiring removal and reinsertion of the metallic endoprostheses, and loosening of the prosthesis at the interface with bone requiring recementing. One patient who did not have a bone replacement had sciatic nerve compression after an internal hemipelvectomy which required reoperation.

Overall survival rate

The overall disease-free survival rate for patients with osteosarcoma was 47% at a follow-up of one to eight years (mean three years)(Fig. 1). The three patients with low-grade intraosseous osteosarcoma are free of disease, as are two of the three patients with parosteal sarcoma. One of the two patients with osteosarcoma developing in Paget's disease is free of disease, and two of the five patients with Stage III disease or metastatic disease on initial presentation remain free of disease. The mean follow-up for all these patients is three years. Forty-seven of the 98 patients with high-grade osteosarcoma, all Stage IIB of the Enneking system, or 48%, have remained disease free at three years. Fifty percent of the patients with recurrence had the recurrence in the lung and the remaining 50% were divided equally between recurrences in a second bony site, with no lung involvement, and a combination of lung and bone involvement. Two cerebral metastases to the liver occurred.

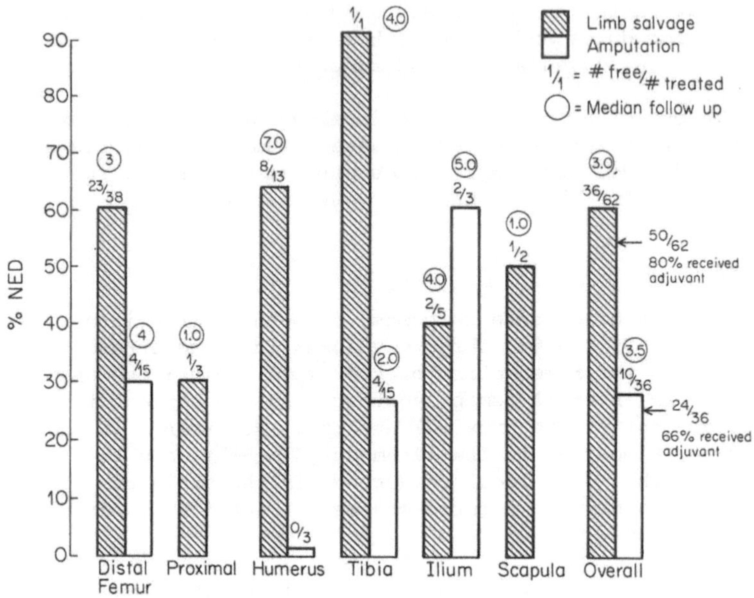

Fig. 1. 1973-1983 high-grade osteosarcoma.

Adjuvant chemotherapy

The survival rate varied for patients who received adjuvant chemotherapy. Of the 55 patients who received doxorubicin and high-dose methotrexate, 24 (45%) remained continuously free of disease from two to seven years after primary treatment (mean follow-up of four years). Of the patients receiving the T-7 protocol, 15 of the 19, or 79%, remained continuously disease free from one to three years (mean of one year). Only seven of the 24 patients, or 30%, who received no adjuvant chemotherapy between 1973 and 1983 remained continuously disease free. The disease-free interval for these patients ranged from one to three years, at a mean follow-up of 18 months.

Tumor necrosis

Seventy-eight specimens were evaluated for necrosis after preoperative chemotherapy. Fifty-nine were high-grade intraosseous osteosarcomas, and necrosis varied from a low of 5% to a high of 100% (mean 90%). The mean percentage necrosis was 20% for parosteal osteosarcomas, 40% for the low grade, and 40% for the osteosarcoma in Paget's disease.

Discussion

The overall disease-free survival rate at a mean of three years' follow-up for this group of patients with osteosarcoma is approximately 50%, which is comparable to that achieved by other investigators. Our overall limb salvage rate, however, of 66% appears to be higher than that reported in other studies. Of the 111 patients with osteosarcomas, 36 had primary amputation and 75 were entered into the limb salvage protocols. If the distal femur alone is considered, 80% of these patients had limb salvage and only 20% required amputation. Experience has taught us that primary lesions of the tibia require amputation because a functional limb replacement for these patients is inadequate. Therefore, of the 37 amputations, 16 (43%) were required because the tumor was located in the tibia. The low local recurrence rate (three) for this series of patients is comparable to most other series. The local recurrence rate for 37 patients who were treated by primary amputation was also three. Therefore, the local recurrence rate for the limb salvage protocol is no higher than that for amputation.

We cannot offer any reason for the high limb salvage rate in our series except to say that our goal was to attempt limb salvage in all patients. Preoperative therapy resulted in a high percentage of tumor cell necrosis and, therefore, the ability to come closer to the tumor may have allowed us to salvage a limb, whereas this procedure may be impossible without preoperative therapy. Since our study was not randomized, we cannot be certain that the preoperative treatment was responsible. It would appear, however, that limb salvage can be achieved in patients with high-grade intraosseous osteosarcoma and that the local control rate is comparable to amputation. If local control is the primary purpose of the primary tumor treatment, then our results appear to be equal.

The functional results of limb salvage have not been completely evaluated. In our experience, cadaver allograft replacement in patients receiving pre- and postoperative chemotherapy was totally unsatisfactory. The high fracture rate and reoperation rate appear to us to be prohibitive. The results achieved with metallic endoprostheses, however, particularly the replacement of the distal femur with a kinematic rotating hinge, appear to be excellent. The problem of loosening at the bony interface is certainly apparent and will have to be solved. The functional results for these patients appear to us worth the effort, at least in terms of the relatively short follow-up. These patients have a remarkable psychologic advantage and require no pain medication. Furthermore, their overall survival rate does not seem to be jeopardized by this procedure.

Our nonrandomized results suggest an advantage to patients who receive adjuvant chemotherapy, both for the group of patients that received doxorubicin and high-dose methotrexate and for those on T-7 protocol. These patients had a markedly improved survival rate compared with those who did not receive any chemotherapy at all. Again, ours was not a randomized prospective study to evaluate the effectiveness of adjuvant chemotherapy, but it is remarkable that only 28% of the patients who did not receive adjuvant chemotherapy remained disease free. The impact of adjuvant chemotherapy on osteosarcoma remains controversial. Mayo Clinic investigators have suggested that the natural history of the disease has changed, and that 50% of the patients will survive whether or not they receive adjuvant chemotherapy (Taylor et al, 1978). The results from this and other series over the last 10 years would not support this view because no patient who did not receive chemotherapy fared well (Harvei and Solheim, 1981). Possibly, an unknown selection process was built into our study, and perhaps these patients had a worse prognosis even though analyses of all factors (age, site, sex) do not support this hypothesis.

Furthermore, we are unable to tell whether the T-7 protocol of Rosen is superior to the doxorubicin and high-dose methotrexate regimen in this study (Rosen and Nirenberg, 1982). The follow-up for these patient groups is not comparable.

Conclusion

Local tumor control can be achieved for most patients with malignant osteosarcoma of the extremities without amputation. Factors which mitigate against limb salvage are location in the tibia and age less than 12 years because of a potentially severe leg-length discrepancy. Only a small percentage of the patients are ineligible for the limb salvage procedure because of tumor size. Overall disease-free survival rates for patients with osteosarcoma of the extremity are the same for primary amputation or a limb salvage. Preoperative intra-arterial doxorubicin and radiation therapy induces tumor cell necrosis in a mean of 90%. Finally, in our opinion, the 57% complication rate from cadaver allografts is too high to warrant continued use of this bony replacement, whereas the 15% complication rate for metallic endoprostheses at four years of follow-up warrants their continued use. Fifty-six of the 111 patients (47%) remain continually disease free, an indication that at least half of the patients with osteosarcoma will live for a long time. Therefore, psychologic and functional benefits of limb salvage certainly warrant continued investigation. Our results indicate, however, that the ideal adjuvant chemotherapeutic regimen has yet to be developed, although our study implies that adjuvant chemotherapy is superior to no adjuvant chemotherapy in terms of disease-free survival. Current efforts are directed towards a more effective adjuvant chemotherapeutic regimen.

References

1. Eilber, F.R., Mirra, J.J., Grant, T.T., Weisenburger, T. and Morton, D.L.: Is amputation necessary for sarcomas? A seven-year experience with limb salvage. Ann. Surg. 192, 431, 1980.

2. Harvei, S. and Solheim, Ø.: The prognosis in osteosarcoma: Norwegian National Data. Cancer 48, 1719, 1981.

3. Rosen, G., Caparros, B., Huvos, A.G., Kosloff, C., Nirenberg, A., Cacavio, A., Marcove, R.C., Lane, J.M., Mehta, B. and Urban, C.: Preoperative chemotherapy for osteogenic sarcoma: Selection of postoperative adjuvant chemotherapy based on the response of the primary tumor to preoperative chemotherapy. Cancer 49, 1221, 1982.

4. Rosen, G., and Nirenberg, A.: Chemotherapy for osteosarcoma: An investigational method, not a recipe. Cancer Treat. Rep. 66, 1687, 1982.

5. Taylor, W.F., Ivins, J.C., Dahlin, D., Edmonson, J. and Pritchard, D.: Trends and variability in survival from osteosarcoma. Mayo Clinic Proc. 53, 695, 1978.

Sarcoma in Paget's Disease of Bone

A. HADJIPAVLOU and J. ZUCKER

Introduction

The development of sarcoma is the most important complication of Paget's disease of bone. These tumors are the most common of the malignant bone tumors in patients over 40 years of age. In the early stage, sarcomas are extremely difficult to differentiate from Paget's disease because of similarities in appearance.

Paget's original case (1877) developed sarcoma at the elbow. His series of 23 patients (Paget, 1889) with osteitis deformans had five patients who died from malignant transformation. Camp (1925) reported two cases of sarcoma in Paget's disease. Bird (1927) collected seven definite sarcomas from four Boston hospitals. Coley and Sharp (1931) described three case reports of 20 sarcomas treated at the Memorial Hospital in New York, and added 51 cases from the Bone Sarcoma Registry. Davie and Cooke (1937) reviewed 49 cases and reported on two patients. They commented on the presence of multicentric foci in 29% of affected patients. Sumney and Pressley (1946) reviewed 73 cases from the literature and added three of their own. The following year, Platt (1947) added 10 cases from the Manchester area. Poretta, Dahlin and Jones (1957) reviewed 1,753 patients with Paget's disease at the Mayo Clinic and collected 16 cases of sarcoma. Greditzer et al (1983) reviewed 4,415 patients with Paget's disease and found 41 cases of sarcoma from the Mayo Clinic. They updated the series by Poretta et al. Both the 1957 and 1983 studies show a relative incidence of 0.9% of sarcomatous transformation. Freydinger et al (1963) collected seven cases of sarcomas from 118 pagetic patients. McKenna (1964) reviewed 250 patients from the English literature and added 33. Barry (1961) reported on an 18-year review of the Australian experience and analyzed 116 cases of sarcomas in Paget's disease of bones. Price (1969) collected 107 examples from the Bristol and Leeds Bone Tumor Registries and reported on 80 of these patients. Finally, Milgram (1977) added 20 more patients (Table I).

The various forms of malignant transformation in Paget's disease of bone are osteogenic sarcoma (50-60%), fibrosarcoma (20-25%), giant cell sarcoma (5-10%), and rarely lymphosarcoma. These sarcomas are very aggressive. Reports vary but a 20% one-year survival rate is typical (Barry, 1969; Price and Goldie, 1969). Sarcoma affects the femur, pelvis, humerus, and skull in decreasing order of frequency (Barry, 1969; Schatzki and Dudley, 1961). Generalized metastasis is found in 30% of the patients at

Current Concepts of Diagnosis and Treatment
of Bone and Soft Tissue Tumors.

Table I. Literature review

Author	Year	Sarcoma	Case population when available
Paget	1889	5	23
Camp	1925	2	
Bird	1927	7	64
Coley et al	1931	3(71*)	
Davie et al	1937	2(49*)	
Sumney et al	1946	3(76*)	
Platt	1947	10	
Poretta et al	1957	16	1,753
Freydinger et al	1963	7	118
McKenna	1964	33(250*)	
Barry	1969	116	
Price	1969	80	
Milgram	1977	20	
Greditzer et al	1983	41	4,415
Hadjipavlou et al	1983	8	1,078

* Review of cases in the literature

the time of diagnosis. The cause of death is most frequently from pulmonary metastasis.

Paget's disease affects certain races and specific geographic populations with a higher prevalence than usual. The familial clustering of sarcoma in Paget's disease has been mentioned by previous investigators (Barry, 1961; Epstein et al, 1970). This pattern is usually horizontal; however, a recent report illustrated one example of vertical transmission affecting identical anatomic sites in a father and his son (Nassar and Gravanis, 1981).

Intranuclear inclusion bodies in the osteoclasts have been reported (Mills and Singer, 1976; Singer and Mills, 1977), and confirmed by electron microscope (Fig. 1). Their ultrastructural and immunologic characteristics are similar to a measles-like virus (Rebel et al, 1980). The prolonged latency period, afebrile course, and the involvement of a single organ indicate a striking resemblance to slow virus infections such as subacute sclerosing panencephalitis (SSPE). These are circumstantial evidences of a viral infection.

It is plausible that the inclusion bodies in some way trigger pagetic bone to behave in a disorganized fashion. Sarcomatous transformation may be related to these inclusion bodies (virus infection). Only knowledge of the etiology of the disease process can lead to prevention and eventual cure.

Material and Methods

A retrospective review was carried out of all patients with symptomatic Paget's disease admitted for therapy at four major institutions in the

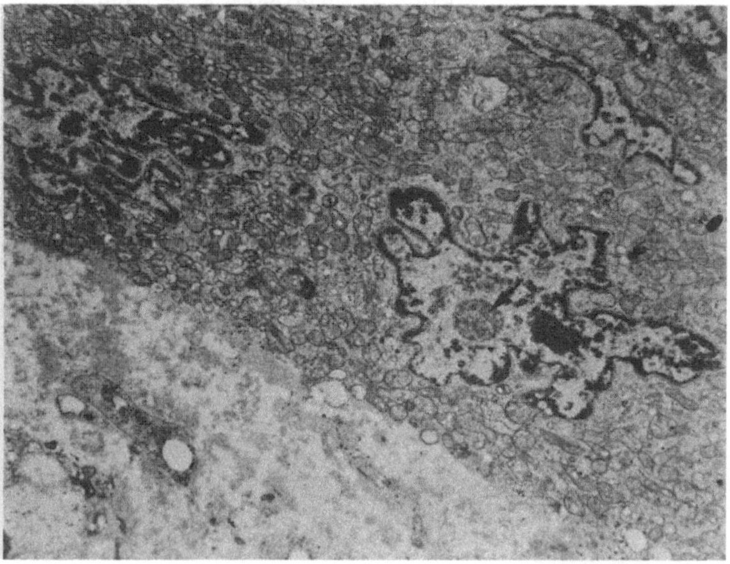

Fig. 1. Electron micrograph illustrating an inclusion body in an osteoclast of Paget's disease.

Montreal area -- The Royal Victoria Hospital, The Montreal General Hospital, the Sir Mortimer B. Davis - Jewish General Hospital, and the Sacré Coeur Hospital -- comprising 1,078 patients from 1970 to 1982. Eight sarcomas and 78 pathologic fractures of long bones were detected; four patients in this study had pathologic fractures associated with sarcoma (Table II).

Two of these eight case reports are described in detail below.

Case report

Case 3 concerns a 74-year old woman who had complained of pain in her knee six months before admission. Radiographs of the distal femur showed a mixed phase of Paget's disease. Her serum alkaline phosphatase level was 2,520 IU on admission.

Six months after the onset of symptoms, she was referred to our center where she underwent an open biopsy which confirmed the diagnosis of an osteogenic sarcoma. She was given calcitonin and then underwent a hip disarticulation in April 1978. Her serum alkaline phosphatase level decreased precipitously to 250 IU four days after the operation. Due to the pagetic focus at the L-1 vertebra, calcitonin administration was continued for one year. In May, 1978 the patient's serum alkaline phosphatase level was 210 IU and had decreased further, one year later, to 100 IU. The patient remained well and the calcitonin was discontinued. In May, 1979 both a clinical examination and a bone scan confirmed that the patient was free of active Paget's disease and that the osteogenic sarcoma had not metastasized.

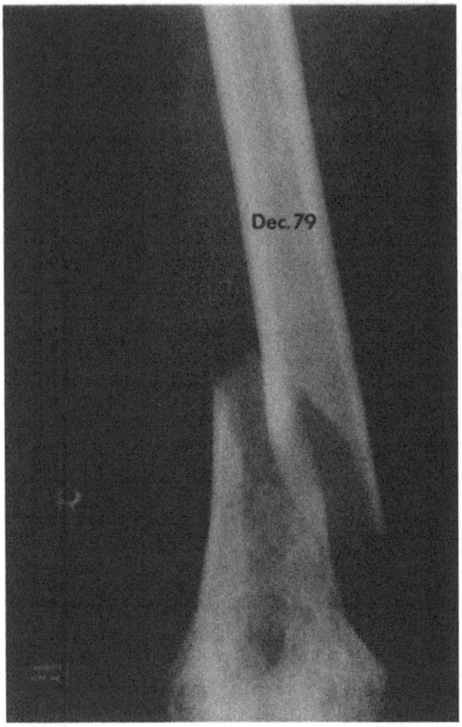

Fig. 2. Radiograph of Case 6. Lymphosarcoma complicating Paget's disease of bone with pathologic fracture.

A few weeks later, however, the patient noted a painless lump in the stump site. She was readmitted to hospital in December, 1979 due to a progressive enlargement of the mass in the stump site, and another mass was noted in the lower abdominal quadrant. An investigation identified metastatic deposits in the cervical spine as well, leading shortly thereafter to paraplegia. Her alkaline phosphatase level had risen to 1,050 IU at that time, increasing to 2000 IU by January 1980. The patient died one month later.

Case 6 concerns a patient with a rare occurrence of lymphosarcoma complicating Paget's disease. This patient had a colonic sarcoma resected in 1974. She was doing well until 1979 when a radiograph of her femur revealed a pathologic fracture through a lytic lesion (Fig. 2). Open biopsy revealed a reticulum cell lymphosarcoma (Fig. 3). She underwent aggressive curettage with cement reconstruction and internal stabilization with a compression plate. Following this procedure, she was given local radiotherapy and systemic administration of doxorubicin, procytox, vincristine, and prednisone. She remains alive 24 months after the operation.

Discussion

The eight patients described in this review came from a highly selected population group with symptomatic Paget's disease. If one considers that

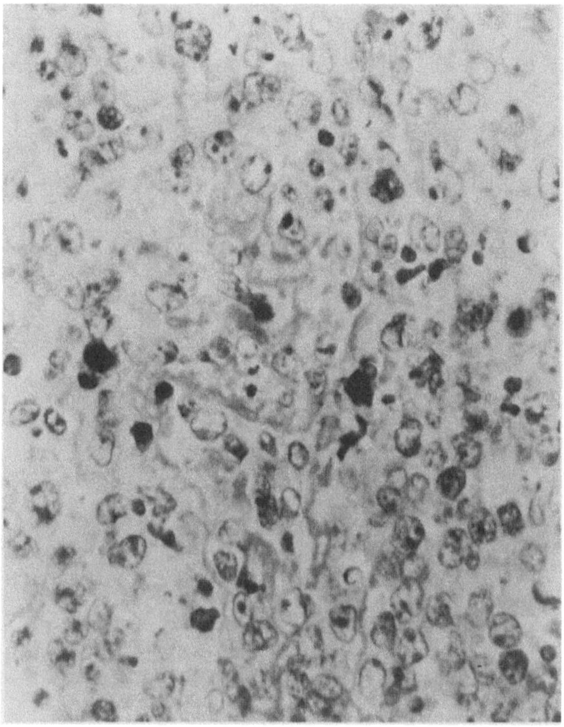

Fig. 3. Biopsy specimen of bone showing reticulum cell lymphosarcoma in pagetic bone.

3.7% of the population over 40 years of age have some form of Paget's disease (Collins, 1956), then our eight cases represent a remote possibility of developing sarcoma in a pagetic patient. One can state that in the Montreal area the chance of developing sarcoma in a patient with symptomatic Paget's disease of bone is 0.8%.

Clinically, acute onset of pain or change in the underlying nature of the pain was common in our patients who developed sarcoma. Lytic pathologic fractures in Paget's disease should alert a physician to sarcomatous transformation (Fig. 4). Radiologically, the tumors in our series were lytic or mixed. Localized focal expansion in Paget's disease of bone with clearly delineated and circumscribed borders of periosteal reaction is usually a benign process -- so-called pumice stone. This should not be confused with malignant transformation. We recall one case where a patient with a pumice stone appearance on the radiograph of a distal fibula was erroneously diagnosed as a sarcoma and underwent radical ablative surgery for this benign lesion (Fig. 5A). When the periosteal borders of an expanding focal pagetic lesion are irregular and interrupted, this should arouse suspicion of malignancy, especially when the lesion is associated with soft tissue swelling (Fig. 5B and 5C).

When malignancy is suspected, biopsy should not be delayed. Since bleeding can be a major complication in Paget's disease of bone, we recommend a bone blood flow before surgery. One of our patients (Case 5 in Table II)

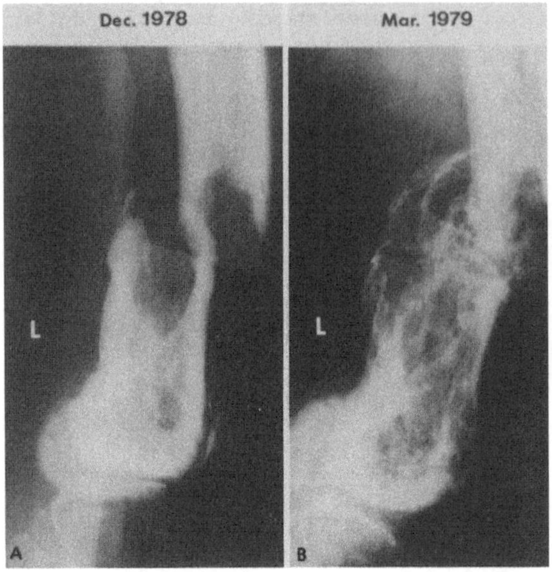

Fig. 4. Radiograph illustrating a pathologic fracture in pagetic bone. Focal osteolysis should arouse suspicion of malignancy.

illustrates an example of a patient hemorrhaging after a needle biopsy of a suspected sarcoma in pagetic bone. Perhaps this could have been prevented with a course of mithramycin before undertaking this biopsy. We have shown that a five-day course of mithramycin (15 μg/kg/day, iv, over 12 h) decreases the vascularity and subsequent bleeding diasthesis (Hadjipavlou et al, 1977). One can confirm the beneficial effect of mithramycin by repeating the bone flow after therapy (Hadjipavlou et al, 1979).

Case 3 illustrates an example of the potentially beneficial and suppressive effects of calcitonin. Metastasis in the soft tissue and elevation of serum alkaline phosphatase levels were observed when the calcitonin was discontinued.

Case 6 illustrates a lymphosarcoma complicating Paget's disease of bone. We are aware of only one other case report in the English literature (Ryan and Schwartz, 1981). Whether the occurrence of lymphosarcoma in pagetic bone was coincidental or secondary to the disease process remains to be proven. Since we do not want to make anecdotal assumptions at this stage, we are only reporting here the occurrence of lymphosarcoma in Paget's disease of bone. **In our series, we had five patients with osteogenic sarcoma, one case of malignant fibrohistiocytoma, one case of epithelioid cell sarcoma, and one case of reticulum lymphosarcoma.**

Paget's disease of bone can also be associated with metastatic tumors of bone (Burgener and Perry, 1977; Epstein et al, 1970). Occasionally, benign focal osteoblastic remodelling pagetoid lesions may be indistinguishable from osteoblastic metastatic lesions (Fig. 6). In our series we had four patients with metastatic osteoblastic lesions in pagetic bone. The diagnosis was made late in each of these cases. These four patients

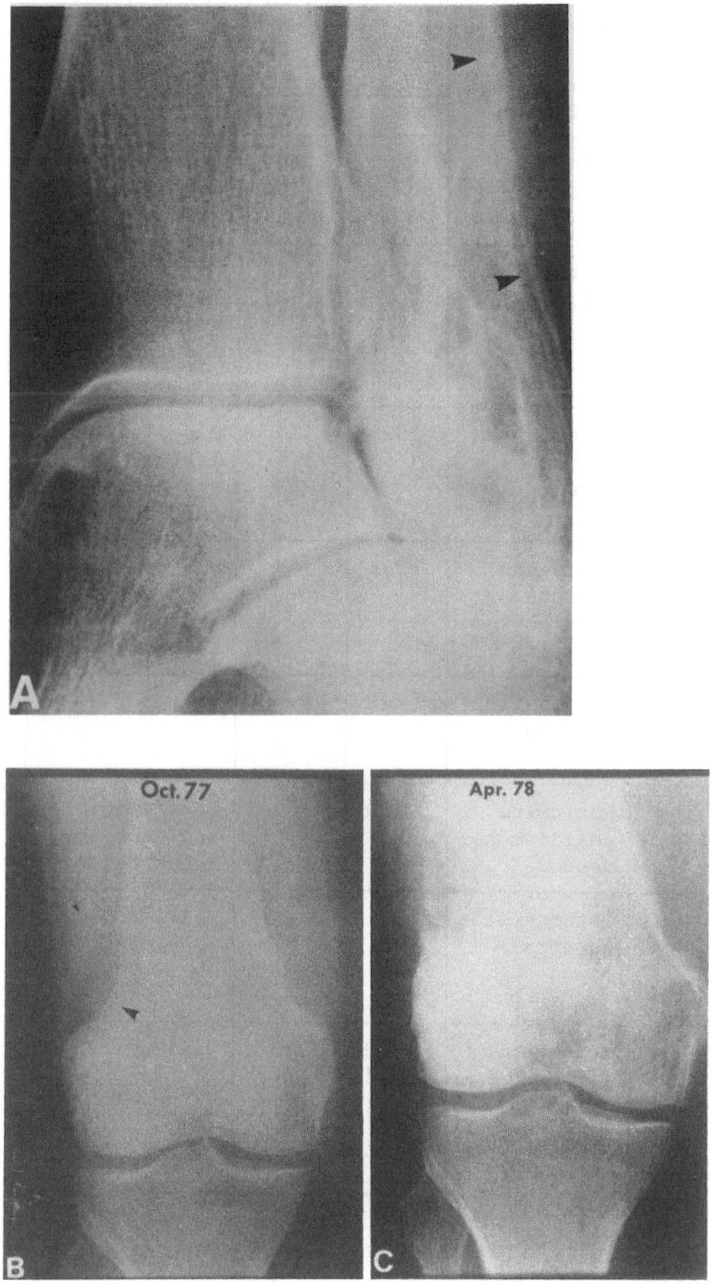

Fig. 5. A) This is a localized periosteal bone expansion: pumice stone. Usually the borders are well defined contrary to sarcomatous transformation: B) the confines of bone are violated; C) progression of sarcoma after six months.

Table II. Summary of cases with sarcomas detected in a retrospective

Patient No/Name Age/Sex	Site	Fracture	Histologic findings	Alkaline phosphatase levels		Treatment
				Preop	Postop	
1 / RM 42 / M	Distal radius	No	Osteogenic sarcoma - lytic	250	110	Above elbow amputation; doxorubicin
2 / RC 76 / M	Proximal femur	Yes	Epitheloid cell sarcoma	225	115	En bloc resection; Moore's arthroplasty; local radiotherapy
3 / DC 73 / F	Distal femur	No	Osteogenic sarcoma - mixed	2520	110	Hip disarticulation; preop and postop calcitonin
4 / AG 64 / F	Middle femur	Yes	Malignant fibrous hystiocytoma - lytic	275	85	Hip disarticulation; preop and postop mithramycin
5 / LM 72 / M	Pubis ilium	No	Osteogenic sarcoma	325	-	Nil
6 / DL 77 / F	Distal femur	Yes	Reticulum cell lymphosarcoma	150	150	Curettage; radiotherapy; chemotherapy
7 / LT 78 / M	Skull	No	Osteogenic sarcoma - mixed	231	-	Nil
8 / TB 55 / F	Humerus	Yes	Osteogenic sarcoma - lytic	168	170	Radiotherapy

review of 1,078 patients with Paget's disease

Metastasis	Survival (mos)		Remarks
	After onset of symptoms	After diagnosis	
Lungs	8	6	Trauma to wrist 2 months before biopsy
Lungs	12	6	Cachectic for 6 months before biopsy
Lungs Spine	29	23	See case report
Lung	16	4	Pathologic fracture one year before diagnosis
Spine	1	1	Extensive hemorrhage after needle biopsy
Inguinal region	36	24	See case report
Lungs	1	1	3 days after admission had massive CVA and died
Unknown	7	1	Nonunion of fractured humerus; no follow-up. Died 1 mo. later

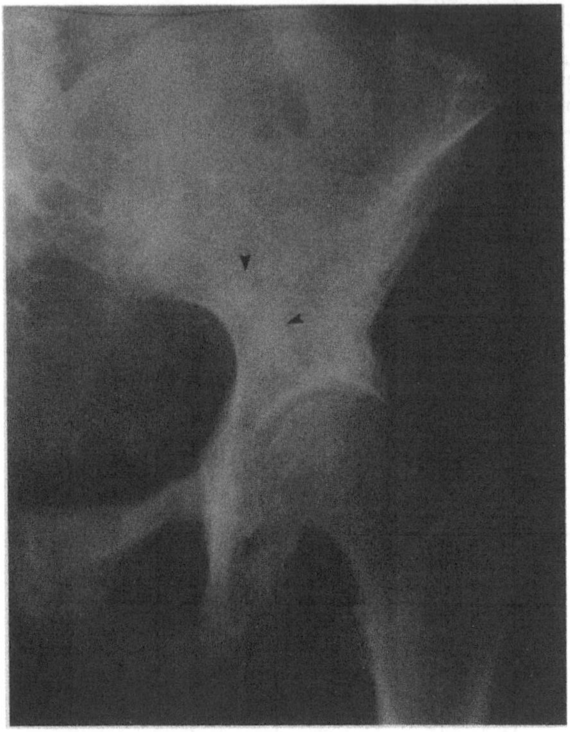

Fig. 6. Focal osteoblastic pagetic lesions can be confused with metastatic lesions (see arrows).

illustrate the difficulty in diagnosing metastatic osteoblastic tumors in pagetic bone. A definitive diagnosis can only be made by tissue biopsy.

Summary

Eight patients with sarcomatous transformation were selected for this review from a series of 1,078 patients with symptomatic Paget's disease from four major Montreal hospitals over a 12-year period (1970-1982). The incidence of sarcoma in symptomatic pagetic patients is 0.8%; however, in asymptomatic patients the incidence is much lower.

The etiology of Paget's disease remains unknown; however, inclusion bodies in the osteoclasts may play a role. Focal expansion of pagetic bone beyond the borders of the periosteum can be either benign (pumice stone-like lesion) or a sarcoma. Biopsy should be performed in suspected lesions. Bleeding can be minimized with a preoperative course of mithramycin. Focal osteolysis in pagetic fractures should arouse suspicion of sarcoma. Focal bone-remodelling abnormalities should not be confused with metastatic blastic lesions.

We are reporting a case of reticulum cell sarcoma in a pagetic lesion. Whether lymphosarcoma is another form of malignant transformation in Paget's disease of bone has not yet been determined. The prognosis of pagetic sarcoma remains grave. One should be aware that metastatic lesions can be found in pagetic bone.

References

1. Agha, F.P., Norman, A., Hirsch, S., and Klein, R.: Paget's disease. Coexistence with metastatic carcinoma. N.Y. State J. Med. 76, 734, 1976.

2. Anderson, J.T., and Dehner, L.P.: Osteolytic form of Paget's disease. J. Bone Joint Surg. 58A, 994, 1976.

3. Barry, H.C.: Sarcoma in Paget's disease of bone in Australia. J. Bone Joint Surg. 43A, 1122, 1961.

4. Barry, H.C.: Paget's Disease of Bone. E. & S. Livingstone Ltd. Edinburgh and London, 1969.

5. Bird, C.E.: Sarcoma complicating Paget's disease of the bone. Arch. Surg. 14, 1187, 1927.

6. Burgener, F.A., and Perry, P.E.: Solitary renal cell carcinoma metastasis in Paget's disease simulating sarcomatous degeneration. Amer. J. Radiol. 128, 853, 1977.

7. Camp, J.D.: Sarcoma complicating osteitis deformans: a case report of two cases. Radiology 5, 495, 1925.

8. Coley, B.L., and Sharp, G.S.: Paget's disease. Predisposing factor to osteogenic sarcoma. Arch. Surg. 23, 918, 1931.

9. Collins, D.H.: Paget's disease of Bone. Incidence and subclinical forms. Lancet ii , 271: 51, 1956

10. Davie, T.B., and Cooke, W.E.: Supervention of osteogenic sarcoma in Paget's disease. Brit. J. Surg. 25, 299, 1937.

11. Epstein, L.T., Bixler, D. and Bennett, J.E.: An incident of familial cancer including three cases of osteogenic sarcoma. Cancer 25, 889, 1970.

12. Freydinger, J.E., Dubig, J.T., and McDonald, I.: Sarcoma complicating Paget's disease of bone. Arch. Path. 75: 496, 1963

13. Greditzer, H.G., McLeod, R.A., Unni, K.K., and Beabout, J.W.: Bone sarcomas in Paget's disease. Radiology 146, 327, 1983.

14. Hadjipavlou, A.G., Tsoukas, G.M., Siller, T.N., Danais, S., and Greenwood, F.: Combination drug therapy in treatment of Paget's disease of bone. J. Bone Joint Surg. 59A, 1045, 1977.

15. Hadjipavlou, A.G., Danais, S., Tsoukas, G., Siller, T.: Use of arterial bone blood flow in assessing pagetic bone lesions. Trans. 25th Annual Meeting of The Orthopaedic Research Society 4, 307, 1979.

16. McKenna, R.J., Schwinn, C.P., Soony, K.Y., and Higinbotham, N.L.: Osteogenic sarcoma arising in Paget's disease. Cancer 17, 42, 1964.

17. Milgram, J.W.: Orthopaedic management of Paget's disease of bone. Clin. Orthop. 127, 63, 1977.

18. Mills, B.G., and Singer, F.R.: Nuclear inclusions in Paget's disease of bone. Science 194, 201, 1976.

19. Nassar, U.H. and Gravanis, M.B.: Familial osteogenic sarcoma occurring in pagetoid bone. Amer. J. Clin. Path. 76, 235, 1981.

20. Paget, J.: On a form of chronic inflammation of bones (osteitis deformans). Med. Clin. Trans. 66, 37, 1877.

21. Paget, J.: Remarks on osteitis deformans. Illus. Med. News 2, 181, 1889.

22. Platt, H.: Sarcoma in abnormal bones. Brit. J. Surg. 34, 232, 1947.

23. Poretta, C.A., Dahlin, D.C., and Jones, J.M.: Sarcoma in Paget's disease of bone. J. Bone Joint Surg. 39A, 1314, 1957.

24. Price, C.H.G., and Goldie, W.: Paget's sarcoma of bone. A study of eighty cases from the Bristol and Leeds Bone Tumor Registries. J. Bone Joint Surg. 51B, 205, 1969.

25. Rebel, A., Pouplard, A., Filman, R., Basle, M., Kouyoumdjian, S., and Lepatezour, P.: Viral antigens in osteoclasts from Paget's disease of bone. Lancet ii, 344, 1980.

26. Ryan, W.C., and Schwartz, T.B.: Mithramycin and malignancy (Letter) Ann. Intern. Med. 94, 279, 1981.

27. Schatzki, S.C. and Dudley, H.R.: Bone sarcoma complicating Paget's disease. Cancer 14, 517, 1961.

28. Singer, F.R., and Mills, B.G.: The etiology of Paget's disease of bone. Clin. Orthop. and Related Res. 127, 37, 1977.

29. Sumney, T.J., and Pressley, C.L.: Sarcoma complicating Paget's disease of bone. Ann. Surg. 123, 135, 1946.

Medical Treatment of Paget's Disease of Bone: Current Status

Z. F. G. JAWORSKI

Introduction

In approximately 1% of patients pagetic lesions undergo a malignant transformation (Barry, 1969; Siris et al, 1980). A discussion of this disorder at this conference devoted to bone and soft tissue tumors is, therefore, appropriate. Moreover, this disease behaves clinically as a low grade neoplasm.

Until some 15 years ago, no effective treatment for Paget's disease of bone was available. Since then, in short succession, mithramycin, calcitonin, and diphosphonate analogues were introduced and proved to be effective. The need to establish criteria for the selection of patients for such treatments revealed considerable gaps in the knowledge of this disease. Research had to be carried out in the etiology, epidemiology, natural history and physiopathology of this disease which affects 1%-5% of the white population after 50 years of age (Barry, 1969; Siris et al, 1980). As a result, we now have a better concept of this disorder. We propose, therefore, to review briefly its pathophysiology and natural history, the mode of its presentation, assessment of patients when first seen, currently available treatments, and developments in this field.

Pathology and Natural History

Pagetic lesions begin as an area of disorganized osteoclastic bone resorption followed sooner or later by equally disorganized osteoblastic activity forming predominently woven bone. Cycles of resorption and formation continue locally although they decrease with time. These processes end in the so-called burned-out phase of the disease, characterized on the roentgenogram by a marked osteosclerosis. They probably evolve differently in compact bone (the diaphysis of the long bones or the vault of the skull) and in spongy bone (Meunier, 1977). In the former, the earliest gross manifestation appears on the roentgenogram as the resorption wedge, so-called osteoporosis circumscripta. In the spongy bone, on the other hand, the coarse and thickened trabeculae are seen as its first manifestation.

Thus, Paget's disease is apparently essentially a disorder of osteoclasts. The number of osteoclasts increases as well as their nuclei, and they show ultramicroscopic evidence of increased activity (Rebel et al, 1980). Most important, the pagetic osteoclast populations escape the factors

Current Concepts of Diagnosis and Treatment
of Bone and Soft Tissue Tumors.

which normally organize them in space, ie, which are responsible for bone remodelling whether within the compacta or on the periosteal or endosteal surface (Frost, 1963). These features remind one, in fact, of a neoplastic type of process. Resorption fronts are supported by highly vascular connective tissue and, as in osteitis fibrosa cystica, associate with the fibrosis of the marrow. The concomitant increase in the formation of predominantly woven bone reflects the usual coupling of bone formation to bone resorption, observed normally in lamellar bone remodelling or in bone repair following injury (Frost, 1963; Meunier, 1977). Interestingly, however, the pagetic bone, even when fairly dense, appears to respond in the usual manner to immobilization.

Disuse osteoporosis in the form of rapidly expanding lytic lesions can be observed on roentgenograms in a limb put into a cast after a fracture (Doyle et al, 1980). Presumably, during such periods, large amounts of mineral released from the pagetic lesions may produce a transient hypercalcemia and hypercalciuria. Nephrolithiasis and sometimes staghorn calculi can develop rapidly. And, conversely, repair of such lesions may increase the demand for mineral and produce a transient hypocalcemia, straining the calcium ion homeostasis in body fluids and parathyroid glands (Posen et al, 1978; Siris et al, 1981) which is reflected in the nonaffected tissue (Meunier et al, 1980). Furthermore, the pagetic osteoclasts appear to be equally if not more sensitive than normal osteoclasts to the parathyroid hormone (Bordier et al, 1974; Siris et al, 1981). The risk of such complications will be greatest in patients with extensive polyostotic disease.

Recently, virus-like particles were identified within the osteoclasts in the pagetic lesions (Mills et al, 1980; Rebel et al, 1980), that is in the cells primarily responsible for the disease. The original designation of the disease -- osteitis deformans (Paget, 1877) -- implying an infectious agent and an inflammatory reaction, is therefore appropriate. The relation, however, between the postulated viral etiology and the definite age, race, and familial predisposition of the disease (Barry, 1969; Cullen et al, 1976; Siris et al, 1980), and a predilection for certain (although asymmetric) skeletal sites, have yet to be adequately explained.

Osteoclasts form by a fusion of mononuclear cells derived from the proliferating precursors related to monocytic series (Hall, 1975). Furthermore, the osteoclasts' nuclei undergo normally a continuous turnover (Jaworski, et al, 1981). The question, therefore, arises which type of cell is primarily infected and what is the fate of the infected nuclei.

Our knowledge of the natural history of the disease comes from retrospective reconstruction rather than prospective studies. It varies greatly in individual patients. It appears that the disease manifests itself at once in all affected areas and while it may spread locally, it does not metastasize to other areas. In most patients, the disease evolves slowly and in a silent fashion. Sites of predilection are the skull, vertebrae, pelvis and proximal bones of the upper and lower extremities, although lesions are asymmetrically distributed.

Clinical Features

The few patients with characteristic textbook deformities (large head, stooping, and bowleggedness) must be viewed as a tip of an iceberg consisting of a much larger population of patients with various forms of the disease, most of them never discovered during life. Even considerable deformities, if asymptomatic, may remain unnoticed for a long time by the victims of the disease themselves or their families.

The pagetic lesion is mostly found by chance on a roentgenogram examined for an unrelated reason. In some individuals, a fracture following a relatively minor trauma may be the first manifestation of the disease (Traver 1936). Many patients are first seen because of bone pain which cannot be always ascribed to the underlying pagetic lesion discovered on the roentgenogram because it is often secondary to Paget's disease or due to unrelated osteoarthritis. In this age group the association of these two diseases is not uncommon.

Metabolic or systemic complications of Paget's disease, almost never a symptom of the disease, include congestive heart failure, secondary hyperparathyroidism, calcium-containing kidney stones, and gout.

In a polyostotic form of the disease, highly vascular pagetic lesions may cause a considerable shunting of blood and increase the demands on cardiac output to the point of precipitating congestive heart failure in predisposed patients (Edholm et al, 1945). As in myeloproliferative disorders, secondary gout may occur because of an increased cell turnover and breakdown products of nucleic acids. With regard to the association of secondary and occasionally primary hyperparathyroidism with Paget's disease of bone (Meunier et al, 1980; Posen et al, 1978; Siris et al, 1981), the question arises whether or not such patients should receive vitamin D to prevent the occurrence of secondary hyperparathyroidism.

As mentioned above, kidney stones may form rapidly in pagetic patients once such patients are immobilized, usually following a fracture.

Assessment and Indications for Treatment

The decision whether or not a patient with symptomatic or asymptomatic Paget's disease should be treated is based on a radiologic skeletal survey (supplemented if necessary by a bone scan) and determination of plasma alkaline phosphatase and urinary hydroxyproline concentrations.

The skeletal survey permits assessment of the extent, the location and the type of lesion, while the bone scan may reveal, due to the vascularity of such lesions, the areas of involvement, especially in spongy bone not yet clearly apparent on the roentgenogram.

The polyostotic but asymmetrical form of Paget's disease is usually observed. It is the location of lesions rather than the number of sites that matters. Polyostotic Paget's disease is more likely to have one or two lesions strategically located, that is, in the vicinity of the joint.

In cranial Paget's disease, the lesion may be close to the cranial nerves and brain stem, or spinal cord and peripheral nerves, if the vertebrae, especially dorsal, are involved.

Occasionally, the disease is limited to one area, for instance one vertebra. A mono-ostotic lesion may present a diagnostic problem, and to differentiate it from other conditions a bone biopsy may be required. Biopsy may also be needed to rule out a malignant transformation (sarcoma) in the pagetic area, manifested either as severe pain or a suggestive radiologic appearance.

Furthermore, if pain is present, roentgenographic examination may determine its nature. As mentioned above, pain in Paget's disease is often due to secondary arthritis or the compression of adjacent structures.

Finally, roentgenographic examination determines whether the predominant type of lesion is lytic, mixed or sclerotic, and may reveal an impending fracture. Predominantly, lytic lesions in a weight-bearing bone, such as tibia or femur, increase the risk of fracture. Occasionally, a typical losei's zone may be noted, such as seen in osteomalacia or a pseudofracture.

The intensity of bone resorption in the whole skeleton is reflected in the levels of urinary hydroxyproline while that of secondary bone formation in elevated levels of plasma alkaline phosphatase. If a small area of the skeleton is involved, for instance one vertebra, such values may be only slightly increased. The main purpose, however, of these determinations, especially serum alkaline phosphatase, is to establish a baseline for monitoring the progress of the disease or to follow the effect of treatment.

The indications for treatment therefore include bone pain when it can be clearly ascribed to the pagetic process; involvement of the joint or compression of the content of the skull and spinal canal; and strategic location of an asymptomatic lesion that may predict potential complications. In the forms of the disease affecting the spine, pelvis, femora and tibia, intervertebral hip and knee joints are often involved, and in forms affecting the skull and the spine compression of cranial nerves and brain stem and cord may occur. Further indications for treatment are the presence of systemic or metabolic complications such as congestive heart failure, gout, or the presence of secondary hyperparathyroidism.

Patients with Paget's disease who do not require treatment are those with asymptomatic lesions distant from joints in nonweight-bearing bones (pelvis, clavicles, ribs, scapulae), or those with minimal radiographic change and little activity in nonstrategic areas, for instance in long bones away from joints.

Available Therapeutic Agents and Their Use

The first effective drug for treatment of Paget's disease introduced in the late 1960s was mithramycin (Ryan et al, 1970); it was displaced shortly afterwards by calcitonin and diphosphonate analogues, the latter being currently preferred therapy.

Among the calcitonins, the synthetic salmon calcitonin is most widely used (Derose et al, 1974). It has to be administered, however, intramuscularly or subcutaneously and is not free from side effects such as nausea and diuresis. It seems to suppress the disease as long as treatment is continued (Derose et al, 1974; Hosking, 1981; Siris et al, 1980), and the side effects generally disappear during the course of treatment. Calcitonin resistance does not usually have an immunologic cause (Haddad and Caldwell, 1972; Hosking, 1981). Like diabetics, patients with Paget's disease are taught to administer the drug subcutaneously themselves. The initial dosage is 100 IU daily for as long as the pagetic process continues to subside. Once the effect is stabilized, the recommended maintenance dosage is 50 or 100 IU twice weekly for several months (Derose et al, 1974). Recurrences, however, are frequent.

The second drug is EHDP or disodium etidronate, a first generation diphosphonate analogue, which has the advantage of being orally effective in one single dose although it should be taken between meals because it is poorly absorbed. This drug produces remissions when given in six month courses, with dosages between 5 to 10 mg/kg of body weight daily (Altman et al, 1973; Douglas et al, 1980; Gunǎaga et al, 1974; Russell et al, 1974; Siris et al, 1981). It should not be given in higher dosages or for a longer period because it will impair the mineralization of newly deposited matrix, producing an osteomalacia-like syndrome (Meunier et al, 1979).

Diphosphonates are rapidly cleared from the blood, and because of a high affinity for calcium phophate crystals concentrate in bone. They are powerful inhibitors of crystal formation and dissolution, but they inhibit the bone resorption probably by affecting intracellular mechanisms (Fleisch, 1980; Rowe and Hausmann, 1944).

Treatment with EHDP at a dosage of 5 to 10 mg/kg/day (according to the extent of the disease) for six months will apply to most cases. If needed, such courses may be repeated following a six-month interval. In resistant cases, a regimen of 20 mg/kg of body weight daily for one month out of every four has been advocated (Siris et al, 1981).

In some cases, however, either calcitonin alone initially, or in combination with diphophonates, may be indicated (Bijvoet et al, 1979). This applies particularly to the large predominantly osteolytic lesions strategically located to increase the risk of fracture. Such treatment should be continued for as long as bone turnover continues to decline; but more experience with a combination of EHDP and calcitonin is obviously required.

Whether or not patients are treated, the nature of the disease should be explained to them and they should be reassured if no major complication is encountered. It is also important to warn the patient that suppression of the pagetic process will not alleviate the pain if it is due to secondary complications, and that the skeletal deformities already present will not recede.

The secondary complications of Paget's disease, such as osteoarthritis and the compression of adjacent structures, should be treated accordingly, either by medical or surgical means. Because of frequent orthopedic and neurosurgical complications, management of advanced Paget's disease requires a coordinated team approach between the surgical and medical treatments. Patients who need a total hip or knee replacement because of pagetic involvement may be given increased dosages of EHDP, for instance 20 mg/kg of body weight daily, for two weeks pre- and postoperatively to prevent the periarticular calcifications to which pagetic patients are particularly prone.

Results of Treatment and Prospects

Judging by objective criteria, the available treatments suppress the pagetic process. The diminished activity of osteoclasts and osteoblasts is reflected in the fall of urinary hydroxyproline and serum alkaline phosphatase levels, occasionally in changes on the bone scan, a decrease in the size of the lytic lesions as seen on the roentgenograms (Doyle et al, 1980), and the reduction of the skin temperature overlying the lesions. As to the pain, double blind studies have demonstrated that both forms of treatment, that is calcitonin and diphosphonates, produce more than a placebo effect in approximately 40% of patients with Paget's disease (Siris et al, 1981).

Nevertheless, it would be unrealistic to expect under those circumstances that the deformities and significant structural alterations could recede, although, as mentioned above, some normalization of bone structure can be observed grossly on the roentgenograms (Doyle et al, 1980) and by microscopic examination (Meunier et al, 1980).

Since the pagetic process can be suppressed, does it mean that treatment applied in the early stages of the disease could eradicate the disease and thus prevent the complications? Will newer and better diphosphonate analogues currently undergoing clinical trials (Douglas et al, 1980; Meunier et al, 1979) offer any promise in this respect? In fact, currently available treatments appear to deal only with the cellular responses to the postulated viral agent rather than with the cause; hence, we can now obtain only a temporary suppression of the disease rather than a permanent cure.

It should be remembered that all these therapeutic agents were introduced at the time when our knowledge of the disease was sketchy. They were designed to control the locally disorganized bone turnover rather than the etiologic factor. In fact, these agents appear to be specific only because of the increased recruitment and turnover of mononuclear cells forming the pagetic osteoclasts. They affect also normal osteoclasts, however, and cells not involved in the pagetic process (Kallio et al, 1973); thus they do affect bone remodelling and renewal in nonaffected areas as well, and insofar as diphosphonate analogues are concerned, they may affect the mineralization process of newly formed normal bone.

Assuming that the disease is due to the invasion of osteoclasts or their precursors by a virus (Rebel et al, 1980), and that only susceptible patients develop the disease (Cullen et al, 1976), one can envisage new forms of therapy which would stimulate natural immune defences against the infecting agent and not only suppress but possibly cure the disease.

Therefore, any progress in the management of patients with this disease will depend first on the availability of even more effective and specific therapeutic agents than currently available and, second, on the earlier detection of candidates for such treatment. Selection of patients to be treated is determined by their symptoms, thus excluding those whom we should be treating if the prevention of complications is our goal. Assuming that the effective treatments become available, the problem of the detection of early cases may not be easy to solve.

Nevertheless, experience with the treatments now available has allowed us to establish reasonable criteria for the selection of patients for treatment, which will probably apply to patients with less advanced forms of the disease if and when specific therapy becomes available.

Acknowledgement

I wish to thank Dr. Denis R. Desjardins from the Department of Orthopaedic Surgery of the Ottawa General Hospital for his valuable help in managing some of our patients, and Miss Sylvie Smith for preparation of the manuscript.

References

1. Altman, R.D., Johnston, C.C., Khairi, M.R.A., Wellman, H., Serafini, A.N., and Sankey, R.R.: Influence of disodium etidronate on clinical and laboratory manifestations of Paget's disease of bone (osteitis deformans). New Engl. J. Med. 289, 1379, 1973.

2. Barry, H.C.: Paget's Disease of Bone. Livingston, Edinburgh and London, 1969.

3. Bijvoet, O.L.M., Hosking, D.J., Frijlink, W.B., Velde, J.J., and Vellenga, C.J.L.R.: Treatment of Paget's disease with combined calcitonin and diphosphonate (EHDP). Metab. Bone Dis. Rel. Res. 1, 251, 1979.

4. Bordier, R., Rasmussen, H., and Doefmann, H.: Effectiveness of parathyroid hormone, calcitonin, and phosphate on bone cells in Paget's disease. Am. J. Med. 56, 850, 1974.

5. Cullen, P., Russell, R.G.G., Kalton, R.J., and Whitely, J.: Frequencies of HLA-A and HLA-B histocompatibility antigens in Paget's dissease of bone. Tissue Ant. 7, 55, 1976.

6. Derose, J., Singer, F.R., Avramides, A., Flores, A., Dziadiw, R., Baker, R.K., and Wallach, S.: Response of Paget's disease to porcine and salmon calcitonins. Effect of long-term treatment. Am. J. Med. 56, 858, 1974.

7. Douglas, D.L., Duckworth, T., Kanis, J.A., Preston, C., Beard, D.J., Smith, T.W.D., Underwood, I., Woodhead, J.S., and Russell, R.G.G.: Biochemical and clinical response to dichromethylene diphosphonate (CL_2MDP) in Paget's disease of bone. Arthritis Rheum. 23, 1185, 1980.

8. Doyle, H., Banks, L.M., and Pennock, J.M.: Radiologic observations on bone resorption in Paget's disease. Arthritis Rheum. 23, 1205, 1980.

9. Edholm, O.G., Howarth, S., and McMichael, J.: Heart failure and blood flow in osteitis deformans. Clin. Sci. 5, 249, 1945.

10. Fleisch, H.: Experimental basis for the clinical use of diphosphonates in Paget's disease of bone. Arthritis Rheum. 23, 1162, 1980.

11. Frost, H.M.: Bone Remodeling Dynamics. Charles C. Thomas, Springfield, Illinois, 1963.

12. Gunčaga, J., Lauffenburger, T., Lentner, C., Dambecher, M.W., Haas, H.C.; Fleisch, H., and Olah, A.J.: Diphosphonate treatment of Paget's disease of bone. A correlated metabolic, calcium kinetic and morphometric study. Horm. Metab. Res. 6, 62, 1974.

13. Haddad, J.G. Jr., and Caldwell, J.G.: Calcitonin resistance: clinical and immunologic studies in subjects with Paget's disease of bone treated with porcine and salmon calcitonin. J. Clin. Invest. 51, 3133, 1972.

14. Hall, B.K.: The origin and fate of osteoclasts. Anatomical Rep. 183, 1, 1975.

15. Hosking, D.J.: Paget's disease of bone. Brit. Med. J. 283, 686, 1981.

16. Jaworski, Z.F.G., Duck, B., and Sekaly, G.: Kinetics of osteoclasts and their nuclei in evolving secondary Haversian systems. J. Anat. 133, 397, 1981.

17. Kallio, D.M., Garant, P.R., and Minkin, G.: Ultrastructural effects of calcitonin on osteoclasts in tissue culture. J. Ultrastructural Res. 39, 205, 1973.

18. Meunier, P.J.: Disturbance in morphology and dynamics of the remodeling process in pagetic bone. In, Human Calcitonin and Paget's Diseases. I. MacIntyre (ed.), Hans Huber, (publ.), Bern 1977.

19. Meunier, P.J., Chapuy, M.C., and Alexandre, C.: Effects of disodium dichloromethylene diphosphonate on Paget's disease of bone. Lancet ii, 489, 1979.

20. Meunier, P.J., Coindre, J.M., Edouard, C.M., and Arlot, M.E.: Bone histomorphometry in Paget's disease. Quantitative and dynamic/analysis of pagetic and non-pagetic bone tissue. Arthritis Rheum. 23, 1095, 1980.

21. Mills, B.G., Singer, F.R., Weiner, L.P., and Holst, P.A.: Cell cultures from bone affected by Paget's disease. Arthritis Rheum. 23, 1115, 1980.

22. Paget, J.: On a form of chronic inflammation of bones (osteitis deformans). Med. Chir. Trans. 60, 37, 1877.

23. Posen, S., Clifton-Brigh, P., and Wilkinson, W.: Paget's disease of bone and hyperparathyroidism: coincidence or causal relationship. Calcif. Tiss. Res. 26, 107, 1978.

24. Ryan, W.G., Schwartz, T.B., and Northrop, G.: Experiences in the treatment of Paget's disease of bone with mitramycin. JAMA 213, 1153, 1970.

25. Rebel, A., Basle, M., Poulard, A., Mackani, K., Filmon, R., and Lepatezour, A.: Bone tissue in Paget's disease of bone. Ultrastructure and immunocytology. Arthritis Rheum. 23, 1104, 1980.

26. Rowe, D.J., and Hausmann, E.: Specificity of the diphosphonate-induced alterations of osteoclast morphology. J. Dent. Res. 23, Special Issue B, B196, 1944.

27. Russell, R.G.G., Smith, R., Preston, C., Walton, R.J., and Woods, C.G.: Diphosphonates in Paget's disease. Lancet i, 894, 1974.

28. Siris, E.S., Canfield, R.E., Jacobs, T.P., Stodardt, K.E., and Spector, P.J.: Clinical and biochemical effects of EHDP in Paget's disease of bone: patterns of response to initial treatment and to long-term therapy. Metab. Bone Dis Rel. Res. 4:301, 1981.

29. Siris, E.S., Jacobs, T.P., and Canfield, R.E.: Paget's disease of bone. Bull. N.Y. Acad. Med. 56, 285, 1980.

30. Traver, C.A.: The association of fractures and Paget's disease (osteitis deformans). N.Y. State J. Med. 36, 242, 1936.

Chondrosarcoma

F. H. SIM

Introduction

Chondrosarcoma is a malignant tumor whose cells produce hyaline cartilage that is usually lobulated. Chondrosarcoma accounts for approximately 10% of all bone sarcomas and is approximately half as common as osteosarcoma. Most of these lesions are primary, occurring in a previously normal bone, but 10% of the tumors are referred to as secondary chondrosarcomas. In the latter type, the original lesion is generally an osteochondroma, particularly in patients with multiple familial osteochondromatosis; less often, the original lesion is a chondroma. Several different variants of chondrosarcoma should be recognized because of their distinctive histologic appearances and different biologic behaviors.

Ordinary Chondrosarcoma

In the Mayo Clinic series of 470 patients with chondrosarcomas (Dahlin, 1978), the tumor occurred primarily in adulthood and old age. More than three-fourths of the tumors were in the trunk and upper ends of the femora and humeri (Fig. 1). The remarkable rarity of chondrosarcoma in the distal portions of the extremities is noteworthy. Symptoms are generally mild and insidious, with a long clinical evolution. Local swelling or pain or both are usually the presenting symptoms. The presence of pain associated with a central cartilaginous lesion is suggestive of active growth.

Radiographic features

The roentgenogram usually provides pathogenic evidence of chondrosarcoma. Central cartilaginous lesions produce fusiform expansion of the shaft, with thickening of the cortex and endosteal scalloping (Fig. 2A). Areas of calcification and ossification are seen in the region of destruction. The calcification gives the appearance of ringlets or snowflakes (Fig. 2B). In secondary chondrosarcoma, the earliest changes are seen in the cartilaginous cap, which becomes fuzzy and irregular. The presence of lucent and poorly mineralized areas within the osteochondroma is suggestive of malignant degeneration.

Pathologic features

Chondrosarcoma is usually characteristically composed of lobules. The lesion produces a matrix substance that varies in consistency. A myxoid quality is

Current Concepts of Diagnosis and Treatment
of Bone and Soft Tissue Tumors.

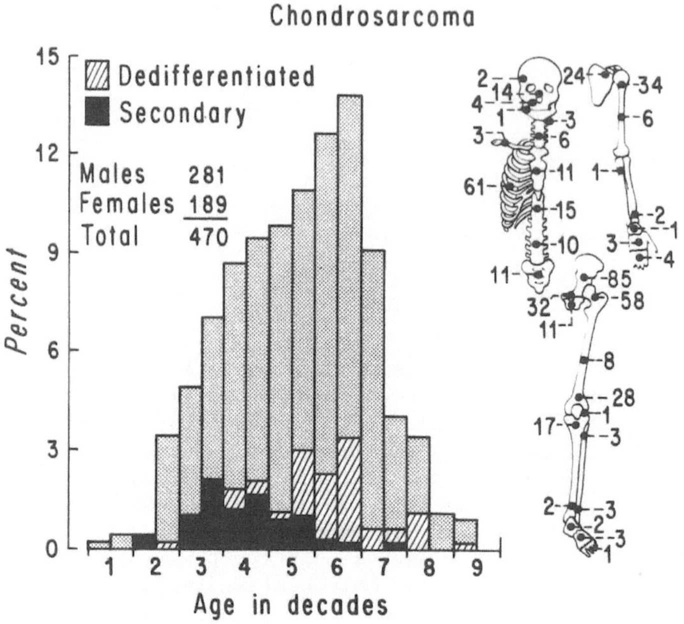

Fig. 1. Age and sex distribution of patients with chondrosarcoma in a Mayo Clinic series. (From Dahlin, D.C.: Bone Tumors: General Aspects and Data on 6,221 Cases, p. 191, Third Ed., Chas. C. Thomas, Springfield, Ill., 1978. By permission)

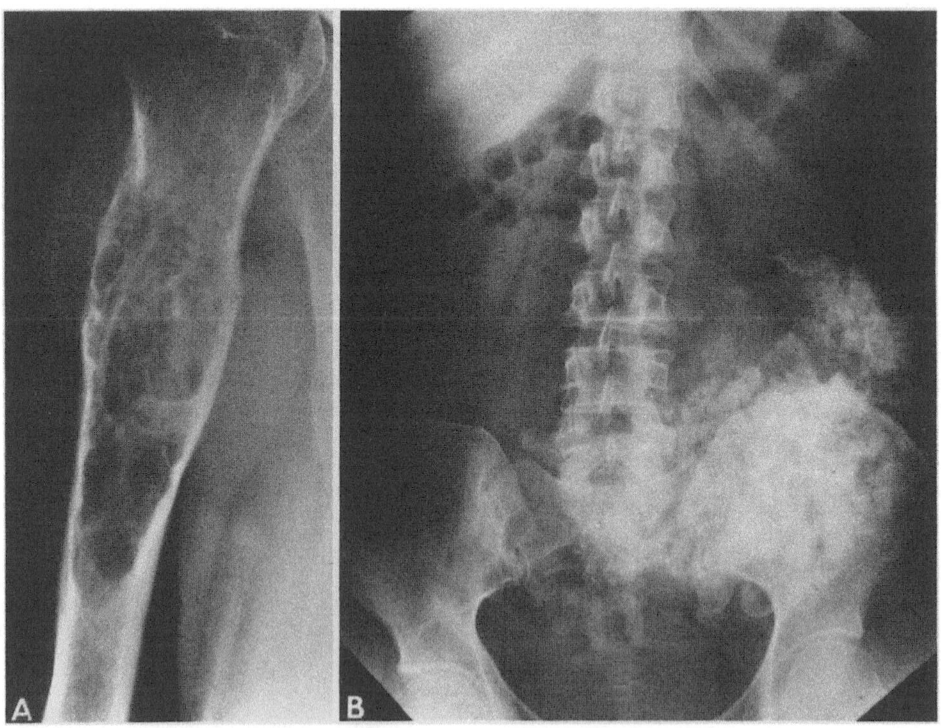

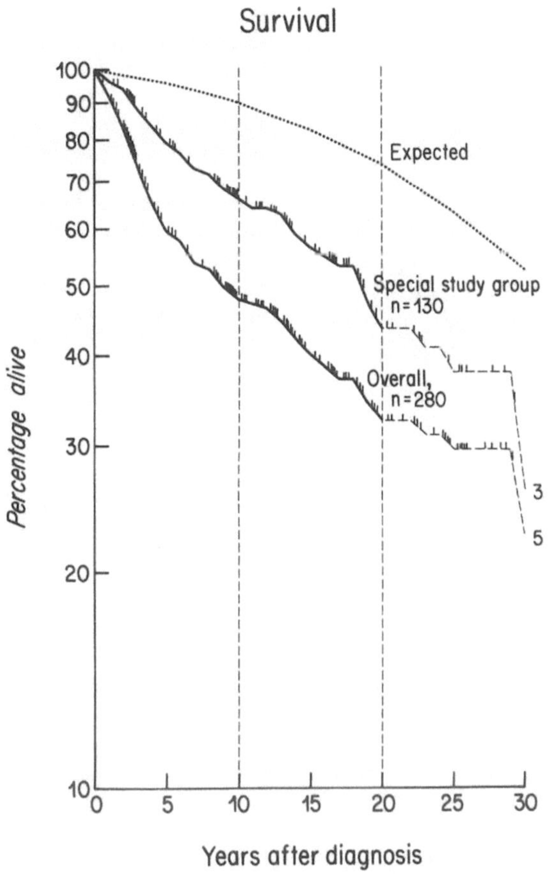

Fig. 3. Survival of patients with chondrosarcoma in a Mayo Clinic series. The 130 patients in the ideal group had no previous treatment and underwent curative procedures.

an ominous sign suggestive of malignancy. Histologically, the lesion may pose diagnostic difficulty for the pathologist. The lesions have a spectrum that ranges from tumors that are difficult to distinguish from chondromas to high-grade poorly differentiated tumors. Important features are cellularity, nuclear anaplasia, and presence of numerous binucleated cells.

Treatment

Treatment of this radioresistant tumor requires surgical ablation (Campanacci et al, 1975; Henderson and Dahlin, 1963). Careful pretreatment assessment

---

◄ Fig. 2. A: Anteroposterior view of right proximal humerus showing typical features of chondrosarcoma. There is expansion of the bone with thinning of the cortex. Areas of calcification are seen within region of destruction. B: Anteroposterior view of pelvis and lumbar spinal column showing extensive chondrosarcoma. Calcification gives the appearance of ringlets.

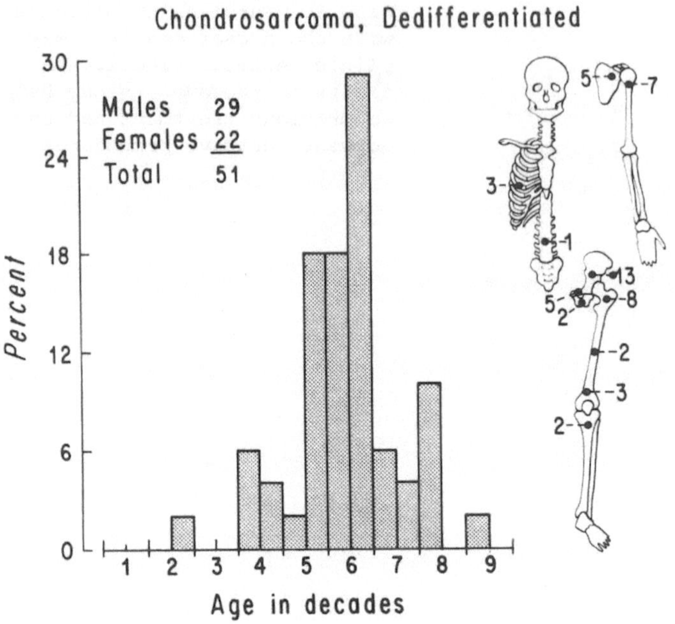

Fig. 4. Age and sex incidence in 51 patients with dedifferentiated chondrosarcoma in Mayo Clinic series. (From Dahlin, D.C.: Bone Tumors: General Aspects and Data on 6,221 Cases, p. 204, Third Ed., Chas. C. Thomas, Springfield, Ill., 1978. By permission)

is necessary to determine the general status of the patient, including evidence of occult systemic metastasis. Computed tomography has been particularly helpful in determining the regional extent of involvement of the tumor and assisting with the decision whether amputation is necessary or whether a limb-saving resection is possible. The aggressiveness of the surgical procedure depends a great deal on the histopathologic grade and the size of the lesion and its location. Control of the primary lesion is of critical importance as local recurrence will adversely affect survival. A long follow-up is necessary because recurrence can develop many years after the initial treatment. The fact that many surgically treated patients subsequently experience a local recurrence attests to the fact that surgery is often inadequate.

The lesion is best controlled by adequate en bloc resection or amputation. While amputation achieves the most complete ablation, provided a limb-sparing procedure can be done and the entire tumor can be resected, the possibility of long-term survival is not altered. Placement of the biopsy incision is important because of the potential for these cartilaginous tumors to implant with subsequent recurrence. The site of the biopsy must be carefully planned and placed where it can be completely circumvented at the time of definitive surgery.

The overall survival is approximately 50% at 5 years (Fig. 3). Our results indicate that one can expect a high cure rate for patients with grade 1 or 2 lesions (Pritchard et al, 1980). The grade of the lesion is also important in terms of survival and in the Mayo Clinic series (Dahlin, 1978) was an important factor in establishing prognostic criteria.

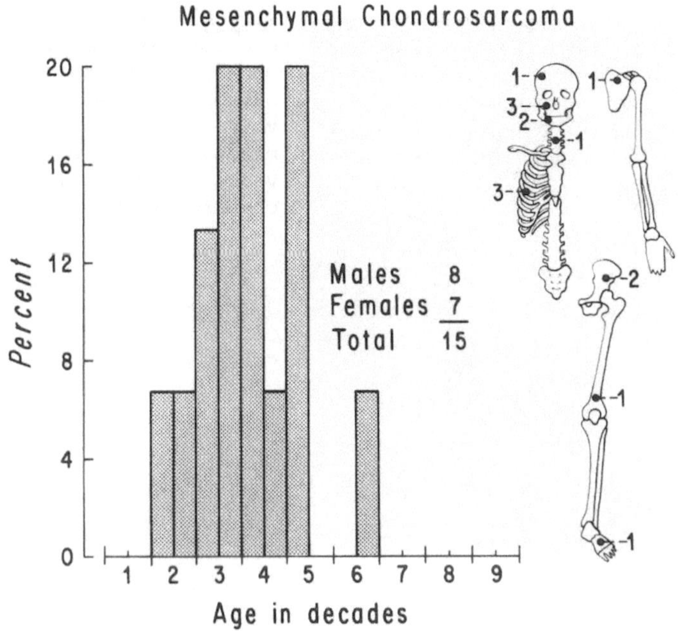

Fig. 5. Age and sex incidence of patients with mesenchymal chondrosarcoma in a Mayo Clinic series. (From Dahlin, D.C.: Bone Tumors: General Aspects and Data on 6,221 Cases, p. 218, Third E., Chas. C. Thomas, Springfield, Ill., 1978. By permission)

Chondrosarcoma Variants

Dedifferentiated chondrosarcoma

Of the 470 patients with chondrosarcoma, 51 had regions of significantly more anaplastic sarcoma (Fig. 4). Twenty-eight of these patients had associated fibrosarcoma, and 23 had osteosarcomas. It is important to recognize these tumors because of their highly malignant nature and correspondingly poor prognosis. In the series reported by Dahlin and Beabout (1971), of 28 patients with a follow-up of two years or longer, 23 had died of the disease. This suggests that treatment must be radical and selected on the basis of the dedifferentiated portion of the tumor.

Mesenchymal chondrosarcoma

In a Mayo Clinic series (Dahlin, 1978), 15 patients had a variant of chondrosarcoma that was characterized by a basically bimorphic pattern consisting of sheets or clusters of highly undifferentiated round cells, which may have a spindling quality, and small or larger chondroid islands (Fig. 5). The cartilaginous component had a well differentiated or benign appearance. In this variant, although the clinical course may vary and often be prolonged, the biologic behavior is usually relentless, with a strong potential to metastasize (Salvador et al, 1971). The most effective treatment probably is radical surgery to achieve complete ablation.

Clear cell chondrosarcoma

Another variety of chondrosarcoma is the clear cell chondrosarcoma, which is so named because it is composed of cells with abundant clear cytoplasm

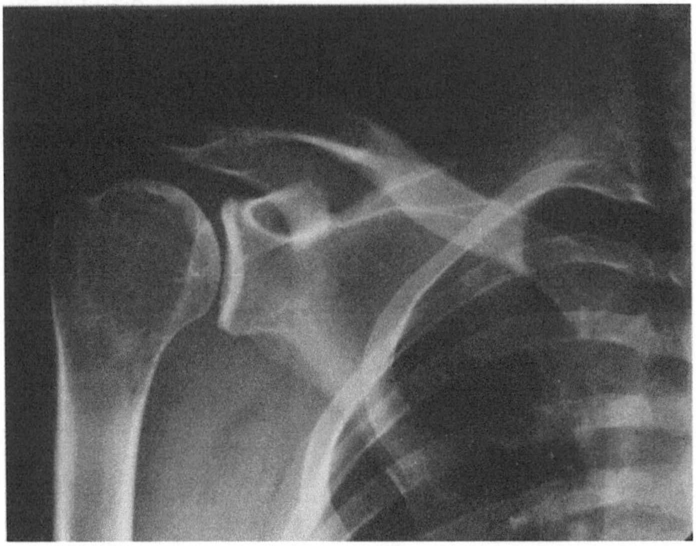

Fig. 6. Anteroposterior view of right shoulder showing osteolytic lesion involving proximal humerus. Biopsy revealed clear cell chondrosarcoma.

and indistinct cell boundaries and associated multinucleated giant cells. Typically, the lesion is seen as an osteolytic expansion at the end of the long bone (Fig. 6) and is frequently mistaken for a benign lesion, such as chondroblastoma or osteoblastoma. Because of the low-grade malignancy of this lesion, en bloc resection is the treatment of choice (Unni et al, 1976).

References

1. Campanacci, M., Guernelli, N., Leonessa, C., and Boni, A.: Chondrosarcoma: A study of 133 cases, 80 with long-term follow-up. Ital. J. Orthop. Traumatol. 1, 387, 1975.
2. Dahlin, D.C.: Bone Tumors: General Aspects and Data on 6,221 Cases. Third Ed. Charles C. Thomas, Springfield, Ill., 1978.
3. Dahlin, D.C., and Beabout, J.W.: Dedifferentiation of low-grade chondrosarcomas. Cancer 28, 461, 1971.
4. Henderson, E.D., and Dahlin, D.C.: Chondrosarcoma of bone: A study of two hundred and eighty-eight cases. J. Bone Jt. Surg. (Am.) 45, 1450, 1963.
5. Pritchard, D.J., Lunke, R.J., Taylor, W.F., Dahlin, D.C., and Medley, B.E.: Chondrosarcoma: A clinicopathologic and statistical analysis. Cancer 45, 149, 1980.
6. Salvador, A.H., Beabout, J.W., and Dahlin, D.C.: Mesenchymal chondrosarcoma: Observations on 30 new cases. Cancer 28, 605, 1971.
7. Unni, K.K., Dahlin, D.C., Beabout, J.W., and Sim, F.H.: Chondrosarcoma: Clear-cell variant; a report of sixteen cases. J. Bone Jt. Surg. (Am.), 58, 676, 1976.

Calcifying Enchondroma of Long Bone

C. D. TELFER and H.-K. UHTHOFF

Introduction

Enchondromas are benign tumors of cartilage, known as the most common benign tumor of the hand (Huvos, 1979; Jaffe and Lichtenstein, 1943; Jewusiak et al, 1971; Takigawa, 1971). In this setting they tend to be innocuous except for their association with pathologic fractures. Malignant transformation of enchondromas in this location is exceedingly rare (Coley and Higinbotham, 1954; Jakobson and Spjut, 1960; Takigawa, 1971). Less commonly, these tumors are found within the metaphysis of long bones such as the femur, humerus and tibia. In these bones the lesions tend to be highly calcified and several authors have shown that they carry a potential for malignant transformation (Coley and Higinbotham, 1954; Sanerkin and Woods, 1979). In this paper we present three cases of enchondromas found within the tibia or femur, and then follow with a discussion of enchondromas of large long bones.

Case Reports

Case 1. A 66-year old woman presented with a three month history of tenderness on the lateral aspect of her right knee. She had no previous trauma or problems with the right knee. Apart from the tenderness she had no other symptoms. Physical examination revealed diffuse swelling anteriorly from her right knee to just above her right ankle. Physical examination was otherwise normal. All laboratory examinations including serum alkaline phosphatase levels were within normal limits. Radiographs revealed an area of dense amorphous calcification at the proximal end of the tibia (Fig. 1, 2). No expansion of bone was noted, the cortices were intact, and no soft tissue involvement was observed. The lesion was approximately 3 cm in diameter. Bone scan revealed increased uptake in the medulla of the proximal portion of the right tibia (Fig. 3). A preoperative diagnosis of calcifying enchondroma was made. This was confirmed at surgery and the lesion was treated by curettage and bone grafting. Microscopic examination revealed a cartilaginous tumor of varying cellularity but no cellular crowding or cellular atypia. Calcification was prominent throughout. Final diagnosis was calcifying enchondroma. A follow-up examination 3 months postoperatively showed no recurrence of her symptoms.

Current Concepts of Diagnosis and Treatment
of Bone and Soft Tissue Tumors.

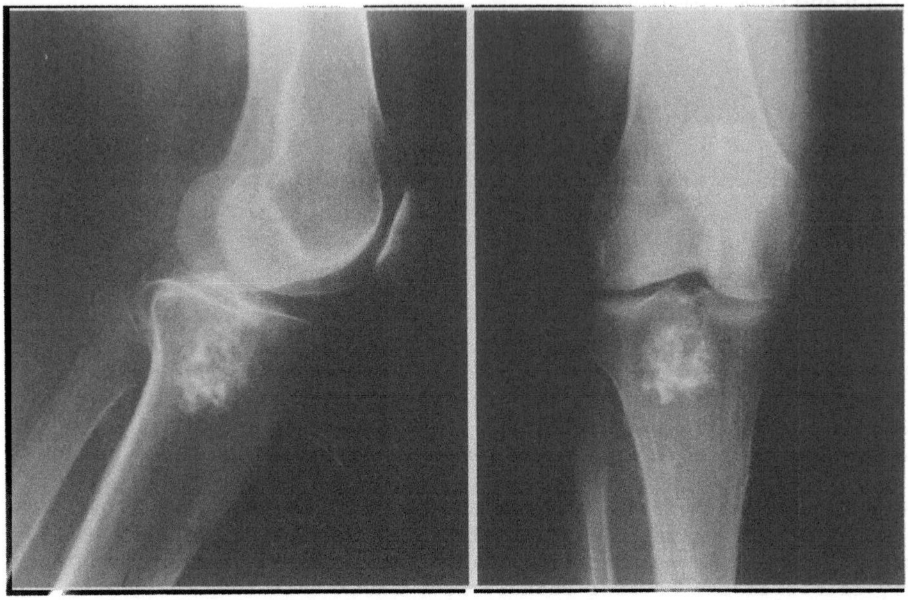

Fig. 1. AP and lateral roentgenogram of a 66-year old woman's right knee showing a circumscribed, highly calcified lesion within the proximal tibia. Biopsy showed calcifying enchondroma.

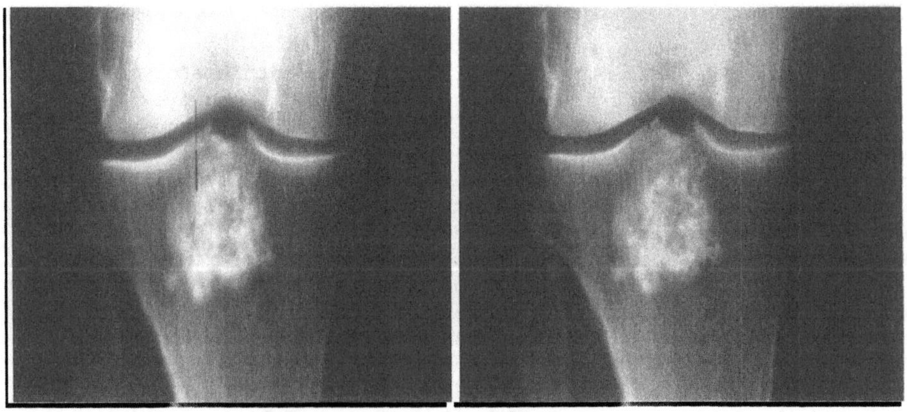

Fig. 2. Bone scan of both knees from patient 1. Posterior view. Note clearly localized area of increased uptake in the right distal tibia.

Case 2. A 53-year old man was admitted for investigation of back pain. During the work-up a bone scan revealed an area of increased uptake in the right distal femur. There was no history of any problem involving the leg and physical examination of the area was normal. Laboratory investigations including serum alkaline phosphatase levels were normal.

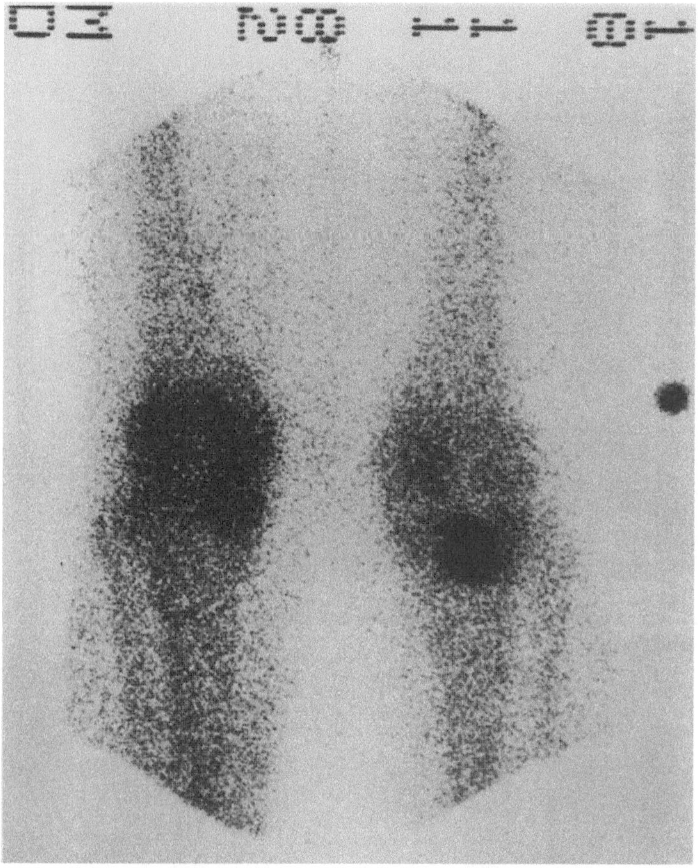

Fig. 3. Tomogram of lesion shown in Fig. 1.

Radiographs revealed a mottled linear calcification extending some 4 cm within the medullary cavity of the distal femur (Fig. 4). Treatment consisted of open biopsy and curettage. Pathologic examination showed a cartilaginous neoplasm with islands of calcification. The lesion was moderately cellular without pleomorphism or mitotic figures. Final diagnosis was calcifying enchondroma.

Case 3. A 51-year old woman presented with polyarticular arthritis. Among other joints involved she complained of poorly localized left knee pain. These symptoms had been present for 3-1/2 years and included periodic swelling and pain which was worse in the morning. During the work-up for this complaint an area of calcification was found within the distal third of her femur. It had a finely calcified appearance on radiographs and was clearly circumscribed within the central portion of the bone. The overlying cortex was unchanged (Fig. 5). The patient was admitted for investigation of the tumor. At the time of admission her

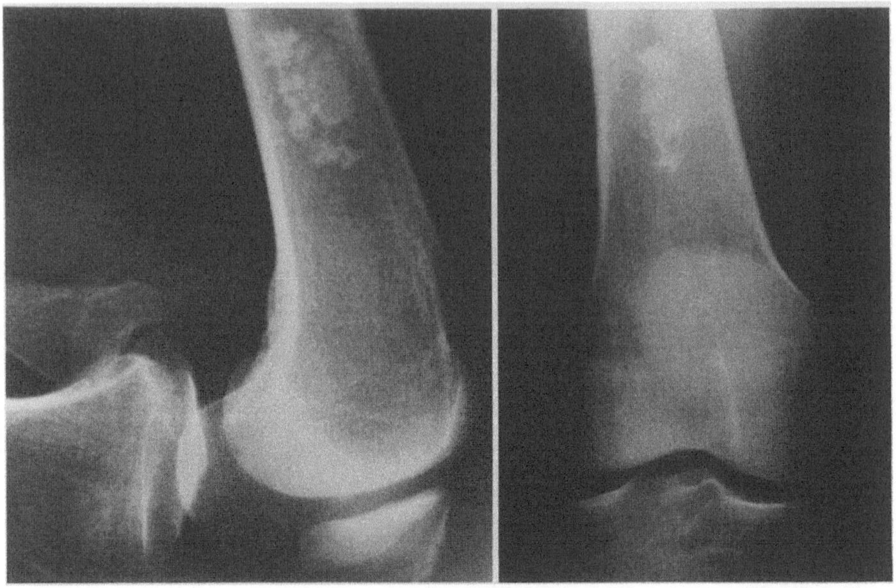

Fig. 4. AP and lateral of right distal femur from patient 2. Biopsy showed calcifying enchondroma.

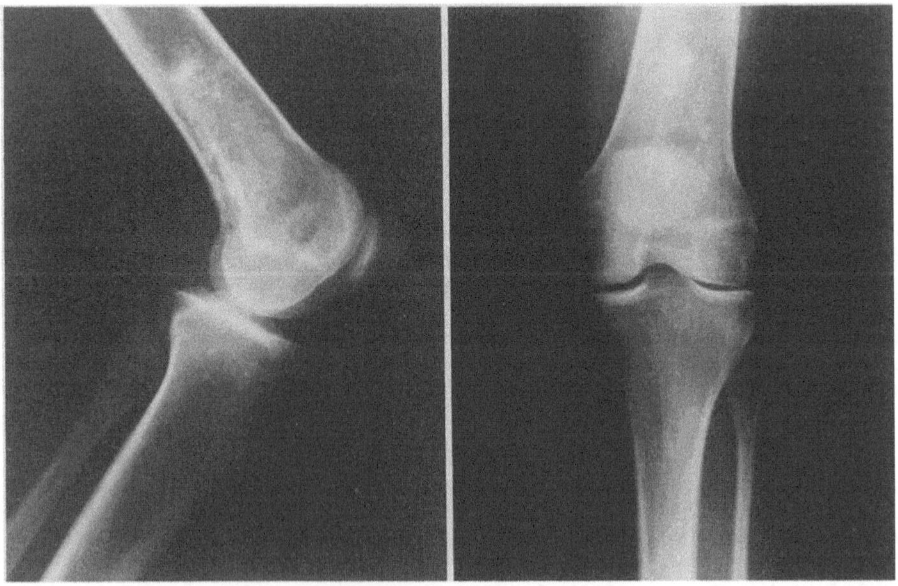

Fig. 5. AP and lateral radiograph from left knee of patient 3. Note finely calcified lesion in metaphysis of distal femur. Overlying cortex is unchanged. Biopsy showed calcifying enchondroma.

symptoms had largely resolved on gold therapy and the physical examination was normal. Bone scan showed increased uptake in the area of the lesion. A biopsy was performed of the lesion without total excision. Pathologic findings revealed an enchondroma. It was believed that the patient's knee symptoms were not caused by the enchondroma and no further treatment was carried out. A follow-up examination one year postoperatively showed that the patient had no symptoms due to the enchondroma and a roentgenogram of the lesion revealed no further growth of the tumor.

Calcifying Enchondroma

Enchondromas are benign tumors of hyaline cartilage which appear as a clearly circumscribed intramedullary lesion in the metaphysis of long bones. These tumors are thought to arise from embryonal rests of chondrocytes left behind in the metaphysis by the advancing physeal plate (Huvos, 1979).

The lesion usually presents in the third or fourth decade with an age range of 5 to 75 years and an equal sex incidence (Huvos, 1979; Jaffe and Lichtenstein, 1943). Enchondromas are the most common tumor of the hand and, according to Huvos (1979), 35% of all enchondromas arise in this region. The femur, humerus, and ribs are the next most common areas with 13% of enchondromas arising in each of these areas. As compared with enchondromas of the hands, enchondromas in these regions tend to show increased calcification as well as an increased incidence of malignant change (Coley and Higinbotham, 1954). Calcifying enchondromas discussed in this presentation occur in large long bones such as the femur, tibia or humerus.

Symptoms

Enchondromas are diagnosed in one of three ways. Often they are asymptomatic and are found during the work-up for other complaints, or they are found secondary to local trauma and a pathologic fracture at the site of the enchondroma. On occasion they may occur as a painful lesion without preceding trauma. This complication may indicate malignant transformation of the calcifying enchondroma (Coley and Higinbotham, 1954; Hamlin et al, 1971; Huvos, 1979; Jaffe and Lichtenstein, 1943; Sanerkin and Woods, 1979).

Radiographic Appearance

Enchondromas usually appear within the metaphysis of long bones as a clearly circumscribed, centrally-located rarefaction with varying degrees of calcification. Alternatively, they may appear as a conglomerate of calcification with no lytic component. Lack of calcification makes the diagnosis doubtful. Typically, no cortical thinning or regional expansion is observed although this may occur. The integrity of the outer cortex is always preserved (Hamlin et al, 1971; Huvos, 1979; Jaffe and Lichtenstein, 1943; Laurence and Franklin, 1953).

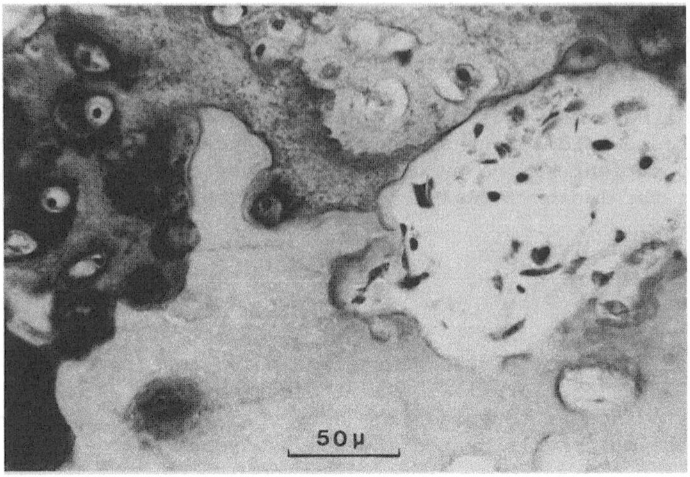

Fig. 6. Enchondroma. Biopsy from patient 1, proximal tibia. Note dense calcification around lacunae in upper left.

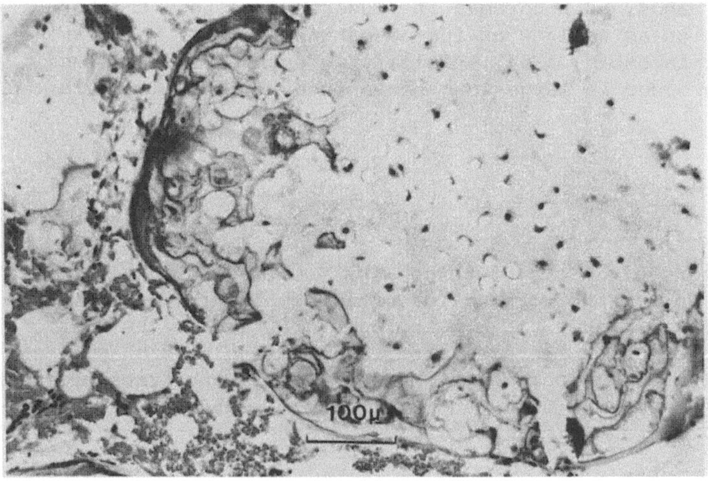

Fig. 7. Enchondroma. Biopsy from patient 2, distal femur. Note hyaline cartilage. Note cells of uniform size with one cell per lacuna. Calcification is at periphery of lobule.

In all of our cases the bone scan showed increased uptake. This may be useful in differentiating an enchondroma from a bone infarct. According to Hamlin et al (1971), signs of malignant change include the following:

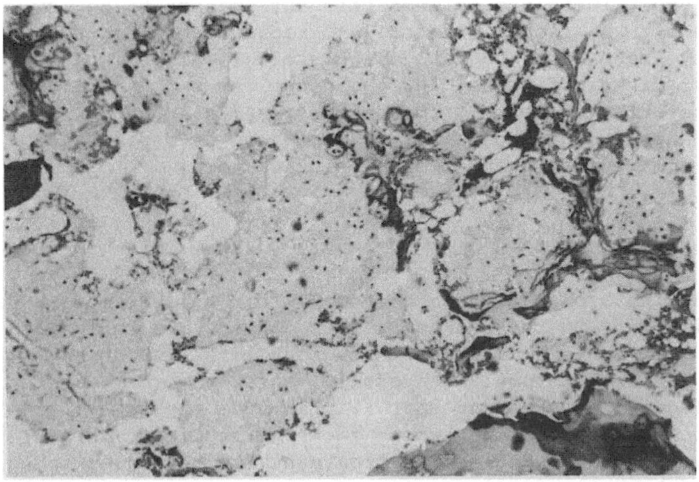

Fig. 8. Enchondroma. Biopsy from patient 2, distal femur. The hyaline cartilage shows a lobular architecture with calcification at the periphery of the lobules.

lack of marginal definition
cortical disruption
local periosteal reaction
growth of a previously known enchondroma
fusiform widening of the shaft associated with cortical sclerosis.

Pathologic Findings

Gross - the tissue tends to be divided into lobules of bluish white, firm or myxomatous hyaline cartilage. Varying degrees of calcification are seen, giving the tumor a yellowish gritty appearance.

Micro - the tissue is divided into lobules. Cellularity varies but usually only one or two cells per lacuna are observed. Calcification tends to appear at the periphery of the lobules or around lacunae. Individual cells are consistently small and have small round and dark staining nuclei. No mitotic nuclei should be present (Huvos, 1979; Jaffe and Lichtenstein, 1943)(Fig. 6, 7, 8).

Malignant Change

Malignant degeneration in enchondromas may occur. This is much more common in enchondromas of long bones than in those of the hand (Coley and Higinbotham, 1954; Jakobson and Spjut, 1960; Sanerkin and Woods, 1979;

Takigawa, 1971). The exact incidence is uncertain but appears to be low and the transition tends to be slow. Clinical and radiographic features of malignant change have been mentioned above. Histologic signs include increasing cellularity, mitotic figures, increased nuclear size and giant cells. The usual transformation is to a chondrosarcoma (Coley and Higinbotham, 1954). Sanerkin and Woods (1979), however, have reported transition to fibrosarcoma or malignant fibrous histiocytoma.

Treatment

The surgeon confronted with a probable enchondroma within a large long bone faces a therapeutic dilemma. On the one hand, these tumors are benign and usually asymptomatic. The patient could go throughout life without any symptoms secondary to his enchondroma. Conversely, these tumors may be associated with pathologic fractures and carry within them the capacity to undergo malignant transformation. Given these facts we recommend curettage and, if necessary, bone grafting as the treatment for these lesions. This treatment provides both tissue diagnosis and eradication of the lesion while complications are rare.

We would suggest follow-up examination at six months and one year to rule out recurrence.

References

1. Coley, B.L., and Higinbotham, N.L.: Secondary chondrosarcoma. Ann. Surg. 139, 547, 1954.

2. Hamlin, A.J., Adler, L. and Greenbaum, E.I.: Central enchondroma - a precursor to chondrosarcoma? J. Assoc. Can. Radiol. 22, 206, 1971.

3. Huvos, A.G.: Bone Tumors. Diagnosis, Treatment and Prognosis. W.B. Saunders Co., Philadelphia, 1979. p 152.

4. Jaffe, H.L., and Lichtenstein, L.: Solitary benign enchondroma of bone. Arch. Surg. 46, 480, 1943.

5. Jakobson, E., and Spjut, H.J.: Chondrosarcoma of the bones of the hand. Acta Radiol. 54, 426, 1960.

6. Jewusiak, E.M., Spence, K.F., and Sell, K.W.: The long bones of the hand. J. Bone Joint Surg. 53(A), 1587, 1971.

7. Laurence, W., and Franklin, E.L.: Calcifying enchondroma of long bones. J. Bone Joint Surg. 35(B), 224, 1953.

8. Sanerkin, N.G., and Woods, C.G.: Fibrosarcomata and malignant fibrous histiocytomata arising in relation to enchondromata. J. Bone Joint Surg. 61(B), 366, 1979.

9. Takigawa, K.: Chondroma of the bone of the hand. J. Bone Joint Surg. 53(A), 1591, 1971.

The Effect of Autoclaving on Normal and Sarcomatous Bone Cells and on Graft Incorporation

C. H. RIVARD

Introduction

The use of autoclaved bone goes back to the early part of this century. Groves, in 1917, used autoclaved bone when he ran out of autogenous bone to complete his grafts. In 1938, Orell used what he calls "Os Purum" in bone grafts. Later, the users of autoclaved bone believed that this procedure delayed graft incorporation (Odell et al, 1951; Reynolds and Oliver, 1950). In 1952, Lloyd-Roberts used autoclaved bone and stated that no immune reaction occurred when an autoclaved allogeneous graft was used. He thus refutes the conclusion that autoclaved bone delays graft incorporation.

Thomson and Steggall (1956) were the first investigators to report the use of autoclaved bone for reconstruction following tumor resection. Smith and Simon, in 1975, published the results of their study in eight patients with chondrosarcoma on whom a local resection and bone replacement was performed using their own autoclaved bone. Sijbrandij reported in 1978 eight patients with seven to 11 years' follow-up on whom bone replacement was performed with autogenous autoclaved bone, including one for chondrosarcoma and two for giant cell tumors. Finally, in 1980, Johnston reported his experience with autoclaved bone in improving the quality of life of older people.

A review of the current literature does not reveal any study of the effect of autoclaving on normal and tumoral bone cells. No experimental study could be located which has investigated the problem of incorporation of an autoclaved autogenous graft. We therefore decided in 1979 to investigate the effect of autoclaving on normal and tumoral (sarcomatous) cell survival in dogs (Rivard et al, 1981).

The second part of our study, in a normal dog, was carried out to observe the effect of autoclaving on the incorporation of an autoclaved autogenous bone graft.

Materials and Methods

It was decided to divide into two parts the study of the bone cell's postautoclave survival. For the first part of the study, we used a histologic approach to attempt a differentiation of the effects of cell

Current Concepts of Diagnosis and Treatment
of Bone and Soft Tissue Tumors.

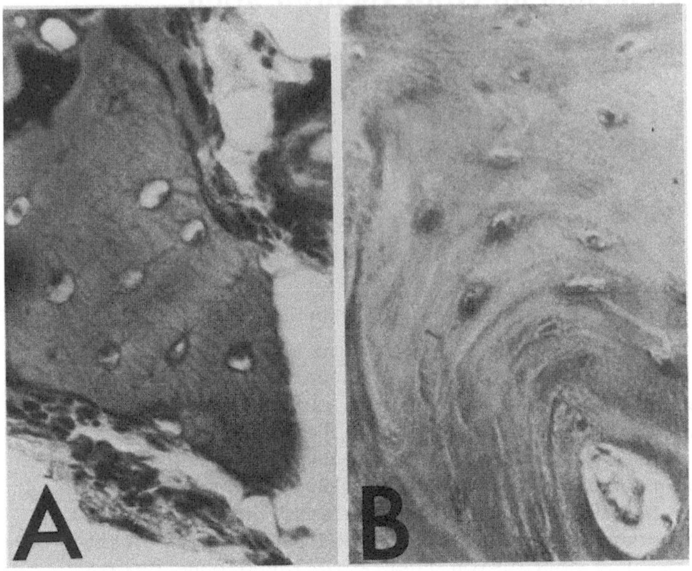

Fig. 1. A) Autoclaved bone fragment. B) Normal bone fragment.

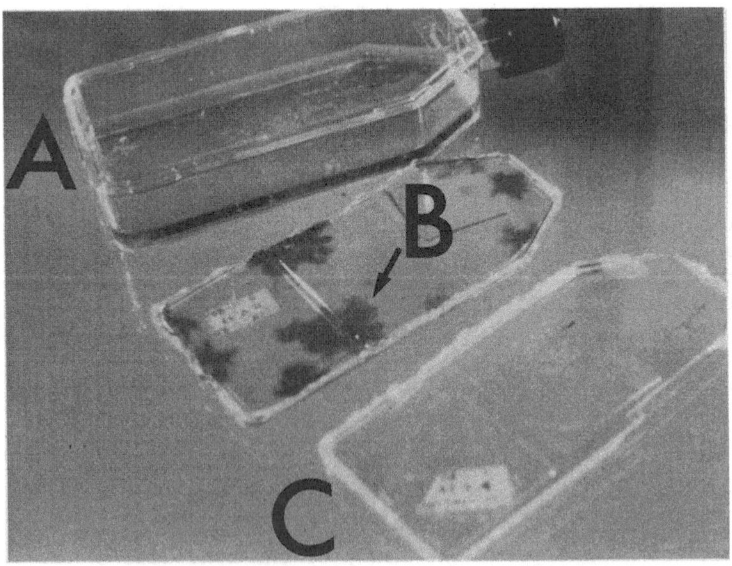

Fig. 2. A) Incubation bottle. B) Colored explant control. C) Treated explant: no cell proliferation.

survival, after autoclaving, on the normal and tumoral cells of the dog. We soon determined that histologic study, even when using the specific nucleus Giemsa stain, did not permit interpretation of the differences between the autoclaved fragment (Fig. 1A) and the control fragment (Fig. 1B). We observed that the nuclei were more lateral in their lacunae in the autoclaved fragment, while in the control bone fragment the nuclei filled their lacunae better; this difference does not adequately confirm the cell survival rate in the autoclaved bone. A more dynamic study was needed. Therefore, later on we used tissue cultures to demonstrate cell death in the autoclaved bone, using the following technique:

Two sterile samples of tumoral bone fragments were taken, one of which was autoclaved at 132°C for 10 minutes without humidity control. The two fragments were then put in a culture medium (F12) followed by a lavage in a Hank solution including antibiotics. The cells were then put in bottles which had been collagenized to better stimulate deep cellular adhesion through the culture medium. The nonautoclaved cells were kept 10 days in the culture medium while the autoclaved cells were kept in it up to 35 days. The culture medium was made up of an Earle solution, bovine fetal serum and penicillin and streptomycin. Incubation took place in a 5% CO_2 controlled atmosphere. At the end of the culture period, the medium was removed and the cells sticking to the collagen coat were fixed and colored using the Wright-Giemsa technique (Fig. 2). The difference between the control fragment and the autoclaved bone was determined under an optical microscope.

To study the incorporation time-factor of the autoclaved fragment versus the control fragment, we used femurs of growing dogs. For the control fragment, we proceeded with an open osteotomy of one femur, removing a 2 cm long fragment of bone and replacing it immediately. Fixation was secured with a 3-hole plate (Fig. 5). We used exactly the same technique on the opposite side, with the difference that the fragment was autoclaved before being replaced in the same fashion. We followed the incorporation of these different fragments by serial radiography, and proceeded with the removal of the plate and screws when fusion was achieved.

Results

Results concerning the cell survival of autoclaved tumoral bone in a tissue culture involve 17 tumoral specimens distributed by diagnosis and differentiating the 12 tumors of the dog and the five tumors of the human (Table I).

After seven days in the culture, all the nonautoclaved tumoral fragments revealed a cellular proliferation phenomenon at the explant periphery (Fig. 3). On the other hand, none of the autoclaved tumoral fragments (17/17 or 100%) showed cellular proliferation (Fig. 4), making it impossible for us to stain them with the Wright-Giemsa procedure. In fact, we found no cellular proliferation even 35 days after the fragments were first put in the culture medium.

Table I. Bone tumor specimens (n = 17)

Tumor	Dog	Human
Angiosarcoma	1	-
Chondrosarcoma	1	-
Fibrosarcoma	1	-
Osteosarcoma	9	3
Neurofibrosarcoma	-	1
Ewing's sarcoma	-	1
Total	12	5

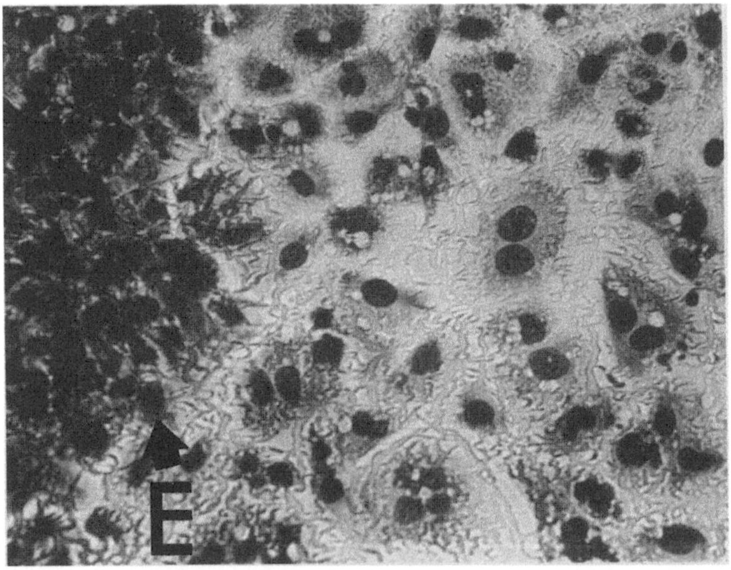

Fig. 3. E: Control explant with cell proliferation at the periphery.

The time factor with respect to the incorporation of the autoclaved graft versus the control femur fragment can be summarized as follows: consolidation was not delayed in the five dogs used in the experiment. Fig. 5A shows the right control femur and Fig. 5B shows the left femur with autoclaved fragment. Fig. 6 shows the same two femora one year after the initial operation and eight months after the removal of plate and screws. Following the removal of the bone samples, torsional testing was performed to pinpoint the degree of strength of the grafted areas. There was no difference between the control and the autoclaved sides.

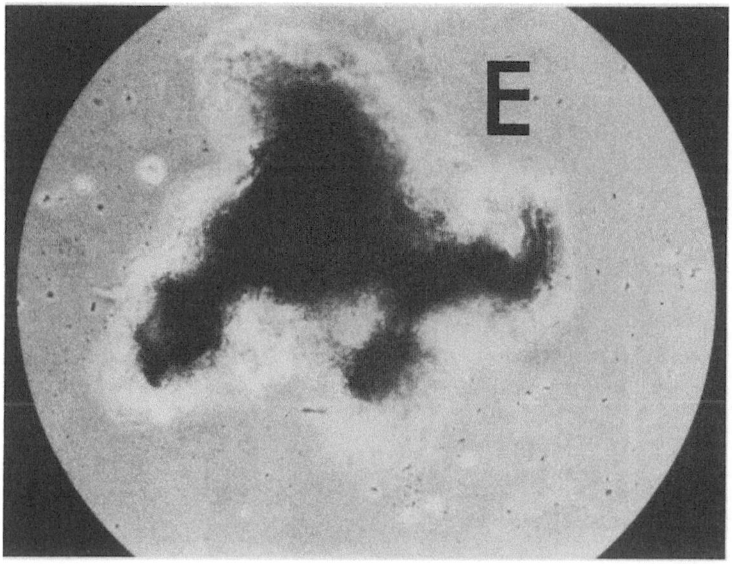

Fig. 4. E: Treated explant under microscope.

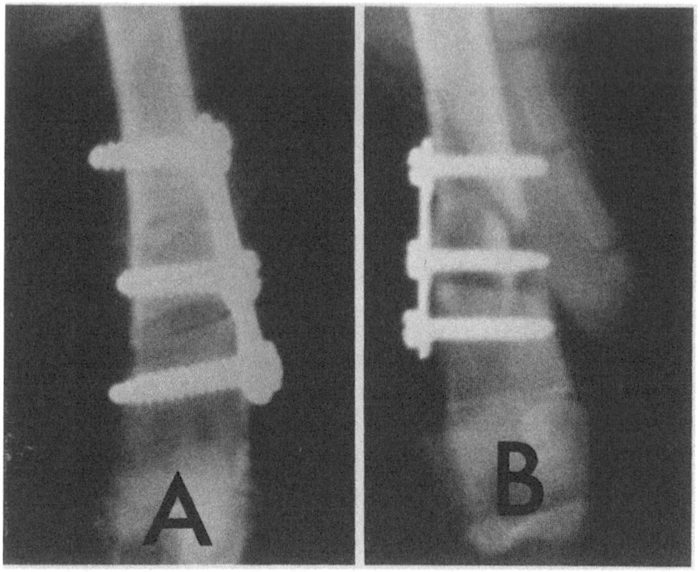

Fig. 5. A) Control femur. B) Treated femur.

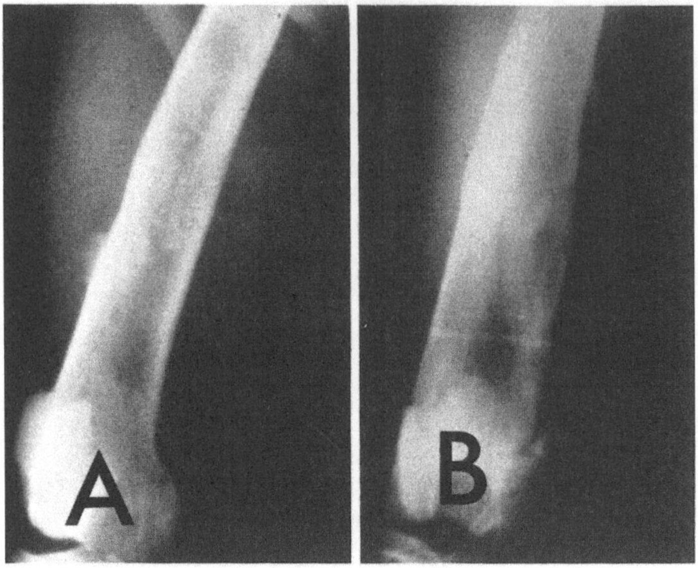

Fig. 6. A) Control femur. B) Treated femur.

Discussion

We believe that autoclaving at 132°C for 10 minutes results in cell death whether tumoral or normal cells are involved. On the other hand, we believe that the shortcoming of this study lies in the histologic interpretation which, per se, cannot be conclusive, even when based on histochemical principles. An investigation of cell metabolism is needed. Therefore, for one year, we have been monitoring cell proliferation in the control and autoclaved fragments, using H^3-thymidine in a culture medium. Using a scintillation counter, we are able to obtain the significant difference between the two tumoral fragments.

We also believe that autoclaving autogenous bone does not delay graft incorporation. We are now monitoring mechanical characteristics of bone which we feel might be affected by autoclaving. We also plan torsional studies three or four months following the removal of the osteosynthesis material.

Conclusions

This study documents that autoclaving leads to cell death, whether tumoral or normal cells are involved. More dynamic studies are needed to put the clinical use of autoclaved tumoral bone on solid scientific foundations.

We therefore come to the same conclusion as Lloyd-Roberts that autoclaving does not produce any delay in the incorporation of an autogenous graft. Finally, we should point out that more sophisticated bone studies, immediately after the autoclaving procedure, could demonstrate that the autoclaved bone readapts more easily than the control fragment.

References

1. Groves, E.W.H.: Methods and results of transplantation of bone in the repair of defects caused by injury or disease. Brit. J. Surg. 5, 185, 1917.

2. Johnston, O.J.: Local resection in primary malignant bone tumors. Clin. Orthop. Rel. Res. 153, 73, 1980.

3. Lloyd-Roberts: Experience with boiled cadaveric bone. J. Bone Joint Surg. 34(B), 428, 1952.

4. Odell, R.T., Mueller, C.B., and Key, J.A.: Effect on bone grafts of radioactive isotopes of phosphours. J. Bone Joint Surg. 33(A), 324, 1951.

5. Orell, S.: The use of Os Purum in bone implantations. Surg. Gynec. Obstet. 66, 23, 1938.

6. Reynolds, F.C., and Oliver, D.R.: Experimental evaluation of homogenous bone grafts. J. Bone Joint Surg. 33(A), 307, 1950.

7. Rivard, C.H., Fallaha, M., and Labelle, H.: The effect of autoclaving in normal and tumoral bone cells in the dog. Orthop. Trans. 5, 474, 1981.

8. Sijbrandij, S.: Resection and reconstruction for bone tumors. Acta Orthop. Scand. 49, 249, 1978.

9. Smith, S., and Simon, M.A.: Segmental resection for chondrosarcoma. J. Bone Joint Surg. 57(A), 1097, 1975.

10. Thomson, V.P., and Steggall, C.T.: Chondrosarcoma of the proximal portion of the femur treated by resection and bone replacement. A six year result. J. Bone Joint Surg. 38(A), 357, 1956.

Subject Index

MIX
Papier aus verantwortungsvollen Quellen
Paper from responsible sources
FSC® C105338

If you have any concerns about our products,
you can contact us on
ProductSafety@springernature.com

In case Publisher is established outside the EU,
the EU authorized representative is:
Springer Nature Customer Service Center GmbH
Europaplatz 3, 69115 Heidelberg, Germany

Printed by Libri Plureos GmbH
in Hamburg, Germany